Marketing and Management Innovations in Tourism

Özgür Yayla, Alper Işın, İrfan Yazıcıoğlu,
Fuat Bayram, Eren Yalçın (eds.)

Marketing and Management Innovations in Tourism

Insights and Impacts on Industry Growth

PETER LANG
Berlin · Bruxelles · Chennai · Lausanne · New York · Oxford

Bibliographic information published by the Deutsche Nationalbibliothek
The Deutsche Nationalbibliothek lists this publication in the Deutsche Nationalbibliografie; detailed bibliographic data are available on the Internet at https://dnb.dnb.de.

A catalogue record for this book is available from the British Library.

Library of Congress Cataloging-in-Publication Data
A CIP catalog record for this book has been applied for at the Library of Congress.

ISBN 978-3-631-93833-1 (Print)
ISBN 978-3-631-93834-8 (ePDF)
ISBN 978-3-631-95345-7 (ePUB)
DOI 10.3726/b23821

Published by Peter Lang GmbH, Berlin, Germany

info@peterlang.com

www.peterlang.com

This publication has been peer reviewed.

Contact for General Product Safety Regulation (GPSR): gpsr@peterlang.com

Contents

Figures

Tables

The Extent of Food Waste in Hotel Management and Proposed Solutions for Prevention

Abdullah AKMAZ[1]*, Metin SÜRME*[2]*, Müge KAYGUSUZ*[3]

Introduction

The term waste refers to the excessive consumption of resources, characterized by inefficiency and futility. The term food waste denotes the excessive and inadvertent eating of produced food or its disposal as refuse due to non-consumption. The discarded and surplus portions of food sourced from gardens, following their production and transformation before reaching our tables, exemplify food waste (Republic of Turkey Ministry of Agriculture and Forestry, 2024). While millions globally suffer from hunger and lack access to food, thousands of meals are squandered due to the open buffet system implemented in luxury hotels. Food waste significantly escalates during vacation periods, reaching critical levels. According to the Turkish Foundation for Waste Prevention (2023), In 2019, the Food and Agriculture Organization (FAO) reported that 1.3 billion tons of food are wasted annually, with an associated cost of around 990 billion dollars (FAO, 2019; Şenel & Çılgınoğlu, 2022:1089). The 2024 Food Waste Index Report by the United Nations Environment Programme (UNEP) indicates that 1.5 billion tons of food are lost globally each year, with a minimum of 8.7 million tons wasted annually in Turkey. In Turkey, each individual wastes 102 kg of food annually. The paper examines the sectoral allocation of food waste, indicating that 28 percent occurs in the service industry and 12 percent in the retail sector (Turkish Foundation for the Prevention of Waste, 2024). This statistic unequivocally illustrates the magnitude of the global issue of food waste. Globally, billions of tons of food waste annually lead to significant economic,

1 Mersin University, Tourism Faculty, aakmaz@mersin.edu.tr
2 Gaziantep University, Tourism Faculty, Department of Tourism Management, surmemetin@gmail.com
3 Gaziantep University, Institute of Social Sciences, Department of Tourism Managementt, muge.cakir@gmail.com

environmental, and social repercussions. Concurrently, this data suggests Turkey's involvement in the global garbage crisis. To mitigate waste issues, it is imperative to implement actions at both individual and societal levels.

Food waste exacerbates hunger issues and indirectly contributes to global warming and climate change through its environmental repercussions (Gönültaş, Kızılaslan, & Kızılaslan, 2020: 23). Consequently, individuals must be informed about these two facets of food waste. Mitigating food waste results in economic savings, enhanced resource efficiency, and diminished adverse environmental effects. Turkey's initiatives to mitigate food waste will advance both national and global sustainability objectives (Girgin, Üzüm, Yılmaz, & Sünnetçioğlu, 2022). To mitigate food waste, it is crucial to precisely define the behaviors that contribute to waste and to enhance awareness regarding this issue. A little awareness and sensitivity about waste-inducing actions can lead to significant global transformations (Sezgin & Ateş, 2020). This study aims to highlight food waste in hospitality enterprises by examining the "Open Buffet" and "All Inclusive" systems in hotels. The objective is to assess the magnitude of food waste in hotel enterprises, identify the factors contributing to this waste, examine the methods of waste utilization, and evaluate the initiatives undertaken by hotel businesses to mitigate food waste. This study, a situational analysis and review, strives to underscore the imperative of minimizing food waste. Identifying the causes and erroneous practices contributing to food waste in hotel facilities and devising solutions to eradicate this issue will be a crucial step towards decreasing and potentially preventing food waste.

Literature Review

The food sector supplies about one-third of global food consumption. Within this sector, hotel establishments serve as prominent examples of places where food waste frequently occurs. The annual number of individuals traveling to various destinations for diverse purposes significantly increases. Consequently, the tourism sector persists in its rapid expansion. It appears improbable that hotel establishments, which cater to essential needs and provide accommodations, can universally attract patrons with their food and beverage services and achieve uniform appreciation. Consequently, the "All Inclusive" concept and "Open Buffet" service, which have gained global popularity and provide extensive options for visitors, emerge as an appealing answer for the hospitality industry. Nonetheless, elements such as the quantity of food allocated only to the guests, unrestricted purchases, and the diversity and appeal of the offerings complicate the visitors' ability to obtain the proper amount of food on their plates. This results in excessive food intake and consequently significant waste (Akmeşe & Ilyasov, 2022: 2). By 2050, the global population is anticipated to approach 10 billion, necessitating a 50% augmentation in worldwide food production to sustain this demographic. According to the Food and Agriculture Organization of the United Nations (FAO), the rapid depletion of existing natural resources significantly hampers the goal of eradicating hunger by 2050, despite ongoing efforts. Moreover, food waste utilizes 250 km^3 of water and generates 3.3 billion tons of carbon emissions globally. Food waste results in economic losses and significant carbon dioxide (CO_2) emissions. FAO data indicates that the average per capita carbon footprint of food waste is roughly 500 kg of CO_2 (Dölekoğlu, Gün, & Giray, 2014: 177).

Dining in hotel facilities transcends just sustenance, evolving into a pursuit of pleasure, flavor, and delight within a festive ambiance. This can lead to increased food waste because guests, while relaxing and enjoying themselves, often consume more than necessary. Hotels contribute by offering large portions and diverse buffet options to showcase their hospitality. Hotel establishments provide a range of services to optimally accommodate their guests, resulting in the generation of solid and liquid waste. Research indicates that minimizing liquid waste, which constitutes roughly 40% of food waste, is essential for cost reduction in the tourism sector, characterized by narrow profit margins. In this regard, it is crucial to address the tourism sector, which can generate significant food waste, by identifying its causes and quantities and implementing essential preventive measures (Akmeşe & İlyasov, 2022: 3).

"All Inclusive System" refers to a certain model of package tour. A package trip encompasses a collection of services, including housing, transportation, meals, beverages, and entertainment, all provided at a unified price. The traveler is presented with a travel itinerary in which all their requirements are prearranged and included at a specified cost (Sezgin & Ateş, 2020: 3491). Hotels operating under an all-inclusive concept have a substantial issue with food waste. For instance, 74% of daily oil waste is generated by hotels utilizing the all-inclusive model (Bekar & Şahin, 2018, p. 1053). This circumstance has necessitated that enterprises implement strategies to mitigate waste. The substantial quantity of food provided in the open buffet system contributes to an escalation in waste. Consequently, industrial processes must be meticulously regulated inside the comprehensive system. The extensive variety of food offered in open buffets contributes to an escalation in waste. To avert this predicament, it is believed that the diversity of diet should be restricted (Keskin, Dinçer, & Dinçer, 2019, p. 52).

Research by Akmeşe and Ilyasov (2022) indicates that a significant proportion of Russian tourists favor buffet service while on vacation. The "All Inclusive System" significantly contributes to the proliferation of the buffet service model within the Turkish tourism sector. Originating in France, the buffet system has become increasingly prevalent in Turkey and has gained global traction. Environments such as breakfast areas, dining establishments, and the hospitality industry particularly favor this approach. The buffet system facilitates the provision of an extensive array of food and beverage options, from breakfast to dinner, at hotel businesses, owing to its user-friendliness. Customers have the freedom to choose from a variety of food, beverages, and sweets, consuming as much as they like. This autonomy is a significant element that enhances consumer pleasure. In the buffet system, patrons purchase excessive food and abandon uneaten portions on their plates (Sezgin & Ateş, 2020: 3491). This circumstance results in a rise in waste.

In 2016, Bekar and Şahin (2018) conducted a waste analysis of 24 distinct five-star hotels in Turkey, concluding that each hotel guest generates approximately 500 grams of food waste daily, given that the average food waste per guest is 700 grams and food waste constitutes 70% of total waste. The annual food waste generated by a consumer is estimated to be roughly 180 kilos. In their 2018 study, Bekar and Şahin elucidated the variables contributing to food waste generation as articulated by hotel personnel.

- Guests take more food than they can consume
- Change in the amount of waste according to the nationality of the guests
- Increased waste generation due to insensitivity of guests
- Serving meals in large portions
- Not using appropriate equipment in the kitchen
- Preparation of meals with poor quality ingredients
- Long waiting and spoilage of food in open buffets

- Insufficient product follow-up
- Prolonged holding of food at the wrong temperature due to lack of knowledge or lack of interest on the part of kitchen staff
- Suppliers sending poor quality and wrong food
- Excessive food preparation, especially in banquets, due to the uncertainty of the number of customers
- Using fruits and vegetables for decoration and discarding them after the decoration is completed
- In rare cases, expired products were found due to lack of regular warehouse checks.

Upon thorough examination of food waste in hotel establishments, it becomes evident that waste can arise at every stage—from procurement to guest service—including selection, sorting, and preparation, as well as plate and table waste resulting from guests' preferences or over-purchasing. A study of four distinct food and beverage companies in Sweden revealed that around one-fifth of the food was wasted, primarily attributable to plate waste. The analysis of plate waste is considered a crucial metric for assessing the necessity for modifications by evaluating food procurement practices and menu sufficiency, in addition to gauging company performance (Akmeşe & İlyasov, 2022: 4). Research indicates that about 92% of hotel establishments report discarding food waste produced after initial kitchen preparation. Concerning the uneaten meals from the open buffet, 42% of the establishments reported that they discard these meals. Concerning the uneaten food on plates post-service, around 67% of firms said that they dispose of these remnants along with all plate scraps (Bekar & Şahin, 2018: 1053). A primary environmental consequence of food waste is its accumulation in landfills and subsequent disposal. As food waste breaks down naturally in landfills, methane and carbon dioxide gases are produced. Methane, a greenhouse gas, possesses the capacity to trap heat 21 times more effectively than carbon dioxide, contributing to climate change (Kurt, Nazik, & Işın, 2020: 4367).

Hotels implement diverse strategies to mitigate food waste; yet, these strategies are inadequate in fully preventing it. Bilgin and Özen (2021) indicate that certain methods implemented include the application of HACCP (Hazard Analysis and Critical Control Points) principles and the adoption of the FIFO (first in, first out) cost scheme. We repurposed kitchen preparation waste from fruits and vegetables to make soups, sauces, olive oil meals, salads, pâté, spice blends, and marmalade or jam. Animal and fish bones were utilized to prepare stocks or sauces, while leftovers were repurposed as staff meals, offered to other patrons, or transformed into alternative dishes. Staff received training in waste management, and informational brochures were employed in restaurants and cafeterias to enhance customer awareness. HACCP is a safety system that establishes and enforces hygiene standards essential for the production of safe food in food operations while also seeking to identify and mitigate potential health risks during production and service phases. This system implements measures to safeguard consumer health by addressing personnel, equipment, raw materials, and environmental sanitation. Nonetheless, the HACCP system has failed to develop universally acknowledged standards (Ündevli, Kadam, Bekdik, Yılmaz, & Çobanoğlu, 2019: 170). The HACCP system was mandated in Turkey through the "Regulation on the Production, Consumption, and Inspection of Foods," published in the Official Gazette in 1998. ISO is a globally recognized organization that pertains to systems for managing food safety. It enhances operational efficiency and process improvement by providing numerous benefits to enterprises. This decreases expenses and enhances profitability. Moreover, possessing ISO certification fosters consumer trust and provides the enterprise with a competitive edge (Ündevli, Kadam, Bekdik, Yılmaz, & Çobanoğlu, 2019: 170).

Studies on food waste indicate that chefs lack adequate expertise or fail to implement methods related to the utilization of stale food products, their incorporation into other dishes, and suitable storage techniques. To mitigate food waste, consumers must get a deeper comprehension of its underlying causes. Food waste is a matter that necessitates focus, investigation, and intervention, particularly to avert waste production (Kurt, Nazik, & Işın, 2020: 4376). A research study in Germany implemented strategies such as offering smaller portions and reducing food quantities in buffet services to mitigate food waste in hotel buffet services. These procedures resulted in a 64.3% reduction in buffet leftovers and yielded savings of 9,000 euros. Significant savings were achieved in the categories of fruit, cheese, and fish (Leverenz, 2020; Bilgin & Özen, 2021: 359). Research undertaken by the Turkish Grain Board in 2008 and 2012 indicated that daily bread waste rose from 4.9 million to 5.9 million, while the national bread waste rate escalated from 5% to 5.9%. Bread constitutes the most squandered food item. The

"Don't Waste Your Bread" campaign, initiated by the public sector in 2013, is regarded as a significant, albeit largely symbolic, measure in tackling the issue of food waste, particularly in relation to other items that contribute to high levels of waste (Dölekoğlu, Gün, & Giray, 2014; Bekar & Şahin, 2018). The Turkish Foundation for the Prevention of Waste (TISVA) is an organization founded in Turkey with the objective of promoting the efficient and effective utilization of the nation's resources. It also seeks to enhance societal awareness of waste prevention and promote improvements in this awareness (TISVA—Stop Waste!). As per 2012 data from the Turkish Statistical Institute (TurkStat), the average daily trash generation per capita in Turkey is 1.12 kg. Furthermore, the 2019 report by the Turkish Foundation for the Prevention of Trash (TİSVA) indicates that the predominant category of trash comprises fruits and vegetables (Şahin, 2021: 529). Digital strategies to mitigate food waste are also implemented in Turkey. LUSE is a smartphone application designed to minimize food waste and optimize the use of surplus food in restaurant operations, facilitating optimal utilization of excess food by restaurants. LUSE is a Turkish-verified internet platform that signifies "delicious," "healthy," "affordable," and "exquisite." This platform unites food and beverage enterprises with customers. When enterprises are unable to utilize their entire daily production, they may sell the surplus food on luseyemek.com. Consequently, firms can market their surplus inventory throughout the evening by posting an advertisement on the site (Erik, 2019). Unconscious consumption is the primary cause of food loss at hotel establishments, while improper cooking procedures and the cook's lack of expertise and experience are the key contributors to food waste during the production phase. The restaurant segment was recognized as the area with the most food waste in hotels, with the food pre-preparation process being the most significant phase contributing to this waste. Fruits and vegetables were recognized as the most wasteful food category, but cheese variations at breakfast, pastry items at lunch, and vegetable dishes at supper were classified as the most wasted food groupings (Sezgin & Ateş, 2020: 3).

Waste oil in hotels constitutes a significant issue regarding food waste. Numerous hotel facilities improperly dispose of waste oil in refuse and fail to engage with contracted entities over this matter. Research in Gökçeada, Turkey, indicated that certain hotel establishments claimed to provide waste oils to the municipality; nevertheless, interviews with municipal authorities disclosed the absence of a process for collecting waste oil (Girgin, Üzüm, Yılmaz, & Sünnetçioğlu, 2022: 760). A proficient strategy to mitigate food waste involves maintaining appropriate temperatures for hot and cold buffet items in hotels, providing visitors with smaller plates for their meals, and decreasing food quantities during the final 30 minutes before the buffet concludes (Turan & Türkay, 2024: 714). Research has examined the influence of gender on food waste, revealing that female visitors have a greater propensity for food waste in categories such as "food-related waste," "attraction-related waste," and "personal-related waste" in comparison to their male counterparts. Taking into account these dimensions, the quality, look, aroma, presentation, sanitary condition, temperature, and psychological variables, particularly personal worries such as weight gain, indicate that women are more susceptible to food waste (Akmeşe & Ilyasov, 2022: 10). Numerous nations, companies, and NGOs are undertaking studies on consumable food waste. We recognize that disposing of consumable food waste raises significant environmental, economic, and ethical issues. Food waste can be categorized into two classifications: consumable and inedible. Edible food waste can be categorized into kitchen waste (produced during food preparation) and customer waste (leftovers and remnants on plates) in food and beverage establishments. Inedible food waste refers to things that are not consumable by humans, including bones, eggshells, and coffee grounds (Turan & Türkay, 2024: 702). The management of food waste entails implementing steps during the preparation, presentation, and consumption phases, strategically structuring operations to reduce waste formation, and repurposing the unavoidable trash that arises. The primary objective of the food waste hierarchy is to avert waste generation at the source or to diminish its quantity if it occurs. If this is not feasible, we should distribute edible food to those in need. Subsequently, food waste appropriate for animal use (e.g., plate remnants) has to be provided to animals. An alternative is to harness energy from food waste or to compost it. Ultimately, non-recyclable food waste should be burned or securely kept to prevent environmental impact (Bekar & Şahin, 2018: 1044).

Initiated in 1967 in the United States and effectively adopted globally, food banks have been formed in our nation to address food waste. Food banks are entities created by associations, foundations, or municipalities that supply donated or surplus food deemed safe for consumption, maintain these products under suitable conditions, and distribute them to individuals experiencing hunger and those impacted by disasters, either directly or through various aid organizations, without profit (Turkish Foundation for the Prevention of Waste, 1016). Kitchen cooks bear significant duties to eradicate or mitigate food waste. Furthermore, staff must be informed about food waste, and business owners should meticulously analyze the quantity and origins of food waste. Furthermore, biological treatment strategies mitigate food waste by minimizing portion sizes, monitoring waste, evaluating delivery systems, donating surplus food to charitable organizations, and transforming organic waste into fertilizer (Turan & Türkay, 2024: 703). To promote environmental protection, enhance environmental awareness, and advocate for sustainable construction and operational practices in tourist accommodations, the Ministry of Culture and Tourism has been awarding Environmentally Friendly Establishment Certificates since 1993. The Green Star Symbol is conferred upon establishments that fulfill the prevailing scoring criteria in domains such as energy and water usage, waste management, and environmental awareness. Businesses that meet these standards are awarded an "Environmentally Friendly Accommodation Facility Certificate." The green star sign, awarded to firms for one year, is subject to renewal upon re-evaluation at the conclusion of this term (Sabancı & Onur, 2024: 60–06).

As nations advance in development, waste challenges, particularly food waste, become increasingly prominent across all sectors. The proliferation of food waste cannot be ascribed to a singular or limited number of sources. Numerous variables, including contemporary food systems, industrialization, economic expansion, urbanization, globalization, cultural influences, and consumer behaviors, contribute to food waste. In light of this circumstance, the Food Waste Prevention and Awareness Platform was formed to raise awareness and mitigate the escalating food waste nationwide, with the objective of coordinating operations across all sectors. The "Orange Flag" initiative encompasses criteria designed to enhance awareness among business personnel and consumers regarding waste reduction in buffet and all-inclusive systems (Çetinoğlu & Ünlüönen, 2020: 323). The Orange Flag application is an extensive system that actively aids in the prevention of food waste and the preservation of food, enhancing societal awareness and providing enduring solutions to the problem of food waste. This system includes all services related to food production, processing, storage, sale, and consumption (turuncubayrak.org, 2018). The Orange Flag application is designed for businesses, including hotels, restaurants, cafes, markets, food wholesalers, shopping malls, food production, processing, packaging, and storage facilities, in addition to ministries, municipalities, schools, public dormitories, NGOs, universities, and other institutions. This application seeks to mitigate food waste in Turkey and worldwide by enabling facilities having a Tourism Operation Certificate to acquire an "Orange Flag Certificate" upon fulfilling designated criteria. This application comprises 45 requirements categorized under 14 principal topics that firms must satisfy (turuncubayrak.org, 2018). This project has been a significant undertaking designed to mitigate food waste. Hotels implementing the Orange Flag program have observed a substantial reduction in food waste following the attainment of this certification. Data from 2019 indicates that 3 million travelers have been informed about food waste via the Orange Flag application, resulting in a notable reduction in buffet waste. In 2020, the objective was to elevate the number of motels utilizing the Orange Flag program to 120, train 10,000 tourism professionals, and attract 20 million vacationers (Çetinoğlu & Ünlüönen, 2020: 332). Hotels utilizing the Orange Flag program exhibit a heightened level of awareness compared to conventional establishments. The objective of these hotels extends beyond merely offering accommodation services; it also includes the reduction of food waste. Hotel enterprises can aid in resource conservation by composting organic waste and using it as fertilizer in their gardens. This strategy satisfies the hotel's fertilizer requirements while facilitating the effective utilization of garbage. Furthermore, by employing alternative technologies like biogas production from organic waste, hotel enterprises can satisfy their electricity requirements at reduced costs and enhance energy efficiency (Şahin, 2021: 535). Certain hotel enterprises overseas employ biodigester systems that produce energy from organic waste. These systems generate gas from organic waste to provide the energy required for culinary operations in hotel kitchens. If executed on a broader scale, hotels can fulfill their electricity requirements with this approach (Şahin, 2021). Given Turkey's fertile terrain, it can be asserted that this approach, which offers multifaceted savings through the utilization of fertilizer derived from organic waste, is exceptionally appropriate.

Results

Food waste in the tourism sector highlights the necessity of preventing, reducing, and recycling waste at every stage, from production to consumption. Food waste management is of great importance in this context and plays a critical role for businesses. Research shows that food waste can be significantly prevented when necessary precautions are taken and effective food management practices are implemented. Studies and literature reviews on food waste indicate that there has been insufficient research conducted in Turkey on this topic, as waste has always been considered one of the significant problems. The knowledge levels, experiences, workloads, and whether kitchen staff in "Buffet System" and "All-Inclusive" hotels have received culinary training have a significant impact on food waste. There are gaps in knowledge regarding the environmental harms of food waste. Food waste is often a result of carelessness and unintended actions. Awareness-raising campaigns targeting the public should be organized in this regard. Considering the current situation of food losses, population growth, and the impacts of agriculture on the ecosystem, immediate and concrete steps are deemed necessary to prevent food waste and loss. The first step in the food waste hierarchy is aimed at preventing or minimizing waste generation at the source. In the last stage, food waste should be recycled by making it usable (e.g., composting). It is also important to store or manage non-recyclable waste in a way that minimizes its negative effects on the environment (Tekiner, Mercan, Kahraman, & Özel, 2021: 126). Based on the research study, it is deemed necessary for hotels, especially those where waste occurs most, to implement certain strategies in their restaurants to prevent or reduce food waste. These strategies include regulating portion sizes, displaying only the necessary amount of food in buffets, stock management, menu planning, utilizing leftover edible food, recycling and composting, raising awareness through training, and monitoring food waste.

Portion sizes for the food served in businesses must be determined correctly, and the prepared dishes should be served to customers in accordance with the recipes (Bekar & Şahin, 2018: 1043). In businesses, customers sometimes request buffet breakfasts simply because they look visually appealing on the table. However, buffet breakfasts are the type of breakfast with the highest waste. In addition to raising public awareness, the government should impose restrictions on such options to ensure that customers become more conscious (Gemeç & Gürbüz, 2022: 173). During the cooking process, it should be ensured that the ovens and stoves used are properly calibrated and that the kitchen staff has the necessary knowledge about the features and functions of these devices. Otherwise, errors from the ovens or stoves, negligence by employees, or lack of knowledge can lead to situations like burning food, resulting in both material and moral losses (Bekar & Şahin, 2018: 1043). Businesses can place food waste bins in appropriate areas to ensure the recycling of packaging and organic waste generated during food production in the restaurant section. It is deemed necessary to create designated collection areas for used oils. Municipalities and relevant institutions can work together to create awareness campaigns and collection programs regarding used oils to prevent the waste oil problem.

Research shows that when individuals are presented with various food options, they tend to take more food, leading to plate waste, even when they are full. It has been determined that food variety encourages people to overindulge and creates a strong attraction effect. Therefore, transitioning from a buffet to à la carte service is considered to be more effective in reducing and preventing waste. The kitchen service equipment in hotel operations plays a significant role in the formation of food waste. In Akmeşe and İlyasov's (2022) research, it is noted that Wansink (2004: 468) conducted a study showing that kitchen tools such as ladles, bowls, glasses, and spoons affect the amount of food taken by 71%. Based on this finding, it is believed that determining appropriate plate sizes would be beneficial for hotels to better control their guests' food servings. It is crucial to provide regular training by experts to staff on waste management and to ensure that practices are monitored. The awareness of separating and collecting waste at the source must be instilled in business employees. To raise customer awareness about waste management, showing visual promotional videos or short films related to waste management on screens at hotel entrances could also be a good idea. Effective coordination and cooperation should be established with relevant public authorities regarding waste management. Additionally, properly and regularly sorting the types and quantities of waste sent to recycling firms is considered important in reducing waste. Determining food production quantities, adjusting portion sizes, menu planning, efficient stock management, and applying proper serving methods are key elements for sustainably managing food waste in hotel operations and reducing costs early on. A detailed examination of these elements and addressing the issue of food waste from this perspective in different service sectors could provide significant guidance for future research.

References

Akmeşe, H., & İlyasov, A. K. (2022). Food Waste in the Buffet Service System: A Study on Russian Tourists. International Journal of Tourism and Technology Research, 3(1), 1–14.

Bekar, A., & Şahin, S. K. (2018). A Global Issue: "Food Waste": Its Dimensions in Hotel Enterprises. Journal of Tourism and Gastronomy Studies, 6(4), 1039–1061.

Bilgin, E. Z., & Özen, C. (2021). Principles of Food Waste Management Applied in 4 and 5 Star Hotels. Journal of Tourism and Gastronomy Studies, 9(5), 355–370.

Çetinoğlu, D., & Ünlüönen, K. (2020). Orange Flag Application to Prevent Food Waste in Hotel Enterprises. Journal of Social, Human and Administrative Sciences, 3(5), 318–335.

Dölekoğlu, C. Ö., Gün, S., & Giray, H. (2014). The Vicious Cycle of Poverty and Food Waste. XI. National Agricultural Economics Congress (pp. 172–182), Izmir, Turkey.

Erik, U. (2019). Development of a Mobile Application Model for Preventing Food Waste in Restaurants and Evaluating Excess Food: LUSE. Social Sciences Institute, Necmettin Erbakan University, Konya.

Furkan Musa ŞENEL, & ÇILGINOĞLU, H. (2022). Trends in Food Waste among Hotel Enterprises: The Case of Kastamonu. Journal of Tourism and Gastronomy Studies, 10(12), 1088–1114.

Gemeç, E., & Gürbüz, İ. B. (2022). Determination of Attitudes and Behaviors towards Preventing Food Waste. Balkan and Near East Social Sciences Journal, 8(Special Issue), 164–175.

Girgin, C. A., Üzüm, Ş. N., Yılmaz, E., & Sünnetçioğlu, A. (2022). Food Waste in Food and Beverage Enterprises. Turkish Journal of Tourism Research, 6(3), 744–764.

Gönültaş, H., Kızılaslan, H., & Kızılaslan, N. (2020). Behavioral Determinants of Food Waste: The Case of Tokat Province. Gaziosmanpaşa Scientific Research Journal, 9(3), 14–24.

Keskin, E., Dinçer, F. İ., & Dinçer, M. Z. (2019). Food Waste in All-Inclusive Accommodation Enterprises. International Gastronomy Tourism Studies Congress (pp. 48–55), Nevşehir, Turkey.

Kurt, Y., Nazik, M. H., & Işın, A. (2020). The Impact of Food Waste Behaviors on the Profitability of Hotel Enterprises: The Case of Eskişehir. Journal of Business Research, 12(4), 4365–4379.
Sabancı, A. E., & Onur, N. (2024). Waste Food Practices in Green Star Hotels: The Case of Istanbul. Journal of Tourism and Gastronomy Studies, 12(1), 604–625.
Sezgin, A. C., & Ateş, A. (2020). A Study to Determine the Dimension of Food Waste in Hotel Enterprises. Journal of Tourism and Gastronomy Studies, 8(4), 3489–3507.
Şahin, S. K. (2021). A Model Proposal for the Evaluation of Food Waste in Hotel Enterprises. International Journal of Social and Humanities Research, 8(66), 527–538.
T.C. Ministry of Agriculture and Forestry. (2024). General Overview of Food Waste and Food Waste. Retrieved on December 15, 2024, from https://www.tarimorman.gov.tr/Sayfalar/Detay.aspx?SayfaId=77
Tekiner, İ. H., Mercan, N. N., Kahraman, A., & Özel, M. (2021). A General Overview of Food Waste and Loss in the World and Turkey. Journal of the Institute of Science and Technology, Istanbul Sabahattin Zaim University, 3(2), 123–128.
TİSVA - End Waste! (2024). Retrieved on December 12, 2024, from https://israf.org/sayfa/VAKIF-SENEDI/244
Turan, F., & Türkay, O. (2024). Waste Food Management in Hotels: The Case of Trabzon. MANAS Social Studies Journal, 13(2), 700–717.
turuncubayrak.org. (2024). Retrieved on December 12, 2024, from https://turuncubayrak.org/sayfa-nedir-2
Turkey Waste Prevention Foundation. (2024). Food Banking. Retrieved on December 10, 2024, from https://israf.org/
Turkey Waste Prevention Foundation. (2024). 2023 Waste Report in Turkey by Numbers. Retrieved on December 21, 2024, from https://www.tgmp.net/eyayin/israf_raporu_2023.html
Turkey Waste Prevention Foundation. (2025). Waste in Turkey by Numbers. Retrieved on January 10, 2025, from https://www.tgmp.net/eyayin/israf_raporu_2024.html
Ündevli, A., Kadam, G., Bekdik, Y. L., Yılmaz, H. İ., & Çobanoğlu, F. (2019). Identification of Food Waste: The Case of Aydın Province. Agricultural Economics Journal, 25(2), 169–184.

A Literature Review on the Complaints of Package Tour Buyers in Turkey

Gözde OĞUZBALABAN[1]

Introduction

The contributions of travel agencies and tour operators are very important in the development of tourism. These businesses not only connect the touristic product providers with the customers, but also contribute to the marketing and promotional activities of the countries. Tourists who participate in package tours offered by travel agencies and tour operators mostly want to be satisfied, and therefore a deficiency or error in any process of the package tour they have purchased can lead to dissatisfaction of tourists. However, the quality of a package tour requires the coordinated work of many businesses that differ from each other, from airline companies to bus companies, from accommodation businesses to restaurants (Güven, 2018).

Travel agencies may resort to "price war" tactics to gain a competitive advantage and create customer satisfaction. However, price tactics alone are not enough to create customer satisfaction. In order to gain an advantage over their competitors, travel agencies should be customer-oriented and act according to their wishes and expectations in order to achieve quality superiority (Yu & et al, 2005).

In this study, the concepts of travel agency and package tour, customer satisfaction and customer complaints are examined. In addition, the results of the studies conducted to determine which types of complaints travel agencies face about package tours are included.

Travel Agency and Package Tour

Travel agency is defined in the Law No. 1618 on Travel Agencies and the Association of Travel Agencies as "a commercial organization that is authorized to provide tourists with tourism-related information for profit, to

1 Assoc. Prof. Zonguldak Bülent Ecevit University, Karadeniz Ereğli Tourism Faculty, Department of Tourism Guidance, oguzbalaban@beun.edu.tr

create package tours and tours, to provide accommodation, transportation, sightseeing, sports and entertainment services for tourism purposes, and to market the product it creates through its own and other travel agencies" (Türsab,2019). One of the important and profitable activities for travel agencies is package tours. A package tour is "a commercial activity organized by a travel agency within the scope of a predetermined program; transportation, accommodation and at least two of the other touristic services that are not considered auxiliary to them, sold at an all-inclusive price or sales commitment is made and the service covers a period longer than twenty-four hours or includes overnight accommodation (T.C. Ministry of Culture and Tourism, 2025).

Package tours consist of a combination of many products and services. This feature of package tours makes it very difficult to ensure customer satisfaction. As challenging as this situation is, it is extremely important for agencies to ensure customer satisfaction in order to survive and grow. (Unur, Çakıcı & Taştan, 2010).

Customer Satisfaction and Complaints

Customer satisfaction can be defined as "a short-term emotional state that occurs as a result of the customer's evaluation of a product or service, whether the goods and services offered meet customer expectations" (Selvi, 2007).

In order to be successful in customer satisfaction, it is necessary to determine the complaint of the dissatisfied customer and the factors affecting this complaint. Because bringing new customers to the business requires more effort. However, the energy to be spent to reconnect the dissatisfied customer to the business is less. Kotler expressed this situation as follows: "In the past, the sole purpose of businesses was to acquire new customers and they put all their efforts into acquiring new customers, not retaining existing customers. Today, however, the first goal of businesses is to ensure that existing customers remain loyal to the business and to try to maintain this bond." (Kotler, 2000).

Customer complaints are negative behaviors or comments that arise as a result of customers' wishes, desires, needs or expectations not being met and that need to be resolved as soon as possible (Saydan, 2008).

Customers who complain are the customers that businesses want to retain because if an unsatisfied customer complains, they want to continue doing business with the business. If the customer does not complain, they have either said goodbye to the business or are about to do so. Answering customer complaints and eliminating problems benefits both businesses and customers (Barış, 2006).

Although businesses aim to provide error-free service, it is not possible to completely eliminate errors or problems. Therefore, customer complaints are an opportunity for continuous improvement of the service provided. Businesses that manage the complaint process well and satisfy their customers strengthen the image they create in the minds of their customers (Limon & Arpacı, 2015).

Tourism services are an area where instant complaints can occur, uncontrollable situational effects are experienced and complaint management is very difficult due to the difficulty of intervention. For example; even if the weather is very windy or raining during the tour, it can be a cause of dissatisfaction and complaint for the customer. At this point, the priority for businesses is to fully fulfill the elements that can be controlled in service delivery. These are; the attitudes and behaviors of the staff, ensuring the cleanliness and hygiene of the venues, good organization of the tour program, quality guidance and consultancy services, etc. Providing these elements in the best way will prevent the formation of complaints and will contribute positively to the business reputation and future sales (Zorlu & Erkuş, 2018).

Considering all these reasons, a travel agency should take into account customer complaints and establish and operate a good complaint management system in order to improve itself, increase its profitability and ensure its continuity. Determining customer complaints in package tours organized by travel agencies will contribute to travel agencies to encounter fewer problems in package tours, to take precautions about the possible causes of complaints and to develop a more effective management style (Unur, Çakıcı & Taştan, 2010).

According to Table 1, the main complaint elements that travel agencies and tour operators should consider and pay attention to when preparing a package tour program are; elements of complaint against the guide, complaints about

Table 1. Complaints of Customers Buying Package Tours in Turkey

Elements of complaint against the guide	References
Insufficient professional knowledge of the guide	Arpacı (2022), Dalgıç et al., (2021), Ay (2019), Tanrısever (2018), Arpacı (2016).
Guide's indifference	Arpacı (2022), Kanlı (2020), Zorlu & Erkuş (2018), Tanrısever (2018), Arpacı (2016).
Rude behavior of the guide	Kanlı (2020), Zorlu & Erkuş (2018), Tanrısever (2018), Arpacı (2016).
Failure of the guide to lead the group	Kanlı (2020).
Inadequate explanation by the guide	Kanlı (2020), Zorlu & Erkuş (2018)
Pressure from the guide for shopping and extra tours	Kanlı (2020), Tanrısever (2018)
Failure of the guide to comply with the tour program	Ay (2019)
Complaints about the content of the sightseeing tour program	**References**
Shortening tour times	Arpacı (2022), Tanrısever (2018)
Changing the tour program	Arpacı (2022), Arpacı (2016).
Tour canceled	Arpacı (2022), Kanlı (2020), Tanrısever (2018)
Tight tour schedule	Dalgıç vd., (2021)
Failure to follow the itinerary in the tour program	Kanlı (2020)
Failure to comply with the promises in the tour program	Kanlı (2020), Arpacı (2016).
Bad tour organization	Kanlı (2020)
Fast realization of the tour	Kanlı (2020)
Spending too much time on the bus	Kanlı (2020)
Insufficient free time	Kanlı (2020)
The route of the tour is longer than necessary	Zorlu & Erkuş (2018)
Elements of complaint about the tour bus	**References**
Old vehicles	Arpacı (2022)
Vehicle uncomfortableness	Dalgıç et al., (2021), Ay (2019)
The vehicle is not of good quality	Kanlı (2020), Ay (2019), Tanrısever (2018)
Vehicle breakdown	Kanlı (2020), Tanrısever (2018)
Pollution of the vehicle	Zorlu & Erkuş (2018)

Elements of complaints against the agency	**References**
Incorrect and incomplete information	Arpacı (2022), Kanlı (2020), Tanrısever (2018)
Indifference of agency officials and employees	Arpacı (2022), Kanlı (2020), Tanrısever (2018)
Agency employees do not find solutions to problems quickly.	Kanlı (2020),
The rudeness of agency employees	Kanlı (2020), Tanrısever (2018)
Lack of knowledge of agency employees	Kanlı (2020), Tanrısever (2018)
Elements of complaints regarding accommodation	**References**
Poor quality of accommodation facilities	Arpacı (2022), Kanlı (2020), Tanrısever (2018)
Hotel food is bad	Kanlı (2020)
The hotel is far from the destination center	Kanlı (2020)
Negative behavior of hotel staff	Kanlı (2020)
Hotel food is limited	Arpacı (2016)
Hotel rooms are dirty	Arpacı (2016)
Products such as sheets, towels in the hotel room are dirty	Arpacı (2016)
Elements of complaints regarding fees	**References**
Extra fees requested	Arpacı (2022), Kanlı (2020), Tanrısever (2018)
Non-refundable cancellation of tour fees	Arpacı (2022), Ay (2019), Tanrısever (2018)
Failure to pay the difference in changes made	Kanlı (2020), Tanrısever (2018)
Contradiction between the fees in the advertisement and the fees requested afterwards	Arpacı (2022), Tanrısever (2018)
High fees demanded by shopkeepers	Dalgıç vd., (2021)
Expensive entrance fees to museums and archaeological sites	Dalgıç vd., (2021)
Elements of complaints regarding rest areas	**References**
Food at the break areas is bad	Kanlı (2020)
Food is expensive at the break points	Kanlı (2020)

(Continued)

Table 1. (Continued)

Poor quality of break areas	Kanlı (2020), Zorlu ve Erkuş (2018)
Spending too much time at break areas	Kanlı (2020)
No breaks	Kanlı (2020), Zorlu ve Erkuş (2018)
Complaint elements regarding shopping	**References**
Lack of shopping time	Dalgıç vd., (2021)
Spending too much time in shopping places	Kanlı (2020)
Expensive contracted shopping places	Kanlı (2020)
Complaint elements regarding museums and historical sites	**References**
Lack of information signs in museums and archaeological sites	Dalgıç vd., (2021)
Waiting in line at museums and archae-ological sites	Dalgıç et al., (2021)
Closure of museums and archaeological sites without information	Dalgıç et al., (2021)
Lack of time allocated to museums and archaeological sites	Dalgıç et al., (2021)
Faulty electronic devices used in museums	Dalgıç et al., (2021)
Elements of complaint about the driver	**References**
Rude behavior of the driver	Kanlı (2020), Arpacı (2016)
Driver's incorrect use of the vehicle	Arpacı (2022)

the content of the sightseeing tour program, elements of complaint about the tour bus, elements of complaints against the agency, elements of complaints regarding accommodation, elements of complaints regarding fees, elements of complaints regarding rest areas, complaint elements regarding shopping, complaint elements regarding museums and historical sites, elements of complaint about the driver.

It is important for travel agencies and tour operators to prepare their package tours by taking these factors into consideration in order to increase customer satisfaction and ensure that customers remain loyal to the business.

Conclusion

Travel agencies aim to ensure customer satisfaction by considering service quality in the tours they organize, and the best way to achieve this is to manage the complaint process well. Dissatisfied customers express that they are still in a relationship with the business by complaining to the business, and each complaint means that the business gives the other party opportunities to be better. Businesses that do not care about customer complaints, ignore them, make misleading advertisements to the customer, and fail to provide the expected service quality will face extinction in the market (Limon ve Arpacı, 2015).

Determining the complaint elements in Table 1, which was created within the scope of this study on customer complaints in package tours, gives important clues about improving the quality of tours. The recommendations recommended to be implemented in this context are as follows;

- The guide is the most powerful representative of the agency and therefore the agency should choose the guide well. The guide should be knowledgeable about the issues that the guide needs to know due to his/her profession, the guide should avoid rude and indifferent behavior towards the customer, and the guide should avoid pressuring customers to shop and buy extra tours.
- One of the most common complaints of customers is the change of tour content. This should not be done unless it is very mandatory, and in mandatory changes to be made, the situation should be explained in a way to satisfy the customer, and if necessary, alternative options should be offered to the customer.
- The main complaints about tour buses are old vehicles, breakdowns and pollution. Since these situations can cause serious problems for customers, especially on long journeys, the service life of the vehicles to be used in tours should be limited to 5 years.
- The most important factor that agencies should pay attention to is to guide the customer correctly. It is especially important that the promotional tour prices and content they share on their web pages are accurate. Staff service quality should be increased by organizing regular training programs.
- Considering that customers who purchase package tours will receive accommodation services throughout the tour, accommodation quality is

one of the most important factors in terms of preferring the tour and ensuring satisfaction with the tour. Poor quality of accommodation facilities and bad food will cause the customer to leave the tour seriously unhappy. For this reason, agencies should pay special attention to and research the selection of accommodation facilities when creating package tours.

- Extra fees demanded after the tour sale and the non-refund of the money paid as a result of any canceled tour cause customers to feel cheated. Therefore, misleading advertisements and campaigns should be avoided.
- Optimization of stopover points in package tours and service quality (infrastructure and superstructure facilities) in these selected places should also be taken into consideration.
- Drivers should be selected from those who are experts and conscious in their field.

References

Arpacı, Ö. (2016). Seyahat acentalarının tur hizmetlerine yönelik müşteri e-şikâyetlerinin örgütsel öğrenme aracı olarak değerlendirilmesi. *İstanbul Ticaret Üniversitesi Sosyal Bilimler Dergisi*. 2016 (2), 203–219.

Arpacı, Ö. (2022). Karadeniz turu satın alan müşteri şikâyetlerinin analizi. *Balkan & Near Eastern Journal Of Social Science,* 2022 (8), 25–32.

Ay, S. (2019). *Seyahat Acentasından Paket Tur Satın Alan Müşterilerin Şikâyetleri Üzerine Bir Araştırma.* İstanbul Arel Üniversitesi, Sosyal Bilimler Enstitüsü, İşletme Anabilim Dalı, Yüksek Lisans Tezi, İstanbul, 2019.

Barış, G., (2006). *Kusursuz Müşteri Memnuniyeti İçin Şikâyet Yönetimi*, İstanbul: Mediacat Yayınları.

Barlow, J. & Moller, C. (1996). *A complaint is a gift: using customer feedback as a strategic tool.* San Francisco: Berrett-Koehler Publishers.

Dalgıç, A., Toksöz, D., Ekici Çilkin, R., & Güvenç, A. (2021). Tur katılımcılarının şikâyet unsurları ve izlenen şikâyet telafi stratejileri: turist rehberleri açısından bir değerlendirme. *Turist Rehberliği Nitel Araştırmalar Dergisi*, 2(2), 130–149.

Güven, Ö. (2018). Paket tur kapsamında sunulan hizmetlerin kalitesinin davranışsal niyet üzerine etkileri: Yurt dışı paket tur satın alan yerli turistler üzerinde bir araştırma. *MANAS Sosyal Araştırmalar Dergisi*, 7(2), 657–674.

Kanlı, K. (2020). *Karadeniz Kültür Turuna Katılan Yerli Turistlerin E-Şikâyet Davranışlarının İncelenmesi.* İzmir Katip Çelebi Üniversitesi, Sosyal Bilimler Enstitüsü, Turizm İşletmeciliği Ana Bilim Dalı, Yüksek Lisans Tezi, İzmir, 2020.

Kotler, P. (2000). *Kotler ve Pazarlama.* İstanbul: Sitem Yayıncılık.

Limon Ş. & Arpacı Ö. (2015). Halkla ilişkiler faaliyetleri açısından tüketici şikâyetlerinin değerlendirilmesi: konaklama işletmeleri şikâyetlerine yönelik bir araştırma, *1. Türk Dünyası Turizm Sempozyumu Bildiriler Kitabı*, 545–555.

Saydan, R. (2008). *Müşteri Memnuniyeti.* Varinli, İnci & Çatı, Kahraman (ed.). Güncel Pazarlama Yaklaşımlarından Seçmeler içinde 105–125. Ankara: Detay Yayıncılık.

Selvi, M. S. (2007). *İlişkisel Pazarlama Stratejiler ve Teknikler.* Ankara: Detay Yayıncılık.

Tanrısever, C. (2018). Paket tur satın alan müşterilerin şikâyet analizi. *Turizm Akademik Dergisi*, 5(1), 114–123.

T.C. Kültür ve Turizm Bakanlığı (2025) Seyahat Acentaları Yönetmeliği. Retrieved from https://teftis.ktb.gov.tr/TR-14578/seyahat-acentalari-yonetmeligi.html (Accessed: 03.03.2025).

Türsab (2019). Seyahat Acentası Tanımı ve Faaliyetleri. Retrieved from https://www.tursab.org.tr/assets/pdf/turizme-giris-seyahat-acentaligi-ve-tur-operatorlugu.pdf (Accessed: 03.03.2025).

Unur, K., Çakıcı, C., & Taştan, A. G. H. (2010). Paket turlardaki müşteri şikayetleri ve seyahat acentalarının bu şikayetlerle karşılaşma sıklıkları üzerine bir araştırma. *Çukurova Üniversitesi Sosyal Bilimler Enstitüsü Dergisi*, 19 (3), 391–408.

Yu, C.J., Wu, L.W, Chaio, Y. & Tai, H. (2005). "Percieved quality, customer satisfaction, and customer loyalty: the case of lexus in Taivan," *Total Quality Management*,16 (6), 707–719.

Zorlu, Ö., & Erkuş, G. (2018). Kültür turlarında turist şikâyetleri ve şikâyet davranışları. *Turist Rehberliği Dergisi*, 1(1), 31–40.

Cross-Border Tourism Partnerships Between Türkiye and Greece

Batuhan AKTEPE[1], *Saadet Pınar TEMİZKAN*[2]

Introduction

Türkiye and Greece are two neighboring countries that have had historical, cultural, and geographical relations with each other for a long time (Keridis, 1999). This unity, which started with the migration of Turks to Anatolia, continued with the Ottoman Empire controlling Greece and the Balkans for a long time (Constantinides, 1996). Events such as the Treaty of Lausanne after the Greek War of Independence constitute key points in relations (Öztürkmen, 2002). Although both countries' accession to NATO after the Second World War strengthened relations, the Cyprus issue and the ongoing disputes over the Aegean Sea continental shelf negatively affected bilateral relations (Mann, 2001). However, the cultural similarity of the two countries and regional and economic cooperation are considered common interests. This has ensured that the two countries interact from time to time. Cross-border tourism agreements between Türkiye and Greece bring economic benefits to both countries. They also strengthen the cultural and historical ties between the two countries. In recent years, visa facilitation, visa exemptions, joint tourism activities, and specific projects have been carried out between the two countries. In this way, the number of people participating in tourism activities between the two countries has increased day by day. Despite the positive developments, there are border problems, migrant crises, and historical disputes between Türkiye and Greece. These problems hinder cross-border tourism agreements (Aydin & Ifantis, 2004).

1 Lecturer, Bartın University, Vocational School, Department of Hotel, Restaurant and Catering Services, baktepe@bartin.edu.tr

2 Prof. Osmangazi University, Tourism Faculty, Department of Gastronomy and Culinart Arts, sptemizkan@ogu.edu.tr

In this context, it is important to understand the tourism dynamics between Türkiye and Greece.

This study examines the cross-border tourism activities between Türkiye and Greece and reviews the partnerships from economic, cultural, and historical perspectives. The study will examine the historical unity of the two countries, as well as joint tourism activities and projects. In addition, evaluations will be made about the economic and political situations that may affect cross-border tourism between the two countries. In this way, it will be revealed how the two countries cooperate despite being competitors in the same geography. Finally, the problems that Türkiye and Greece may face in cross-border tourism activities and future cooperation opportunities will be discussed, and recommendations will be given.

Historical Background and Tourism Relations Between Turkey and Greece

The Aegean Sea is central to the geographical and cultural connection between the two countries. This sea, which runs along the coastline of both countries, is vital for trade, culture, and tourism (Athanasopulos, 2001). Both countries have made significant investments to revitalize the tourism sector by exploiting the opportunities and historical values offered by the region (Koberidze, 2023). The islands of Greece and the Aegean coast of Türkiye attract tourists not only from the region but also from many parts of the world (Ballis, Paravantis & Moschovou, 2018). This strong tourism potential and economic impact of the Aegean region have led the two countries to compete, sometimes interdependently and sometimes confrontationally.

Tourism is a good tool for strengthening relationships between communities (Nunkoo, 2017). Such contacts also enhance the industry by providing personal relationships (March & Wilkinson, 2009). There have been attempts to organize joint tours between the two countries, especially for foreign tourists. However, visa exemption, immigration, and economic problems have created problems for such organizations (Ulusoy & Battjes, 2017).

In 2004, the number of tourists traveling from Türkiye to Greece was 185 thousand, while the number of tourists traveling from Greece to Türkiye was 491 thousand. These numbers have increased steadily over the years (Tsarouhas & Yazgan, 2018). In the pre-COVID-19 period, the number of tourists traveling from Türkiye to Greece reached 789 thousand, while the number of tourists traveling from Greece to Türkiye reached 820 thousand. The number of tourists coming from Greece is predominantly higher. This is because Greek citizens have been subject to visa exemption to enter Türkiye since 1985 (Euronews, 2024).

Tourism provides significant economic benefits but has certain limitations. It is susceptible to events in the host country (Tosun, 2000). Two countries are expected to offer each other alternative tourist destinations. Especially after the terrorist attacks, refugee crisis, and economic problems in Türkiye in recent years, this benefit has emerged visibly (Dogru-Dastan & Tütüncü, 2024).

Key Cross-Border Tourism Initiatives

As a part of the service sector, the tourism industry is a significant source of income for many countries (Zaei & Zaei, 2013). The 1980 Manila World Tourism Declaration recognized tourism as an essential activity for the life of nations, directly influencing the social, cultural, and economic sectors of national communities (Var & Ap, 2013). For these reasons, tourism also generates substantial economic revenue through payments for goods and services (Sinclair, 1998).

Cross-border tourism refers to activities conducted between two or more countries based on geographical and cultural proximity. This type of tourism often allows communities living in border regions to mutually explore each other's cultural, historical, and natural riches. Cross-border tourism not only provides economic benefits but also promotes social and cultural interactions (Gelbman & Timothy, 2011).

The cross-border tourism agreements between Greece and Türkiye have been developed to enhance the tourism potential of both countries and strengthen economic cooperation in border regions. In cross-border agreements, knowledge transfer between regional administrations and innovative collaborations ensure the long-term sustainability of these agreements (Makonnen et al., 2018). In this context, tourism collaborations between Greece and Türkiye not only offer economic benefits but also have the potential to increase cultural interactions. Several cross-border agreements exist between the two countries, which are listed sequentially.

a) İzmir-Thessaloniki Ferry Services Protocol (2016): İzmir and Thessaloniki, two of the most important and developed port cities in the Aegean Islands, are among the most significant cross-border agreements between Türkiye and Greece. The İzmir-Thessaloniki Ferry Services Protocol, signed between Türkiye and Greece in 2016, aimed to promote maritime transportation between the two cities, thereby increasing tourism and trade (Ege Postası, 2016).

Initially, the project was planned with the Greek-origin Ropax ship carrying passengers and tourists from Türkiye three times a week. However, the project could not be implemented for years. Factors such as the unavailability of a suitable ship, operational challenges, and the pandemic delayed the process. In 2022, the first trial voyage was conducted, but due to low demand and high costs, the services could not be regularized (İzto, 2022).

b) İzmir-Thessaloniki Cooperation (2024): In 2024, Dr. Cemil Tugay, the Mayor of İzmir Metropolitan Municipality, visited the city of Thessaloniki, Greece, to explore cooperation opportunities in logistics, smart city applications, trade, and tourism. During these meetings, discussions focused on strengthening economic and cultural ties between the two cities, developing mutual projects in sustainable urban practices, and enhancing trade relations.

Smart city applications and environmental sustainability projects were also highlighted as key topics during the meetings. It was assessed that İzmir and Thessaloniki could collaborate on green energy usage, waste management, and digital urban projects (Ege Telgraf, 2024).

c) Türkiye-Greece Tourism Forum and Joint Commission Meetings: The Türkiye-Greece Tourism Forum and Joint Commission Meetings, organized to enhance tourism relations and strengthen cooperation between the two countries, have served as an essential platform for evaluating the current state of the tourism sector and implementing joint projects. However, the fluctuating nature of bilateral relations over the years has caused interruptions in these forums and commission meetings during certain periods. These meetings address tourism's economic, cultural, and political dimensions, facilitating direct dialogue between sector representatives and government officials.

During these meetings, topics such as visa policies, maritime and air transportation, border crossings, and sustainable development in tourism are discussed. Additionally, Turkish and Greek tourism authorities have agreed to create joint tourism packages to attract more tourists from distant markets (e.g., the USA, China, and Japan) (Dış Ekonomik İlişkiler Kurulu, 2024).

d) European Union (EU) Cross-Border Cooperation (CBC) Programs: The European Union (EU) has developed Cross-Border Cooperation (CBC) Programs to strengthen economic and social ties between member and candidate countries. Within the scope of these programs, the two countries have created economically important tourism projects, especially to be sustainable (Ohtamış, 2008).

In this context, the INTERREG Europe Program is a plan designed to realize the economic goals of the two countries. With the projects implemented in the two countries' border regions, local governments and academic institutions are working on developing tourism and increasing environmental sustainability (Yıldırım et al., 2016). In addition, with these cross-border agreements, studies are also carried out in terms of eco-tourism and gastronomy tourism. These collaborations accelerate economic development and strengthen diplomatic relations between the two countries.

e) On-Arrival Visa Application and Ease of Access to Islands (2024): Visa on arrival allows Turkish citizens to enter certain Greek islands such as Kos, Rhodes and Lesbos without a visa. Initially launched as a pilot program in 2012, the Greek government reintroduced the scheme for 10 islands in 2024. This regulation facilitates Turkish tourists' access to Greek islands, revitalizing regional tourism and contributing to the economy. Ferry services from Turkish cities such as Ayvalık, Çeşme, Kuşadası, Bodrum, and Marmaris to the Greek islands demonstrate the impact of this initiative on the tourism sector. Greek authorities expect a threefold increase in Turkish tourists (Ministry of Foreign Affairs, 2024).

The on-arrival visa strengthens tourism relations between the two countries, promoting economic integration and cultural interaction. However, it is important to note that this facilitation is not an alternative to the Schengen visa and is a temporary measure applicable only to specific islands. The continuity and expansion of this initiative would further enhance cooperation between the two countries. Additionally, the interest of other Aegean islands in joining the project indicates its potential for future expansion (Kalogirou & Velissariou, 2024).

In light of the examples provided, cross-border agreements, projects, and conferences have been organized between Türkiye and Greece in the field of tourism. These efforts are significant for the development of tourism in both countries.

Economic and Political Factors Affecting Tourism

In 2023, the travel and tourism sector contributed 9.1% to global GDP and generated 1.7 trillion dollars of revenue from tourism worldwide. This ratio shows the size and economic impact of tourism worldwide (World Travel Tourism Council, 2023).

Cross-border tourism agreements are an indicator of economic and political relations between countries. Türkiye and Greece are significantly affected by tourism agreements in terms of their historical ties and proximity. Economic factors are one of the most important issues in these tourism agreements. According to studies, cross-border agreements reveal that economic factors contribute to the growth of the tourism industry. It also shows that tourism activities generate a high-income increase (Maratovna et al., 2021).

Political situations have a significant impact on cross-border agreements between countries. The characteristics of cross-border crossings in terms of tourism are among the main factors that determine agreements in terms of border security and diplomatic relations. How the political relations between Türkiye and Greece affect tourism mobility is an issue that needs to be emphasized. The political situation of the two countries and their mutual agreement on tourism issues affect the viability and sustainability of cross-border agreements (Weidenfeld, 2013).

Cross-border tourism agreements increase cultural interactions and social convergence. Cultural relations between Türkiye and Greece have a decisive influence on the progress of these agreements. By strengthening tourism cooperation, the two countries can develop economically, politically, and culturally. Expanding these partnerships will further deepen relations by increasing mutual understanding.

Challenges and Opportunities in Cross-Border Tourism

Although cross-border cooperation is important in strengthening economic, social, and political ties between neighboring countries, Türkiye and Greece have not developed comprehensive cooperation due to historical conflicts and political tensions. In particular, cross-border tourism projects require building trust between the two countries, a process made difficult by past hostilities. In addition, deficiencies in tourism infrastructure, transportation, and communication inadequacies are among the important factors limiting touristic mobility (Rumelili, 2005).

Historical enmity and nationalist discourses affect national identity formation and the public's perspective on cross-border cooperation (Güven, 2018). Educational systems complicate reconciliation efforts by positioning the other side as rivals, while political leaders use historical disputes to gain popular support during times of domestic political crisis (Aydın & Ifantis, 2004). This hinders the development of cross-border cooperation not only in tourism but also in areas such as trade, security, and environmental management.

In addition, Türkiye's non-membership in the European Union (EU) and exclusion from the Schengen Area limit the feasibility of cross-border projects. Furthermore, security concerns and ongoing territorial disputes between the two countries complicate the establishment of joint border management mechanisms. Instead of fostering cross-border cooperation, Türkiye and Greece continue to prioritize national security interests, adopting policies that restrict regional integration.

Cross-Border Tourism Agreements That Can Be Made Between the Two Countries

Türkiye and Greece, with their historical, cultural, and geographical ties, have great potential for cooperation in the field of tourism. Despite the competition-oriented tourism policies of the past, cross-border tourism agreements and joint ventures offer an opportunity to strengthen economic and cultural relations between the two countries. Sustainable tourism models, joint marketing strategies, and regional cooperation can increase tourism mobility and contribute to sectoral growth (Tosun et al., 2005).

Gastronomy and cultural heritage tourism are among the most important areas that can promote cross-border cooperation. The similarities between Turkish and Greek cuisine can be utilized for gastronomy tourism through joint events and restaurant investments (Temizkan and Aktepe, 2023). In addition, the preservation of Ottoman and Byzantine heritage and the promotion of destinations on the UNESCO World Heritage List through joint routes will contribute to the development of cultural tourism. A more integrated cruise route between Turkish ports such as Istanbul, Kuşadası and Bodrum and Greek ports such as Piraeus, Thessaloniki, and Crete and the integration of maritime transportation is another important factor that can strengthen the port economy of both countries (Kapsomenakis et al., 2023). Alternative tourism types, such as health tourism, can be supported by the two countries' joint efforts and taken to a different level.

Although there are many areas for joint tourism activities between the two countries, specific problems also stand out. The difference in visa processes between the countries is one of the biggest obstacles to tourism. The historical conflicts and political developments between Türkiye and Greece may prevent possible cooperation in the tourism sector. In addition, the fact that the two countries develop strategies based on competition instead of making joint decisions in the Aegean Region causes difficulties in tourism activities (Naniopoulos et al., 2016).

Results

There have been many cross-border cooperation agreements between Türkiye and Greece in the field of tourism from the past to the present. This has strengthened the economic, cultural, and political relations between the two countries. However, due to historical tensions, political conditions, and bureaucratic obstacles, cross-border agreements between the two countries have not reached the desired level. Especially in the coming years, gastronomy tourism, health tourism, and special studies on the Aegean Sea can improve the tourism potential of the two countries. However, for cross-border cooperation to be permanent, the two countries need to solve visa problems, migration crises, and fundamental transportation problems.

However, factors such as political instability, maritime jurisdiction disputes in the Aegean Sea, and the Cyprus issue sometimes hamper tourism cooperation between the two countries. Therefore, it is crucial to treat tourism as an area of collaboration independent of political developments and to develop long-term strategic partnerships in the sector. Türkiye and Greece should build on existing cross-border cooperation and use tourism for economic gain and as a tool to improve cultural and diplomatic relations. Sustainable and mutually beneficial tourism agreements can strengthen ties between the two countries and promote regional stability.

References

Athanasopulos, H. (2001). *Greece, Turkey and the Aegean Sea: A case study in international law*. McFarland.

Aydin, M., & Ifantis, K. (Eds.). (2004). *Turkish-Greek relations: the security dilemma in the Aegean*. Routledge.

Ballis, A., Paravantis, J. A., & Moschovou, T. (2018, July). Assessing the tourism potential of the Greek islands of South Aegean. In *2018 9th International Conference on Information, Intelligence, Systems and Applications (IISA)* (pp. 1–7). IEEE.

Constantinides, S. (1996). Turkey: the emergence of a new foreign policy, the neo-Ottoman imperial model. *Journal of Political & Military Sociology*, 323–334. https://www.jstor.org/stable/45294324

Dış Ekonomik İlişkiler Kurulu. (2024). *Türkiye-Yunanistan 6. Dönem karma ekonomik komisyon (kek) toplantısı vesilesiyle Türkiye Yunanistan iş forumu.* Retrieved from 07.02.2025 https://www.deik.org.tr/etkinlikler-turkiye-yunanistan-6-donem-karma-ekonomik-komisyon-kek-toplantisi-vesilesiyle-turkiye-yunanistan-is-forumu

Dış İşleri Bakanlığı. (2024). *Greece* [Yunanistan]. Retrieved from 07.02.2025 https://www.mfa.gov.tr/yunanistan.tr.mfa

Dogru-Dastan, H., & Tütüncü, Ö. (2024). Impacts of the terrorist attacks and political incidents in major cities on tourism industry: Evidence from the tourism managers in Turkey. *Cities*, *153*, 105255. https://doi.org/10.1016/j.cities.2024.105255

Ege Postası. (2016). *Here are the details of that agreement* [İşte o anlaşmanın detayları]. Retrieved from 07.02.2025 https://www.egepostasi.com/haber/Iste-o-anlasmanin-detaylari/127635

Ege Telgraf. (2024). İzmir, Selanik ile ticaret ve turizmde yeni bir dönem başlatıyor! Retrieved from 07.02.2025 https://www.egetelgraf.com/izmir-selanik-ile-ticaret-ve-turizmde-yeni-bir-donem-baslatiyor

Euronews (2024). *Yunan adalarına giden Türk turistlerin sayısı üçe katladı; diğer adalar da aynı uygulamayı istiyor* Retrieved from 07.02.2025 https://tr.euronews.com/gezi/2024/04/13/yunan-adalarina-giden-turk-turistlerin-sayisi-uce-katladi-diger-adalar-da-ayni-uygulamayi-.

Gelbman, A., & Timothy, D. J. (2011). Border complexity, tourism, and international exclaves: A case study. *Annals of Tourism Research*, *38*(1), 110–131.

İzmir Chamber of Commerce (İZTO). (2022). *Years of effort paid off: Thessaloniki-Izmir ferry services start tomorrow* [Yıllar süren çaba sonuç verdi: Selanik-İzmir feribot seferleri yarın başlıyor]. Retrieved from 07.02.2025 https://izto.org.tr/tr/e-oda/haberlerimiz/yillar-suren-caba-sonuc-verdi-selanik-izmir-feribot-seferleri-yarin-basliyor

Kalogirou, C., & Velissariou, E. (2024, December). Promoting Cross Border Tourism in Europe. The case of Greece–Turkey. In *Proceedings of the International Conference on Contemporary Marketing Issues*.

Kapsomenakis, J., Douvis, C., Poupkou, A., Zerefos, S., Solomos, S., Stavraka, T., ... & Zerefos, C. (2023). Climate change threatens cultural and natural heritage at UNESCO sites in the Mediterranean. *Environment, Development and Sustainability*, *25*(12), 14519–14544.

Keridis, D. (1999). Political Culture and Foreign Policy: Greek-Turkish Relations in the Era of European Integration and Globalization. *A NATO fellowship final report*, 97–99.

Koberidze, G. (2023). Analyzing the Aegean Sea Crisis between Greece and Turkey and Its Implications for Regional Stability. *Journal of Politics and Democratization*, *5*(5), 43–53.

Makkonen, T., Williams, A. M., Weidenfeld, A., & Kaisto, V. (2018). Cross-border knowledge transfer and innovation in the European neighbourhood: Tourism cooperation at the Finnish-Russian border. *Tourism Management*, *68*, 140–151.

Mann, S. (2001). *The Greek-Turkish dispute in the Aegean Sea: its ramifications for NATO and the prospects for resolution* (Doctoral dissertation, Monterey, California. Naval Postgraduate School). http://hdl.handle.net/10945/9725

Maratovna, M. A., Tatyana, I., Bekzadaevich, K. M., Yessengalievna, Y. A., Dinara, Z., Kalievna, Z. S., & Shakizada, N. (2021). Management of sustainable development of tourism in cross-border territories. *Academy of Strategic Management Journal*, *20*, 1–9.

March, R., & Wilkinson, I. (2009). Conceptual tools for evaluating tourism partnerships. *Tourism Management*, *30*(3), 455–462.

Naniopoulos, A., Tsalis, P., & Nalmpantis, D. (2016). An effort to develop accessible tourism in Greece and Turkey: the MEDRA project approach. *Journal of Tourism Futures*, *2*(1), 56–70.

Nunkoo, R. (2017). Governance and sustainable tourism: What is the role of trust, power and social capital?. *Journal of Destination Marketing & Management*, *6*(4), 277–285. https://doi.org/10.1016/j.jdmm.2017.10.003

Ohtamış, E. (2008). *Avrupa Birliği sınır ötesi işbirliği programları yönetim yapıları ve Türkiye uygulamaları*. DPT Bölgesel Gelişme ve Yapısal Uyum Genel Müdürlüğü.

Öztürkmen, A. (2002). From Constantinople to Istanbul: Two sources on the historical folklore of a city. *Asian Folklore Studies*, 271–294. https://doi.org/10.2307/1178974

Rumelili, B. (2005). The European Union and cultural change in Greek-Turkish relations. *The European Union and Border Conflicts*, *17*, 1–31.

Temizkan, S. P. & Aktepe, B. (2023). Türkiye ve Yunanistan'daki Michelin yıldızlı restoranların internet sitelerinin incelenmesi. *Journal of Gastronomy Hospitality and Travel* 6(2)787–800. 10.33083/joghat.2023.301

Tosun, C. (2000). Limits to community participation in the tourism development process in developing countries. *Tourism Management*, *21*(6), 613–633. https://doi.org/10.1016/S0261-5177(00)00009-1

Tosun, C., Timothy, D. J., Parpairis, A., & MacDonald, D. (2005). Cross-border cooperation in tourism marketing growth strategies. *Journal of Travel & Tourism Marketing*, *18*(1), 5–23. https://doi.org/10.1300/J073v18n01_02

Tsarouhas, D., & Yazgan, N. (2018). Trade, non-state actors and conflict: evidence from Greece and Turkey. *Cambridge Review of International Affairs*, *31*(3–4), 291–313. https://doi.org/10.1080/09557571.2018.1501346

Ulusoy, O., & Battjes, H. (2017). Situation of readmitted migrants and refugees from Greece to Turkey under the EU-Turkey statement. *VU Amsterdam Migration Law Series*, *15*.

Var, T., & Ap, J. (2013). Tourism and world peace. In *Global tourism* (pp. 63–76). Routledge. 9780080507446

Weidenfeld, A. (2013). Tourism and cross border regional innovation systems. *Annals of Tourism Research*, *42*, 191–213.

Yıldırım, U., Akin, S., Gül, Z., & Çiçek, Y. (2016). A Review on EU Transportation Projects within the Case of Turkey for European Union Regional Development Policy. *International Journal of Sustainable Development*, *5*(1), 1–11.

Zaei, M. E., & Zaei, M. E. (2013). The impacts of tourism industry on host community. *European journal of tourism hospitality and research*, *1*(2), 12–21.

Talent Management in Technology-Oriented Service Businesses

Ela OĞAN[1]

Introduction

The perspective on concepts such as information, people, products and production has changed with the help of technological developments. Despite the increasing intensive use of technology in production, innovation, creativity, talent, agility and the workforce potential that can bring these to the fore have gained significance. Social life is increasingly supported by new-generation technologies such as artificial intelligence, the Internet of Things, big data, and cyber security while cloud computing, and business processes integrated with technology are on the agenda in business life. The digital age requires holistic development and success with technology-human-process collaboration. This also provides a competitive advantage among businesses. Besides having a good technological infrastructure, accessing and maintaining human resources skilled in these technologies has become extremely critical. In other words, emphasis is placed on the identification, selection, acquisition, and sustainability of business talent to enhance performance and contribute to sustainable success in today's conditions. The best way to control this situation in business is through talent management activities.

Service businesses are different in terms of their fields of activity and characteristics. Therefore, they should have their own qualified human resource capital and ensure their sustainability. This study discusses the elements of service businesses that are gaining value today and the processes related to talent management. In this context, topics such as change in human resources management, technology-oriented talent management practices and talent management activities in service businesses are included.

1 Assistant Professor, Artvin Coruh University, Arhavi Vocational School, elaogan40@gmail.com, ORCID ID: 0000-0001-5491-771X

From Human Resources Management to Talent Management

The advancements following the Industry 4.0 revolution have highlighted the significance of foresight and strategic planning across almost all sectors. There have been differences in the perceptions of value on people, environment and products, and the value given to people in every field has increased compared to others (Matt et al., 2015). Although technology has come to the fore, human resources are the most important element that can bring the environment and products to the best condition. The strategic management approach argues that opportunities and threats can be determined by analyzing the environment. Thus, the negativities or opportunities in the environment can be foreseen in advance and the necessary actions can be taken into account. This situation gives the business a competitive advantage (Porter, 1985). Penrose (1956) suggests that a sustainable competitive advantage is determined by the resources of a business. He draws attention to more holistic business processes with the resource-based approach. Following the subsequent developments, the understanding of human resources management has been addressed strategically. Then, efforts to obtain and maintain talented human resources which create a competitive advantage for businesses came to the fore (Bal, 2010: 272).

According to Silzer & Dowell (2010: 18), talent management is among the future goals of the business and has strategic value. Within the framework of this vision, it is necessary to have a good talent management capability to identify, acquire, adapt and maintain the required talents. The concept of "talent management," expressed as the new vision of the field of human resources management, is being carefully implemented in all kinds of organizations (Atlı, 2010; Çevik, 2014; Öktan, 2019). The concept of talent management refers to the process of recruiting, directing, evaluating, and maintaining the human resource capital of the business (Khatri et al., 2010: 39). Talent management has an extremely comprehensive nature in business and it includes organizational development, learning and innovation. Therefore it fulfils functions such as planning, recruiting, career-related guidance and support of the human resources (Yalçın & Kalmuk, 2025).

Talent management practices are strategic and organization-specific (Rožman et al., 2023). Businesses are to create a talent management process depending on their fields of activities, their core competencies, organizational structures and other resources (Guerra et al., 2023). Within the scope of this process, the strategic human resource capital requirements of the business are determined first. Determining the qualifications required for the human resource talent is essential at this stage. Then, it is required to determine in which tasks the talents are taken place in the business. The stages related to the adaptation, training, development and continuity of human resources are to be planned (Karcıoğlu and Atasever, 2024: 2). Among the characteristics of talented employees having the ability to adapt, being able to work flexibly, having high motivation, having a broad perspective, being able to communicate effectively, having a vision, constantly learning, and being able to see and evaluate opportunities are included. (Yalçın & Klamuk, 2025).

Technology-Focused Talent Management

Digitalization has made the necessary transformation in the human resources field like in every field. The operating mission has changed with the introduction of artificial intelligence, the Internet of Things, big data and autonomous systems, which are referred to as new-generation technologies, in business processes. Physical labour has been replaced by mental labour. Although working life is technology-oriented, the presence of human creative thinking has become a critical success factor for more innovation (Oğan & Çetiner, 2024). The important elements of technology-oriented and dynamic talent management are listed below.

Digital Systems and Automation Systems

Adapting technology to business systems and processes has become a basic requirement (Oğan & Wolff, 2022). In other words, technological infrastructure is a necessity for future goals. Systems which do not adapt themselves to technology or resist it fail. Therefore, the tasks expected from the employee have changed and it is necessary for employees to adapt to digitalization. Human resources management must determine the skills needed by determining the new tasks and positions in a question. At this stage, it can perform stages such as identifying, selecting and starting talents with automation technologies. In this way, effectiveness and efficiency are achieved in business and transactions.

Flexible Working Models

Information and communication technologies have introduced mobile and remote working. Thus employees do not have to be in the office environments physically. It has become possible to work independently of place and time. Flexible jobs can be designed in businesses and can be turned into opportunities. Flexibility, trust, technology use and coordination ability come to the fore in the management of human resources. It can be said that the ability of managers and employees to communicate with each other instantly and provide feedback increases transparency, trust and interaction in the organization.

Big Data Analysis Applications

Big data is used in the selection of human resources and other processes. At the same time, effective and efficient results can be obtained by analyzing the information obtained. Depending on the human resource needs of the business, big data can be used in the search, identification of talented employees, and the evaluation of their performance and motivation.

Global Talent Management and Diversity

Digitalization has eliminated borders all over the world. Just as trade is carried out on a global scale, businesses also seek and obtain talented human resources within global borders (Bonneton et al., 2019). In addition, being able to select the workforce from a wider area increases human diversity in organizations. This situation creates important opportunities for business talent.

Digital Culture

Digitalized business processes reveal an organizational culture supported by digital technology and digital talent. In order to achieve digital technology-oriented business processes in the organization, it is crucial to ensure human-technology-process adaptation (Oğan & Wolff, 2022). Thus, committed and happy employees are created with an organizational culture adapted to technological work processes.

Talent Development and Learning Technologies

Mobile technologies have made access to information quite practical. Thanks to many digital applications, libraries, and various databases, education, and seminar activities have become accessible. Interactive learning technologies provide opportunities for employees who prefers acquiring different skills or constantly improving their skills.

Artificial Intelligence (AI) Applications

In business life, artificial intelligence applications greatly relieve the physical burden of the employee. In addition, AI supports cognitive development, creativity and innovation (Oğan, 2024). Artificial intelligence tools are used in the development of business talents as well as in their selection, recruitment and adaptation processes. Thanks to health-related artificial intelligence tools, various data are obtained about the psychological and physical health of the employee. In addition to providing feedback on the situation, it allows for health-enhancing actions. Thus, it contributes to increasing the motivation levels of the employees and creating a peaceful environment in the workplace.

Service Businesses and Talent Management

Service Businesses and Their Characteristics

Service businesses unlike manufacturing businesses, do not produce products or finished goods. Such businesses offer various services to customers in any field. Businesses focus on creating value for their customers with the services they offer, mostly through knowledge, skills and experience (Aksu et al., 2024). Services are intangible and synchronous. Additionally, it cannot be stored and is not standard. The characteristics of service businesses are as follows (Uyar and Akar, 2024):

Service Businesses Are Intangible: Services are not physical. Therefore, services cannot be perceived with the 5 senses. For example, education service, guidance service, or the experience of accommodation service received at a hotel are intangible.

Simultaneous Production and Consumption: Services are generally consumed at the same time they are produced. For example, food service received from a restaurant is instantaneous. This shows that the people who receive the service and the people who provide the service interact.

Non-Standard: Services are specific to the person. Therefore it is difficult to receive or provide the same quality service from everyone at all times. Thus, service businesses need to make continuous efforts to improve quality constantly.

Cannot be stored: Since services are produced instantly, they cannot be stored in order to be presented later. Plans regarding the presentation of services must be managed realistically parallel to its capacity.

Service businesses provide services in various areas in order to meet customer needs. The areas that service businesses serve can be listed as Marketing/Sales Services, Health Services, Education Services, Banking/Financial Services, Transportation Services, Tourism/Accommodation Services, Consultancy Services, and Entertainment Services (Uyar and Akar, 2024).

Service businesses differ from businesses that produce concrete products in terms of their fields of activity and characteristics. For this reason, it can be said that the factors affecting the success of service businesses are more concentrated in the field of needs and expectations specific to the person. The quality expectation regarding the education and skills of employees for the activity in which the person wants to receive service is considered very important. In addition to quality, the price expectation of the customer regarding the service they receive is important. The customer wants to get more than the price he pays for the service. The place and time conditions where the customer receives the service are at the forefront. Since customer needs and desires change new service expectations are constantly in question. All of these factors affect customer satisfaction and create positive or negative situations regarding the continuity of business preference. Therefore, all the capabilities of service businesses are focused on maintaining that quality.

Talent Management in Service Businesses

In service businesses, talent management can be addressed within the scope of the management of human resources, infrastructure resources and core talents (Alayoğlu, 2010). With digitalization, technology-supported services are integrated into talent management activities. While management of online services, e-commerce, mobile applications and artificial intelligence-supported activities come to the fore, management of human resources with the ability to use digital technologies and create innovation comes into question. In this context, talent management activities integrated with technology play a critical role in service businesses improving customer experience and increasing process efficiency (Mesci & Biçer, 2024) since the basis of talent management is employee talent compatible with organizational goals (Aytaç, 2014: 4). It can be said that all these talents together develop the core talent of the business. The talent management process refers to obtaining, developing and using the right talents effectively in order to achieve organizational goals (Bonneton et al., 2019). This process is one of the important stages of human resources management. It includes strategic practices to ensure the development of existing talents in the organization as well as the participation of talents that are likely to be needed in the future (Kumar, 2022). Important elements of talent management can be listed as; talent acquisition, training and development, performance management, career planning, talent protection, analysis and evaluation (Yıldız & Esmer, 2023).

Talent Acquisition: It can be said that technological developments are the main factors determining the need for talent in businesses. Determining the talents needed in the long term within the framework of strategic goals is critical in terms of sustainable activities. For this reason, it is necessary to hire talents who have the skills to achieve business goals (Kock & Burke, 2008: 464). Businesses develop various strategies in this process. Methods such as job postings, career fairs, social media, and big data analysis can be used in the selection of new talents (Sever, 2024).

In talent selection for the service sector, employees are expected to have advanced communication, empathy, emotional intelligence and other social skills as well as the ability to apply digital technologies. It is an important skill for employees to integrate their technical skills with their personal skills (Sever, 2024).

Learning and Development: Competition conditions make continuous learning and development a necessity. This situation is valid for all jobs, businesses, processes and resources. If human resources do not develop their skills in accordance with their current conditions, they will not be preferred or will not be able to ensure continuity in the job. Employees in service businesses need to be able to use new technology well and adapt to innovations. Thus developing leadership ability is considered extremely important (Betchoo, 2014; Haromszeki, 2024). Similarly, business managers need to offer training and development opportunities to employees (Postacı & Polat, 2017).

Performance Management: The performance of business capabilities is monitored and evaluated (Kumar, 2022). Feedback is received regarding whether business goals have been achieved or not. In line with this feedback, talent selection and development plans are repeated. At this stage, successful talents are required to be rewarded.

Career Planning: As a result of feedback on the achievement of goals, the career planning phase is initiated from an individual and organizational perspective. Organizational opportunities increase employee loyalty and ensure continuity (Ekhsan et al., 2023). This is a method of retaining talent within the company.

Talent Retention: In order to retain talents that add value to the company, it is necessary to implement certain strategies. Achievement awards, flexible working conditions and career development opportunities can be accepted as examples of these strategies. Various events and meetings can be held to increase organizational commitment. These types of strategies increase motivation and direct talents to development (Yıldız & Esmer, 2023).

Monitoring and Analytics: In order to evaluate the attainability of the goal, the performance of business talents is monitored and adaptation to organizational change is analyzed.

Conclusion

Customer satisfaction is prioritized in service businesses. In the digital age, customer expectations are shaped in line with the developing technology. This situation requires business capabilities to become more digitalized. Technology-supported services provide practical, agile, flexible and fast solutions. Additionally, the customer expects more quality. Businesses are developing their capabilities to meet these expectations. Human resource talent in particular, as well as the infrastructure that supports this talent, basic activities and business process management is gaining importance.

Talent management in service businesses performs important functions in terms of identifying, supporting, developing and protecting the skills which are prioritized in digitalized working conditions. In this context, the "value" element that affects organizational success and provides competitive advantage emerges. The ability to create this "value" in the organization is the human resource capital capability. For this reason, there has been a recent focus on talent management regarding human resources.

Today's competitive environment is directed towards talents that can achieve business goals and add value to the organization. Human abilities are critical in order to design the future in the best and most beneficial way. It is known that the main key to overcoming competition is to innovate. It is necessary to evaluate and improve the innovation performance of human resource capital. Studies show that talent management activities affect innovation performance (Örücü and Akyüz, 2018: 20). Therefore, it can be said that talented human resource capital is the main element that creates a significant advantage in competitive conditions (Chen and Huang, 2009). The ability to correctly identify the talents needed by the business increases efficiency and effectiveness (Rauter et al., 2019). After determining the business talents, innovation is to be encouraged to ensure continuity. Therefore, human resource talent is required to be supported by training and development opportunities. This is also necessary for service sustainability and quality.

Talent management is an important part of strategic human resources management that adds value to the business. It can produce smart, practical and efficient solutions with technology-focused applications. It aims to evaluate talented human resources at the most appropriate time and position for the business goals.

References

Alayoğlu, N. (2010). İnsan kaynakları yönetiminde yeni dönem: Yetenek yönetimi. *Ticaret ve Turizm Eğitim Fakültesi Dergisi, 1*, 68–97.

Aksu, N., Ata, N.S., & Oğrak, A. (2024). Yetenek yönetiminin otel çalışanlarının performansına etkisi: Van ilinde bir araştırma. *Sosyal, Beşeri ve İdari Bilimler Dergisi*, *7*(2), 117–133.

Atlı, D. (2010). *İnsan kaynakları yönetiminin yeni vizyonu yetenek yönetimi ve basın işletmelerinde bir uygulama* (Doctoral dissertation), Marmara Universitesi, Turkiye.

Aytaç, T. (2014). Okullarda bütünleştirilmiş yetenek yönetimi modeli: Öğretmen görüşleri. *Eğitim Bilimleri Dergisi*, *39*, 1–24.

Bal, Y. (2010). Rekabet avantaji yaratmada kaynak temelli yaklaşim bağlaminda insan kaynaklarinin rolü. *Sosyal Ekonomik Araştırmalar Dergisi*, *10*(20), 267–278.

Betchoo, N.K. (2014). The need for effective leadership in talent management in Mauritius. *International Letters of Social and Humanistic Sciences*, (16), 39–48.

Bonneton, D., Schworm, S.K., Festing, M., & Muratbekova-Touron, M. (2022). Do global talent management programs help to retain talent? A career-related framework. *The International Journal of Human Resource Management*, *33*(2), 203–238.

Chen, C.J., & Huang, J.W. (2009). Strategic human resource practices and innovation performance: The mediating role of knowledge management capacity. *Journal of Business Research, 62*(1), 104–114.

Çevik, B. (2014). *Turizm sektöründe yetenek yönetimi uygulamaları: Antalya yöresinde beş yıldızlı otellerde bir araştırma.* (Master's thesis), Akdeniz University, Turkiye.
Ekhsan, M., Parashakti, R.D., & Perkasa, D.H. (2023). The impact of talent management on employee performance mediated by employee engagement. *East Asian Journal of Multidisciplinary Research*, *2*(4), 1821–1834.
Guerra, J. M. M., Danvila-del-Valle, I., & Méndez-Suárez, M. (2023). The impact of digital transformation on talent management. *Technological Forecasting and Social Change*, *188*, 122291.
Haromszeki, Ł. (2024). The effects of leadership talent management on the company performance results. *Zeszyty Naukowe Politechniki Śląskiej. Organizacja i Zarządzanie*, (191).
Karcıoğlu, F., & Atasever, S. B. (2024). Üniversitelerde stratejik insan kaynakları yönetiminde yetenek yönetimi uygulama düzeyi ile bireysel kariyer yönetimi ilişkisi: Atatürk Üniversitesinde Bir Uygulama. *Academic Knowledge*, 7(1), 1–18.
Khatri, P., Gupta, S., Gulati, K., & Chauhan, S. (2010). Talent management in HR. *Journal of Management and Strategy*, *1*(1), 39–46.
Kock, R., & Burke, M. (2008), Managing talent in the South African public service. *Public Personnel Management*, *37*(4), 457–470.
Kumar, S. (2022). The impact of talent management practices on employee turnover and retention intentions. *Global Business and Organizational Excellence*, *41*(2), 21–34.
Matt, C., Hess, T., & Benlian, A. (2015). Digital transformation strategies. *Business & Information Systems Engineering*, *57*, 339–343.
Oğan, E., & Wolff, R.A. (2022). Fortune 500 işletmelerinin dijital dönüşüm performansında dijital kültürün aracılık rolü. *Fiscaoeconomia*, *6*(3), 1282–1307.
Oğan, E., & Çetiner, N. (2024). Yönetici asistanının dijital yetenekleri ve kariyer sürdürülebilirliğine etkisi. *Pamukkale Üniversitesi Sosyal Bilimler Enstitüsü Dergisi*, (64), 117–139.
Oğan, E. (2024). Artificial intelligence in restaurant businesses: A systematic review on service robots. *Worldwide Hospitality and Tourism Themes*, *16*(2), 150–162.
Öktan, S. (2019). *İnsan kaynakları yönetiminde yeni bir yaklaşım olarak yetenek yönetimi: İşletmelerin yetenek yönetimi birimlerine ilişkin keşifsel bir araştırma.* (Master's thesis), Dokuz Eylul Universitesi, Turkiye.

Örücü, E., & Akyüz, A.N. (2018). Algılanan yetenek yönetimi uygulamalarının yenilik performansı üzerine etkisi. *Dumlupınar Üniversitesi Sosyal Bilimler Dergisi*, (57), 20–36.
Porter, M.E. (1985). Technology and competitive advantage. *Journal of Business Strategy*, *5*(3), 60–78.
Postacı, B., & Polat, D.D. (2017). Liderlik tarzı ile yetenek yönetimi arasındaki ilişkinin analizi: Beş yıldızlı otel işletmelerinde bir araştırma. *Researcher*, *5*(1), 21–37.
Rauter, R., Globocnik, D., Perl-Vorbach, E., & Baumgartner, R.J. (2019). Open innovation and its effects on economic and sustainability innovation performance. *Journal of Innovation & Knowledge*, *4*(4), 226–233.
Rožman, M., Tominc, P., & Štrukelj, T. (2023). Competitiveness through development of strategic talent management and agile management ecosystems. *Global Journal of Flexible Systems Management*, *24*(3), 373–393.
Sever, M.M. (2024). The relations between digitalization, service innovation and service value creation capability: A model adaption in the service industry. *Verimlilik Dergisi*, *1*(58), 61–72.
Silzer, R., & Dowell, B.E. (2010). Strategic talent management matters. *Strategy-Driven Talent Management: A Leadership Imperative*, 3–72.
Uyar, E., & Akar, V. (2023). Hizmet İşletmelerinde Hizmet Hatası ve Telafisi. Kayabaşı, A. (Ed.), *Güncel İşletmecilik Araştırmaları-4*, (133–144). Ankara: Eğitim Yayınevi.
Yalçın, A., & Kalmuk, G. (2025). Z kuşağında yetenek yönetim algısının iş tatmini ve işten ayrılma niyeti üzerine etkisi. *Yönetim Bilimleri Dergisi*, *23*(55), 1–32.
Yıldız, R.O., & Esmer, S. (2023). Talent management strategies and functions: A systematic review. *Industrial and Commercial Training*, *55*(1), 93–111.

Awareness of Activities Conducted in the Context of Event Tourism and Their Importance for the Hotel Sector

Mevlüt UYAR[1], Emrullah CANSU[2]

Introduction

According to the World Tourism Organization, tourism is defined as the activities that people carry out for various purposes such as work, health, leisure and sports from their place of residence to a different place (UNWTO, 2025). Looking at its history, tourism dates back to 4000 BC. Tourism, which started with the need for accommodation and purchasing materials while people travel from one destination to another, has reached the dimension of intercontinental travel and getting to know different cultures in a short time (Çakmak, 2021).

In line with these travels, the trade routes that have been used extensively until today have a long history and date back to Central Asia. In the past, world trade between eastern and western cultures was carried out through two different trade routes. One of these routes is known as the Silk Road and the other as the Spice Road (Yener, 2018). The history of the famous trade route called the "Silk Road" started in 138 BC with the Chinese traveler Chang Chien, but it is estimated that its history actually goes back to the Bronze Age (Yılmaz, 2017). This road starts in China, crosses Asia and extends to Europe via Anatolia and the Mediterranean (Günel, 2010). The spice route was one of the safest routes used for thousands of years and was used to transport hundreds of kinds of spice products from the East to the West for commercial purposes. It was especially aimed to take the products from countries with abundant spice flavors such as India and Egypt to western countries such as Europe (Türkhan, 2025). These two important trade routes have an important place in the beginning of tourism.

1 PhD., Alanya Alaaddin Keykubat University, ALTSO Tourism Vocational School, Hotel, Restaurant and Catering Services Department, mevlut.uyar@alanya.edu.tr

2 PhD., Alanya Alaaddin Keykubat University, ALTSO Tourism Vocational School, Travel, Tourism and Entertainment Services Department, emrullahcansu58@gmail.com

Post-Covid-19 tourism is trying to establish a balance within itself and catch the momentum of 2019 and before. For this, hotels, which are among the most important stakeholders in the tourism sector, are used intensively. Especially with events such as sports, etc. organized in hotels, it is aimed to create a driving force such as "the resumption of tourism" (Borovcanin, Cuk, Lesjak, & Juvan, 2020). Within these intensities, hotels that satisfy their customers can gain an advantageous position in subsequent tourism data. For this, they need to increase their existing services and quality (Avcı, 2015). They can do this by organizing activities such as sporting events. Because tourists use accommodation places such as hotels not only for sleeping and eating but also for participating in activities and getting to know different types of entertainment (Maglovska, 2020: 6). Therefore, in order to compete with the global market, they should have the ability to constantly change their existing services. Because the sector in which it exists is changing day by day and keeping up with innovations. For this reason, stakeholders who cannot adapt to the sector increase the risk of leaving the customers' preference lists over time (Güripek, 2022).

Although accommodation establishments are temporary accommodation places, they are among the most important tourism components that directly affect people's vacation processes. For this reason, there is a constant control and change in the service (Özdemir & Akpınar, 2002). Because the main purpose of hotels is to maintain the purpose of existence by providing customer satisfaction and profit (Davras & Gülmez, 2013).

Today, due to the economic effects of events, the demand for creating these events is increasing. For this reason, the number of events with different criteria and criteria is increasing in destinations and accommodations. These events become an attraction for tourists and contribute to the development of tourism (Jacob, Payeras, Florido & Domínguez-Quintero, 2024: 1). Resort hotels are one of the accommodation businesses that offer tourists the most activity opportunities along with accommodation.

Resort Hotel Concept

In ancient Greek and Roman times, holiday settlements, derived from bathing and spa culture, initially emerged as accommodation units built for recreation and entertainment in thermal centers. Since then, it has developed and progressed until today (Yener, 2018). Especially after World War II, the transfer of existing technologies to tourism has revealed the importance of this sector. The use of warplanes for touristic purposes in terms of transportation can be given as an example. In this sense, the sea, sand, sun trio, which emerged from mass tourism, is one of the promotion methods used especially by hotels, where they try to meet more requirements to keep tourists within the hotel boundaries (Demir & Demir, 2015).

Resort hotels can be defined in many ways, but in short, they are defined as "accommodation areas that offer a wide range of tourism activities and services, especially organized for recreation, rest, health and experience, and include relatively all kinds of facilities and activities" (Özdemir, 2006). In resort hotels, the services offered must be of a certain quality and standard. For this reason, these hotels have accommodation, recreation and cultural activity units, health facilities, conference and meeting areas of different sizes, cinemas, casinos, night clubs, theaters and entertainment facilities, sports halls, shopping centers, children's playgrounds and child care units, SPA centers, restaurants with international cuisines, hairdressers, beauty centers and dry cleaning services (Güncan, 2022).

Today, due to the increasing competition, resort hotels mostly use the all-inclusive system. This system is a sales policy that includes many elements such as accommodation, activities, services, etc. and offers them at a single price. Especially in the last 20 years, it has emerged as a very popular system in Turkey. Among the most intense reasons for this, the organization of customers' demands by a single place is considered to be the most important part of their holiday requests. For this reason, this "all-inclusive" system, which is also used extensively in resort hotels, has a direct or indirect effect on the holiday purchases of customers (Çakıcı & Çetinsöz, 2010). In addition to this system, the "Ultra All Inclusive" system has emerged in recent years. In this system, in addition to providing everything free of charge for 24 hours, food and drinks in the minibar in the rooms in many resort hotels are also included. It is also known that this system has a direct impact on satisfaction. Among the reasons is that the customer does not pay any more fees in the hotel in any way other than the first fee paid (Dağ & Karaatlı, 2020).

Hotels, especially in holiday regions where the sea, sand and sun trio are located, play an active role in the planning processes of tourists by offering fun activities including sporting activities as well as activities such as eating and drinking. These hotels strive to differentiate their services and try to leave a positive impact on customers (Uslu & Aysal, 2021). Satisfaction levels in resort hotels can be analyzed in many ways. In some studies, these levels have been analyzed from 6 different perspectives: physical facilities, comfort, design and environment, professionalism, interest and communication between guests.

In the research conducted by Uşaklı and Öztürk (2021), it was revealed that these 6 levels have a direct or indirect effect on tourists' loyalty. The importance of the concept of "loyalty," which is a very important concept especially for the hotel sector, emerges here. Because just as every business owes its existence to the people who may prefer it later, hotels can be preferred again after those who prefer them are satisfied. One of the places where this loyalty should be provided most intensely is definitely resort hotels. Because the all-inclusive system not only includes many services, but also stands out with its activities. Therefore, it should be constantly updated.

Event Practices and Their Importance for the Hotel Industry

Events have a special place in tourism. Generally, tourists want to participate in different types of activities that are different from the activities that take place in the places where they live outside of situations such as accommodation. Not only foreign tourists but also local people living in the city want these activities to be differentiated (Jung, Draper, Malek, Padron, & Olson, 2024: 1583). The concept of "event tourism," which generally includes these activities, was used in 1987 by the New Zealand Tourist and Publicity Department in their report (Getz & Page, 2016:597; Getz, 1989)

Events in hotels are seen as a stimulating social activity not only by hotel customers but also by local people. Especially pool parties and concerts can be examples of these. Pine and Gilmore's (1998) examine customers' leisure activities in hotels in 4 dimensions. These are categorized as aesthetics, entertainment, escape from reality and education. These dimensions are classified according to the purpose of people's participation in these activities. Participants in special events also have a social impact on each other. Therefore, it creates a social environment directly among customers. This leads to loyalty and customer satisfaction. Hotel events are not only profitable for the hotel itself, but also for the suppliers it uses as markets etc.

A well-organized experience leads to a positive attitude and creates a good memory. Customers' emotional involvement in the experience of a product or service creates a new consumption experience and is embedded in their memory for the future.

Events can serve as a strategic platform where consumers can directly experience the cognitive, emotional, relational and sensory values offered by the product or organization. As a consumer is not only a logical decision-maker but also an emotional human being, they seek enjoyable consumption experiences. The emotional connection with the experiences of intangible hotel products and services tends to influence customers' purchasing decisions (Boo and Park, 2019).

It is known from research that sporting events are directly related to hotel revenues and preference. Resort hotels that want to strengthen economically increase their revenues, especially by giving importance to sports competitions (Borovcanin, Cuk, Lesjak, & Juvan, 2020). Therefore, activities such as sports events are associated with changes in accommodation prices of hotels (Jakar & Binesh, 2024: 237)

When looking into the activity diversity of hotel businesses servicing places where mass tourism activities are prevalent, they can find that more than one activity option is available to guests. These activities, which are shared especially through websites, cover tourists of all ages. Among these, the most intensely offered activities are sports and music activities. Club activities, fitness, yoga, golf, cycling, cultural tours, live music, concerts, massages, hammams, water sports, parties, theme parks, festival celebrations, adventure park activities, animations and shows are examples of all these activities. Special games and animation shows are also organized for children in their gardens. These activities, which are offered free of charge to the majority of resort hotels, are updated every week and made attractive for every tourist. The most intense effect of these activities is the socialization process. Especially for guests from abroad, it is an excellent opportunity to mingle with different cultures and learn about them. Bringing local people and tourists of different nationalities together in these activities increases the socio-cultural levels between people. After the end of the holiday process, since tourists will leave with a happy expression when they return to their countries, it will support the continuous preference of those businesses and increase the customer portfolio with reference, as it will affect the sense of loyalty in the next holiday plan.

In tourism destinations, congress tourism emerges as an important alternative in order to spread tourism to 12 months and to sustain tourism activities in dead seasons. It is known to come from French origin and the meetings held with the participation of managers, delegates, etc. from various countries are called congresses, but it also means "kurultay" (TDK, 2025).

Internationally, congresses, seminars, etc. that organizations or individuals attend to discuss or monitor certain issues are called congress tourism. This type of tourism also has positive aspects such as diversifying the tourism of the country. It provides regional income increase especially in low tourism seasons such as winter. The daily expenditures of people attending the congress affect tourism income through purchasing activities such as touristic expenditures. It is known that especially these people have high levels of education. Therefore, their spending levels will also be higher (Cömert & Azbaz, 2021). Especially the participants of congress tourism, which has an important place in the world, mostly come from universities and the health sector. According to 2022 data, the United States of America is one of the countries where the most congresses are organized. Turkey ranked 31st with 82 internationally organized congresses. When analyzed on a city basis, a total of 3 cities were among the most preferred ones. 58 international congresses were organized in Istanbul, ranked 23rd, 9 in Antalya, ranked 202nd, and 5 in Izmir, ranked 326th (ICCA, 2022). People come together for many reasons such as globalization, networking, and industry meetings for the formation of new markets. For this reason, due to the growth of many sectors, congress tourism is growing at the same rate and becoming widespread all over the world. The most important thing in these congresses, which are gathered for many purposes, is the realization of the business process. However, people usually want to visit the city and country where the congress is held at the end of the congress, whether included in the congress or not. This reason is also known to be an incentive for people to attend congresses (Altunyüzük, 2022). The "All Inclusive" system offered by high-tech hotel businesses is also one of the factors that encourage congress tourism. Because the participants will be able to benefit from many activities and events of the hotel and meet their rest and vacation needs after the congress (Gogitidze, 2021).

International research shows that hotels are mostly preferred for congresses. Therefore, hotels will be most affected by congress tourism. In this context, the proliferation of congresses is directly related to the increase in the congress halls of hotels and the proliferation of congress centers (Delice, 2012). It is an alternative tourism type that is expected to affect concepts such as "image building" in tourism. In this context, creating a positive image is directly related to convention centers and hotels. Leaving a positive image will continue to contribute to the country economically and culturally in addition to the growth of this type of tourism in the sector (Ersun & Aslan, 2009).

Conclusion

Looking at 2024 tourism revenues, it is seen that Turkey generated 61 billion dollars in revenues. 62 million people visited a destination in Turkey in 2024 for many reasons (TUIK 1, 2025). However, according to the statistics for the fourth quarter of 2024 (October-December) obtained by TUIK, 64.9% of tourists who visited Turkey in Chart 1 visited for "sightseeing, entertainment, sportive and cultural activities" (TUIK 2, 2025). This confirms the conclusion

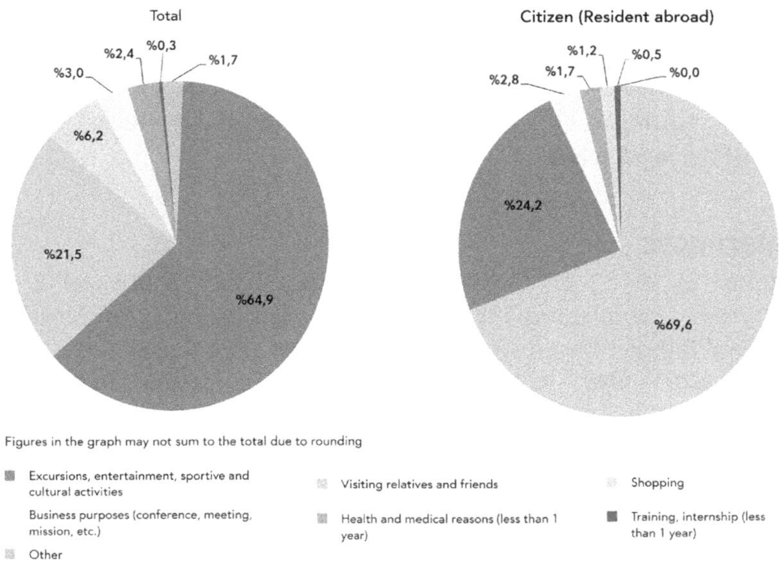

Figure 1: Purpose of Arrival of Tourists Entering the Country in 2024

Source: https://data.tuik.gov.tr/Bulten/Index?p=Turizm-Istatistikleri-IV.-Ceyrek:-Ekim-Aralik-ve-Yillik,-2024-53660#

that the most important purpose of people traveling is entertainment. 2.4% traveled for "Business purposes (conferences, meetings, missions, etc.)." It is one of the undeniable facts that especially visitors to the country extend their vacations for rest after congresses and meetings.

In addition, it is understood from the researches that there is a relationship between the education level of tourists and their consumption preferences. It has also been proven that the consumption preferences of people with higher levels of education increase during vacation (Öner & Cansu, 2019). Therefore, it can be concluded that the expenditures of "business" travelers are higher than other tourists. For this reason, events organized by resort hotels, which are important stakeholders in the tourism sector, have a very important place in the tourism sector. This place is not only limited to satisfying tourists, but also contributes to tourists spending money and local people earning income from it. For this reason, the development of hotels, their technological differences and the opportunities they offer directly affect the preferences of tourists.

Therefore, these revenues will improve the livelihoods and employment opportunities of local people. In addition, investments in the regions will be diversified. In this context, increasing the number of social events organized especially within the scope of resort hotels and the hotel facilities offered will have a direct impact on both the country's tourism and the socio-economic level of the local people.

References

Altunyüzük, A.İ. (2022). *Coğrafi Özellikleri Yönüyle Belek'te (Antalya) Kongre Turizmi*. (Yayımlanmış Yüksek Lisans Tezi). Bursa Uludağ Üniversitesi, Sosyal Bilimler Enstitüsü, Coğrafya Anabilim Dalı, Bursa.

Avcı, N. (2015). Otel İşletmelerinde Yönetici Yetkinliklerinin Analizi: İzmir Örneği. *Anatolia: Turizm Araştırmaları Dergisi*, 26(1), 7–16.

Boo, S., ve Park, E. (2019). Locals' Experiences of Hotel Special Event Packages and their Purchase Intention of Hotel Room Nights. *Event Management*, 23(3), 287–301.

Borovcanin, D., Cuk, I., Lesjak, M., & Juvan, E. (2020). The importance of sport event on hotel performance for restarting tourism after COVID-19. *Societies*, *10*(4), 90.

Borovcanin, D., Cuk, I., Lesjak, M., ve Juvan, E. (2020). The İmportance of Sport Event on Hotel Performance for Restarting Tourism After COVID-19. *Societies Journal*, 10(4), 90.
Cömert, M. ve Azbaz, S. (2021). Covid-19 Pandemisinin Dünya ve Türkiye Kongre Turizmi Gelirlerine Etkisinin Değerlendirilmesi. *Journal of Tourism and Gastronomy Studies*, 9(3), 2259–2271.
Çakıcı, A. C. ve Çetinsöz, B. C. (2010). Otel İşletmelerinde Her Şey Dahil Sistemin Satın Alma Politikalarına Yansımaları Üzerine Bir Araştırma. *Gazi Üniversitesi Ticaret ve Turizm Eğitim Fakültesi Dergisi*, (2), 1–19.
Çakmak, S. (2021). *Beş Yıldızlı Resort Otel Menülerinde Yöresel Ekmeklerin Kullanım Düzeyi: İzmir İli Örneği.* (Yayımlanmış Yüksek Lisans Tezi). Dokuz Eylül Üniversitesi, Sosyal Bilimler Enstitüsü, Gastronomi ve Mutfak Sanatları Anabilim Dalı, İzmir.
Dağ, O. ve Karaatlı, M., (2020). Resort Otellerin Kümeleme Analizi ile İncelenmesi: Antalya ili Örneği. *Süleyman Demirel Üniversitesi Sosyal Bilimler Enstitüsü Dergisi*, (36), 200–232.
Davras, Ö., ve Gülmez, M. (2013). Otel İşletmelerinde Çalışan Memnuniyetine Etki Eden Faktörler: Kemer-Lara-Belek-Side-Alanya Bölgelerinde Bir Çalışma. *Anatolia: Turizm Araştırmaları Dergisi*, 24(2), 167–184.
Delice, K. (2012). *Kongre Turizmi ve Türkiye'deki Toplantılarda Katılımcı Algılarının İncelenmesi; Antalya Yöresinde Bir Uygulama*. (Yayımlanmış Yüksek Lisans Tezi). Akdeniz Üniversitesi, Sosyal Bilimler Enstitüsü, Turizm İşletmeciliği ve Otelcilik Anabilim Dalı, Antalya.
Demir, M. ve Demir, Ş. Ş. (2015). Otel Animasyon Hizmetlerinin Yöneticiler ve Turistlerin Bakış Açısıyla Değerlendirilmesi. International *Journal of Social Sciences and Education Research*, 1(1), 35–48.
Ersun, N ve Aslan, K. (2009). Kongre Turizminin Geliştirilmesinde Kongre ve Ziyaretçi Bürolarının Rolü ve Önemi. *İstanbul Ticaret Üniversitesi Sosyal Bilimler Dergisi*, 8(15), 89–114.
Getz, D. (1989). Special events: Defining the product. *Tourism management*, *10*(2), 125–137.
Getz, D., & Page, S. J. (2016). Progress and prospects for event tourism research. *Tourism management*, *52*, 593–631.
Gogitidze, J. (2021). *Kongre Organizatörlerinin Otel Seçimini Etkileyen Faktörler ve Kongre Hizmetlerinin Değerlendirilmesi Üzerine Bir Çalışma: Antalya Örneği*. (Yayımlanmış Yüksek Lisans Tezi). Akdeniz Üniversitesi, Sosyal Bilimler Enstitüsü, Turizm İşletmeciliği Anabilim Dalı, Antalya.

Güncan, Ö. (2022) *Çocuk Rekreasyonu: Oyun Davranışı Bağlamında Resort Otellerde Bir Araştırma*. (Yayımlanmış Doktora Tezi). Anadolu Üniversitesi, Sosyal Bilimler Enstitüsü, Turizm İşletmeciliği Anabilim Dalı, Eskişehir.
Günel, G. (2010). Anadolu Selçuklu Dönemi'nde Anadolu'da İpek Yolu-Kervansaraylar–Köprüler. *Kebikeç*, 29, 133–146.
Güripek, C, U. (2022). İç Girişimcilik Algısı: Alanya Otelleri Örneği, *İşletme Araştırmaları Dergisi*, 14 (4), 3321–3343.
ICCA, (2022). https://www.iccaworld.org/ (Erişim Tarihi: 26.07.2023).
Jacob, M., Payeras, M., Florido, C., & Domínguez-Quintero, A. (2024). The economic impact of a scientific event on an island tourism destination. *Journal of Policy Research in Tourism, Leisure and Events*, 1–21.
Jakar, G. S., & Binesh, F. (2024). College sport, events tourism and the demand for short-term rentals. *Sport Management Review*, *27*(2), 234–253.
Jung, S., Draper, J., Malek, K., Padron, T. C., & Olson, E. (2024). Bridging Theory and Practice: An Examination of How Event-Tourism Research Aligns With UN Sustainable Development Goals. *Journal of Travel Research*, 00472875241231273.
Öner, S ve Cansu, E. (2019). Yerli Turistlerin Tatil Satın Almadan Önce Tercihlerine Etkisi Olan Faktörlerin Belirlenmesi. *Journal of Tourism Intelligence and Smartness*, *2*(2), 110–129.
Özdemir, E., ve Akpınar, A. T. (2002). Konaklama İşletmelerinde İnsan Kaynakları Yönetimi Çerçevesinde Alanya'daki Otel ve Tatil Köylerinde İnsan Kaynakları Profili. *Kocaeli Üniversitesi Sosyal Bilimler Dergisi*, (4), 85–105.
Özdemir, G. (2006). Resort Planlaması: Turizm Planlaması İçindeki Yeri ve Önemi. *Journal of Yaşar University*, 1 (3), 239–253.
Pine, B. J., ve Gilmore, J. H. (1998). Welcome to the Experience Economy. *Harvard Business Review*, 76(4), 97–105
Ristova, C. (2020). Sports event tourism-an evolving business opportunity for the hotel industry. *Activities in Physical Education and Sport*, *10*(1–2), 6–9.
TDK, (2025). https://sozluk.gov.tr/ (Erişim Tarihi: 05.02.2025).
TUİK 1. (2025). https://yigm.ktb.gov.tr/TR-9851/turizm-istatistikleri.html (Erişim Tarihi: 06.02.2025).
TUİK 2. (2023). https://data.tuik.gov.tr/Bulten/Index?p=Turizm-Istatistikleri-IV.-Ceyrek:-Ekim-Aralik-ve-Yillik,-2024-53660# (Erişim Tarihi: 06.02.2025).

Türkhan, M.S, (2025). Baharat Yolu. https://ansiklopedi.tubitak.gov.tr/ansiklopedi/baharat_yolu (Erişim Tarihi: 05.02.2025).

UNWTO (2025). Dünya Turizm Örgütü. https://www.unwto.org/glossary-tourism-terms (Erişim Tarihi: 02.02.2025).

Uslu. A. ve Aysal, O. (2021). *Otel Ambiyansının Otelin İmajına ve Turist Memnuniyetine Etkisi: Alanya'daki Otel İşletmelerinde Bir Araştırma*. Güncel Turizm Araştırmaları Dergisi, 5(2), 425–438.

Uşaklı, A., ve Öztürk, Y. (2021). Resort Otel Deneyiminin Turist Memnuniyeti ve Sadakati Üzerindeki Etkileri. *Journal of Tourism & Gastronomy Studies*, 9(4), 2854–2873.

Yener, E. (2018). *Beş Yıldızlı Şehir İçi ve Resort Otellerdeki Lobi İç Mekan Tasarımlarının İstanbul ve Antalya Örnekleri Üzerinden Karşılaştırmalı Analizi*. (Yayımlanmış Yüksek Lisans Tezi). Mimar Sinan Güzel Sanatlar Üniversitesi, Fen Bilimler Enstitüsü, İç Mimarlık Anabilim Dalı, İstanbul.

Yılmaz, E. (2017). İpek Yolu Hakkında Kısa Bir Değerlendirme. *Şarkiyat İlmi Araştırmalar Dergisi*, 9(2), 672–683.

Ecotourism in the Scope of Sustainable Development: A Conceptual Framework

Onur ATAK[1]

Introduction

With the effect of globalisation, urbanisation and industrialisation create some negative effects on individuals and increase the desire of individuals to spend time alone with nature and in peace. Especially recently, the demand for tourism has been shifting towards natural environments, intangible and tangible cultural heritage elements, and regions away from the negative effects of the environment (Ankaya et al., 2018:69–70). Local, regional or international ecotourism activities have an important role in the sustainable development of underdeveloped regions, in eliminating the problems caused by unbalanced development between regions, and in creating employment opportunities (Sayın, 2022: 136). Ecotourism activities contribute to the economy of the region by attracting tourism demand to unpopular or unpopular regions and mediate the promotion of unrecognised destinations. In addition, ecotourism activities help to market rural areas, raise awareness about cultural, historical and natural attractions, contribute to the development of products or services by including different attraction centres on a certain route, and create entrepreneurial opportunities in this respect.

Anup, Rijal & Sapkota (2015), in their study conducted in the Annapurna conservation region of Nepal, found that ecotourism improves awareness for the protection of the natural environment and contributes to the socio-economic development of the local people living in the region. Jalani (2012) aimed to determine the effects of ecotourism on local people living in Sabang, Palawan, where the world's longest underground river 'Puerto Princesa' is located, and found that ecotourism has positive economic, social and cultural effects on local people, especially creating employment opportunities and contributing to regional development. Elands et al. (2015) conducted a research aiming to determine the perceptions of local people

1 PhD., Mugla Sıtkı Kocman University, Ortaca Vocational School, onuratak@mu.edu.tr

towards the environmental, economic and sociocultural effects of ecotourism. According to the results of the research, it was determined that there is a perception that ecotourism does not provide enough positive economic contribution. In addition, it was revealed that ecotourism positively strengthens the perception towards the protection of biodiversity and the environment. In 1993, the Dominican Government, with the participation of the local people, created a method plan based on ecotourism, which emphasises the natural, cultural and geographical values of the region, contributes to the economy of the region and aims to use resources effectively. With the ecotourism management plan, local people in the region started to earn income from handicrafts and local agricultural products, accommodation services in their own properties, and the ecotourism revenues obtained in the region contributed to the increase in the gross national product of the Dominican Republic. In addition, with the ecotourism management plan, new employment opportunities such as taxi drivers and tour guides have emerged (Demir & Çevirgen, 2006).

The aim of this study is to determine the importance of ecotourism and its tourism potential through a comprehensive review of national and international literature, to propose solutions to increase national and international tourism demand and tourism employment, to emphasise the importance of ecotourism activities in ensuring sustainable development by determining the contribution of ecotourism activities to destination promotion and marketing of rural areas. Considering that achieving sustainable development goals can be possible by using natural and cultural resources efficiently, it is thought that this study will contribute to the literature.

The Concept of Development

The concept of development has come to the fore especially after the Second World War. Today, 3/4 of the countries are in the category of underdeveloped countries and development is seen as the most important means of getting rid of underdevelopment. Although Jacob & Abel (2002) state that it is not possible to talk about a standard determinant as an indicator of development, they emphasise that welfare level is seen as a measure of development. In fact, the concept of development encompasses much more than the level of welfare. Concepts such as the ability of individuals to act with their free will, developments in basic human rights, political freedom are seen as indicators and components of the concept of development.

Economic growth is a necessary but not a sufficient condition for development. The concept of economic growth is basically considered as the expression of the increase in the country's production or per capita income. The concept of economic development, on the other hand, refers to the growth observed in the economy through changes in the economic structure or the distribution of current output. The changes mentioned in this definition include the increase in material welfare in poorer environments, the weakening of the share of agriculture in national income, technological development, and the increase in the level of education and skills of the labour force (Nafziger, 2006: 15).

Sustainable Development

With the 1987 report entitled 'Our Common Future', known as the Brundtland Report, the concept of sustainable development was defined by the United Nations World Commission on Environment and Development. In the report, sustainable development is defined as 'development that meets the needs of the present without compromising the ability of future generations to meet their needs'. In this approach, natural resources and environmental elements, which are considered as free goods, are considered as scarce inputs (Kaynak, 2014: 57). The concept of sustainable development, which is also handled with its ecological dimension, emphasises the protection and transfer of natural resources to future generations and addresses the concept of sustainable development with its economic and social dimensions.

The Role of Tourism Sector in Balanced Development

Tourism sector is one of the sectors that countries are orientated towards in order to create the necessary financial resources for development (İçöz & Kozak, 1998: 159). This is because the resources offered by the tourism sector have an important place in ensuring regional development and effective and efficient utilisation of resources. Tourism sector is a sector that enables balanced development and enables the planned and effective utilisation of destinations with touristic supply data. Especially countries with underdeveloped industrial and agricultural sectors but rich tourism supply are turning to the tourism sector in order to realise their economic development (Braden & Winer, 1980: 37).

The benefit generated by the income obtained as a result of tourist expenditures in the tourism sector is more than the benefit provided by economic aid provided from abroad. In particular, the income obtained from the tourism sector is not affected by political and economic restrictions. The product produced in the tourism sector differs from the product obtained as a result of agricultural production and the product obtained by exporting raw materials. The supply produced in the tourism sector is considered as an element of diversity among the export products of countries (Çeken, 2016: 133).

In order for a region to develop and develop with tourism, certain factors need to be taken into consideration. It is thought that analysing and evaluating these factors will increase the impact of tourism on regional development (Çeken, 2016):

- It is necessary to make an Infrastructure Analysis covering factors such as the natural vegetation of the region, the condition of the lands suitable for tourism, whether the climate is suitable for tourism or not.
- It is necessary to analyse the cultural diversity and values existing in the region.
- It is necessary to determine the status of the existing facilities (hotels, motels, pensions, etc.) that constitute the tourism supply and whether new ones can be added.
- The status of public investments for tourism in the region should be determined.
- The attitudes of the local people living in the region towards tourism should be determined.
- Analyses should be made regarding the international tourism demand for the region.
- It is necessary to determine whether the tourism supply creates the potential to provide job opportunities for the people living in the region.
- It is necessary to determine whether there are non-governmental organisations directly or indirectly related to the tourism sector in the region.
- The relationship between the tourism sector and other sectors operating in the region should be analysed.

From an economic point of view, the tourism sector has an important share not only for undeveloped or developing countries but also for developed countries. Considering that the main resources that constitute the supply in tourism are natural and historical resources, it is thought that the tourism sector has certain advantages over other sectors in terms of development speed. When the tourism supply potential of the regions is supported by effective tourism marketing and promotion, regional development is positively affected and the imbalances observed in regional development are minimised. The development of the tourism sector has an important place in the development of the country's economy by providing foreign currency inflow to the country and increasing the employment rate. In addition to all these important economic effects, the tourism sector plays a leading role in the development of other sectors, as it is a sector that affects other sectors with which it is directly or indirectly related (Akın, 2018: 4).

Ecotourism in the Scope of Sustainable Development

The increase in industrial production in the process of globalisation has created negative effects on the natural environment, which has increased the interest in nature and strengthened the awareness of environmental protection. This situation has changed touristic consumption behaviours and led to the emergence of new concepts such as 'soft tourism', 'eco-tourism', 'sustainable tourism', which observe the balance of conservation-use for the protection of the natural and cultural environment (Tekin, 2017: 379). Ecotourism, which emerged with this understanding, is a type of tourism that is economically efficient, develops social awareness and protects the natural environment. Weaver (1999: 792–816) defines ecotourism as a type of tourism that aims to protect and utilise natural and cultural resources with a sustainable tourism approach.

The objectives of ecotourism are listed as follows (Turoğlu & Özdemir, 2005: 97–116):

- To contribute to the protection of the natural and cultural environment
- To increase the welfare level of the people of the region through eco-tourism
- To introduce the natural environment and cultural heritage elements of the region to the visitors
- To minimise irrecoverable waste of resources and environmental damage.

- To increase the awareness of sustainable tourism in tourists and local people, to ensure that the public takes a participatory role in tourism activities and tourism management
- To inform tourists about the eco-tourism elements of the region
- To provide social and cultural contribution to the local people
- To help ensure equal distribution of income among local people, relevant institutions and organisations by preventing monopoly in eco-tourism activities
- To minimise the negative effects of tourism activities on the natural environment and cultural elements

The most preferred types of ecotourism that allow domestic and foreign visitors to experience natural, cultural and archaeological assets are as follows:

Plateau Tourism: It refers to the visits of individuals who want to spend their free time allocated for tourism in a natural environment by travelling to plateaus with these resources (Tekin, 2016).

Agro-tourism: It refers to visits to farms, agricultural production areas and places, participating in agricultural activities, receiving training on agricultural activities and shopping (Türkben et al, 2012).

Cave tourism: It is a type of tourism that refers to visits to caves for adventure, recreactive, cultural, educational, sports, etc. purposes due to the geological, ecological and cultural values of caves (Yozcu, 2020).

Hiking and Trekking: Trekking is called trekking for walks made in a certain order without using any means of transport in the natural environment. Hiking is defined as short-term daily nature walks without the need for overnight stay or camping (Çiçek, 2012).

Botanical tourism: It is a tourism activity that includes accommodation, eating and drinking or utilising touristic goods and services in order to identify or observe plant diversity (Cesur et al., 2021:1062).

Horse trekking: It is horse-drawn nature walks for recreational purposes with trained horses on specially designed tracks or on a natural route with predetermined standards and boundaries (Topay, 2003: 48).

Ornithological tourism: It is defined as the activities of recognising, observing, photographing and researching bird species in their natural habitats (Şekercioğlu, 2002: 282–289).

Bicycle tourism: It is defined as cycling to routes with a distance of at least 40 km with daily or overnight stays during the time that individuals allocate for leisure or holiday (Lamont, 2009).

Mountain tourism: It includes activities carried out in mountain ecosystems for the purpose of resting, holidaying, participating in mountain sports actively or as a spectator (Çiçek, 2012).

Camping tourism: It is considered as one of the types of nature-based tourism, which takes place in camping or natural environments and includes the use of touristic goods and services (Hewer et al., 2017: 1806).

River tourism: It is an alternative form of tourism that includes sportive activities such as rafting, river skiing, etc. carried out for tourism purposes in rivers (Köroğlu, 2015: 157).

Considering that development is a concept with environmental, economic and sociocultural aspects, economic, socio-cultural and natural environment in the ecotourism region should be coordinated with each other in order to create sustainable benefits in ecotourism activities. The contribution of ecotourism activities in ensuring sustainable development can basically be summarised within the framework of environmental, economic and socio-cultural contributions:

Environmental contributions of ecotourism activities:

- Since ecotourism is based on nature experience, it contributes to the transfer of natural resources to future generations with an understanding of sustainability.
- Ecotourism activities strengthen awareness for the protection of nature and biodiversity (Mosammam et al., 2016: 170, Sayın, 2022: 136).
- Ecotourism activities are concentrated in protected areas. Ecotourism activities encourage governments to declare valuable areas as protected areas and create public opinion on the prevention of environmentally damaging practices (Kuter and Ünal, 2009: 151).

Socio-cultural contributions of ecotourism activities:

- Ecotourism contributes to the control of migration movements with the employment opportunities it provides (Kaypak, 2012: 26).

- Ecotourism contributes to the improvement of the region's health and education services, the increase in regional welfare, and the transfer of cultural assets to future generations (Gaul, 2003).
- Ecotourism introduces the natural environment, tangible and intangible cultural heritage elements of the region, contributes to the production of artistic environments where these values will be exhibited, and the formation of cultural activities such as folklore, theatre, etc. (Benzer, 2006: 40).
- People in the ecotourism region play an active role in the decisions taken about the region.
- Ecotourism activities enable cultural interaction between tourists and local people and develop an environment of tolerance (Sayın, 2022: 136).

Ecotourism creates positive economic impacts in the destinations where it develops. Increased employment opportunities for local people, direct income from ecotourism activities, entrance fees to parks, income from marketing local arts and crafts, income generated in food and beverage businesses, accommodation businesses, recreation businesses in cultural activities such as festivals, shows, shows, and taxes from these businesses (Mosammam et al., 2016: 170) are among the important economic impacts that arise with the increase in tourism demand. In addition, the increase in infrastructure and superstructure investments to meet the tourism demand contributes to the development of the region through ecotourism.

Costa Rica is one of the most important destinations for sustainable development and ecotourism. Especially local production items such as coffee and bananas have an important place in Costa Rica's export revenues. In addition, La Gamba, where agricultural production takes place, and Cerro de Oro, which borders the Corcovaro National Park, are important ecotourism destinations (Braun et al., 2015: 358). In the villages of Kaski Tanahu Syangja, Lamjung and Gorka regions in the west of Nepal, tourism has become an income-generating sector with the impact of ecotourism. In addition, the Annapurna region, which is under protection, contains many trekking routes. Eco-tourism activities in the region contribute to the increase in forest areas, the protection of fauna and flora, and the promotion of renewable energy use (Anup et al, 2015: 252). The USA is one of the most preferred countries by ecotourists in terms of high ecotourism investments and ecotourism resources. National parks and resorts in states such as Hawaii, Alaska, Virginia and Arizona are the most important attraction centres of ecotourism. The ecotourism programmes implemented in these regions include many policies that aim to improve the lives of the local people with the positive economic, environmental and sociocultural effects they create and develop the region in which they are implemented (Simoni, 2011: 171–172). In Cambodia, ecotourism revenues have an important place in the country's gross domestic product. The elements that are the source of ecotourism are protected by state organisations and high penalties are imposed on parties that harm ecotourism. In Cambodia, ecotourism projects are thought to support economic growth and help reduce poverty (Lopez & Toman, 2006: 437).

In Turkey, the concept of ecotourism was officially mentioned for the first time in the Eighth Five-Year Development Plan covering the years 2001–2005. Compared to ecotourism practices in the world, ecotourism projects in Turkey are not sufficient in terms of institutional infrastructure, ecotourism awareness and lack of investment. In Turkey, Nevşehir Avanos, Muğla Fethiye where ecological farms are located, Çandır village, an ecological village in Muğla Ortaca, Dalyan, known as the spawning ground of Caretta Caretta sea turtles, and Kaz Mountains are among the important ecotourism destinations.

Conclusion

Ecotourism is a type of tourism that is closely related to many concepts such as employment, migration, agricultural production, entrepreneurship, tourism promotion, and includes nature and community-based approaches to ensure sustainable development, especially in underdeveloped and developing countries. Ecotourism, which contributes to the prominence of local products, natural environment attractions, tangible and intangible cultural heritage elements, is considered as an important part of development efforts with its economic, environmental and sociocultural aspects. In addition, ecotourism practices that do not comply with the principles of ecotourism can lead to overuse of natural and cultural environmental elements, deterioration and destruction of the ecosystem. For this reason, there is a need for policies that are compatible with the natural, economic and socio-cultural environment in accordance with the objectives and principles of ecotourism, in which the participation of the local community is ensured in ensuring sustainable development. In this framework, some suggestions for increasing the effects of ecotourism activities in realizing sustainable development are as follows:

- Ecotourism activities should be carried out with the coordination of the government, local administrations and non-governmental organizations, with strategic and management plans and processes involving local people.
- The carrying capacity of destinations with high ecological sensitivity for ecotourism activities should be calculated, and emergency precaution plans should be implemented in areas where the capacity is exceeded. In addition, investments should be made in accordance with the carrying capacity of the region, natural and cultural environmental elements, and the use of renewable energy should be supported.
- Ecotourism requires the active participation of local people in planning and implementation processes. Local people should be informed about the impacts of ecotourism and sustainable tourism, and training programs should be organized to ensure voluntary participation.
- Policies that enable local people to use their properties for tourism purposes should be developed, and local people should be supported to become ecotourism entrepreneurs with investment incentives.
- For ecotourism regions, promotional activities should be carried out for the environmental, natural and sociocultural qualities of the region, and activities such as fairs, festivals, festivals, fairs should be organized without causing the carrying capacity to be exceeded and without destroying the natural environment.

References

Akın, A. (2018). Türk turizminin makroekonomik göstergeler üzerindeki etkileri.*Journal of Tourism and Gastronomy Studies*.6(3): 3–23.

Ankaya, Ü. F., Yazıcı, K., Balık, G. & Aslan, B. G. (2018). Dünyada ve Türkiye'de ekoturizm, sosyokültürel ve ekonomik katkıları. *Ulusal Çevre Bilimleri Araştırma Dergisi*, 1(2), 69–72.

Anup, K.C., Rijal, K. & Sapkota, R.P. (2015). Role of ecotourism in environmental conservation and socioeconomic development in Annapurna Conservation Area, Nepal. *International Journal of Sustainable Development & World Ecology*, 22 (3), ss.251–258.

Benzer, A.N.K. (2006). *Bolu-Göynük ve yakın çevresi doğal ve kültürel kaynaklarının ekoturizm açısından değerlendirilmesi*, Yayımlanmamış Doktora Tezi, Ankara Üniversitesi, Ankara

Braden, P. V. & Louse W. (1980). Bringing travel, tourism and culturel resource activities in harmony with regional economic development. *Tourism Marketing and Management Issues*, George Washington Üniversity, Washington.

Braun, Y.A., Dreiling, M.C., Eddy, M.P. & Dominguez, D.M. (2015). Up against the wall: Ecotourism, development, and social justice in Costa Rica. *Journal of Global Ethics*, 11 (3), ss.351–365.

Cesur, E., Özer, Ç. & Çeken, H. (2021). Alternatif turizm kapsamında Aromatik Turizm ve gelişimi. *Abant Sosyal Bilimler Dergisi*, 21(3), 1055–1078.

Çeken, H. (2016). Turizm ekonomisi, (2. Baskı), Ankara: Detay Yayıncılık.

Çiçek, A. (2012). Ekolojinin sürdürülebilir turizm için önemi. Çiçek, A. (Ed.). Ekoloji ve Turizm. Eskişehir: Anadolu Üniversitesi Yayınları.

Demir C. & Çevirgen A. (2006). Ekoturizm yönetimi., ISBN: 975-591-844-2, 222 s, Nobel yayın dağıtım, Ankara.

Elands, B., Islam, W. & Van Der Duim, R. (2015). Ecotourism in Lawachara National Park in Bangladesh: Perceptions of local people and co-management officials. South Asian Journal of Tourism and Heritage, 8(1), 5–20. https://www.researchgate.net/publication/282875638_Ecotourism_in_Lawachara_National_Park_in_Bangladesh_Perceptions_of_Local_People_and_Co-management_Officials [Erişim Tarihi: 17.03.2025].

Gaul, D. (2003). Environmental impacts of ecotourism. http://foris.fao.org/static/pdf/foph/ecotour.pdf, [Erişim Tarihi: 24.02.2025].

Hewer, M. J., Scott, D. J., & Gough, W. A. (2017). Differences in the importance of weather andweather-based decisions among campers in Ontario parks (Canada). InternationalJournal of Biometeorology, 61 (10), 1805–1818.

İçöz, O & Kozak, M. (1998). Turizm ekonomisi, Ankara: Turhan Kitapevi.

Jacob, T., J. & Abel, R. (2002). Sustainable development and health, An Indian perspective, *Lancet*, 360, (24).

Jalani, J. (2012). Local people's perception on the impacts and importance of ecotourism in Sabang, Palawan, Philippines. Social and Behavioral Sciences, 57, 247–254. https://doi.org/10.1016/j.sbspro.2012.09.1182

Kaynak, M. (2014). Kalkınma iktisadı, (gözden geçirilmiş ve genişletilmiş 5. baskı), Ankara: Gazi Kitabevi.

Kaypak, Ş. (2012). Ekolojik turizm ve sürdürülebilir kırsal kalkınma. *Karamanoğlu Mehmetbey Üniversitesi Sosyal ve Ekonomik Araştırmalar Dergisi*, 14 (22): ss.11–29.

Köroğlu, Ö. (2015). Turizmin çeşitli kriterlere göre sınıflandırılması. Z. Aslan (Ed.), Genel Turizm içinde (s. 111–178). Grafiker Yayınları.

Kuter, N. & Ünal, H.E. (2009). Sürdürülebilirlik kapsamında ekoturizmin çevresel, ekonomik ve sosyo-kültürel etkileri. *Kastamonu Üniversitesi, Orman Fakültesi Dergisi*, 9 (2): 146–156

Lamont, M. (2009). *Independent bicycle tourism in Australia: A whole tourism systems analysis* (Doctoral Thesis). Lismore, NSW: Southern Cross.

Lopez, R. & Toman, M.A. (2006). Economic development & environmental sustainability – New policy options, Oxford: Oxford University Press

Mosammam, H.M., Sarrafi, M., Nia, J.T. & Heidari, S. (2016). Typology of the ecotourism development approach and an evaluation from the sustainability view: The case of Mazandaran Province, Iran. *Tourism Management Perspectives*, (18), ss.168–178.

Nafziger, E., W. (2006). Economic development. 4th. Edition, Cambridge University Press, Cambridge.

Sayın, K. (2022). Ekoturizmin yerel kalkınmaya katkısına ilişkin yerel halkın algılarının belirlenmesine yönelik Taşucu'nda bir araştırma. *Journal of Gastronomy, Hospitality and Travel*, 2022, 5(1), 133–144

Simoni, S. (2011). Ecotouristic marketing strategies in the United States of America. Agricultural Management / Lucrari Stiintifice Seria I, *Management Agricol*, 13 (4), ss.171–178.

Şekercioğlu, Ç. H. (2002). Impacts of birdwathcing on human and avian communities, *Enviromental Conversation*, 29(3), s. 282–289.

Tekin, Ö. (2017). Ekoturizm açısından Konya İli doğa yürüyüşü rotaları üzerinde bir araştırma. *Manas Sosyal Araştırmalar Dergisi*, 6(3), s. 377–200.

Tekin, Ö. F. (2016). Alternatif bir turizm türü olarak yayla turizmi ve Bozkır ilçesinin potansiyelinin değerlendirilmesi, Uluslararası Sempozyum, *Geçmişten Günümüze Bozkır*, 6–8 Mayıs, Konya,1093–1107.

Topay, M. (2003). *Bartın-Uluyayla peyzaj özelliklerinin rekreasyon-turizm kullanımları açısından değerlendirilmesi üzerine bir araştırma.* Yayınlanmamış Doktora Tezi, Ankara Üniversitesi Fen Bilimleri Enstitüsü, Ankara, s.48.

Turoğlu, H. & Özdemir, H. (2011). Bartın İlinin ekoturizm potansiyelinin belirlenmesi. *Doğu Coğrafya Dergisi*, 10(13), S. 97–116.

Türkben, C., Gül, F. & Uzar, Y. (2012). Türkiye'de bağcılığın tarım turizmi (Agro-Turizm) içinde yeri ve önemi. *KMÜ Sosyal ve Ekonomik Araştırmalar Dergisi*, 14 (23), 47–50

Weaver, D. B. (1999). Magnitude of ecotourism in Costa Rica and Kenya. *Annals of Tourism Research*, 26(4) s. 792–816.

Yozcu, S. (2020). Türkiye mağara turizmine yönelik bir değerlendirme. *Türk Turizm Araştırmaları Dergisi*, 4 (2), 1493–1508.

The Role of Social Media in the Hotel Selection Process by Potential Travelers and Tourists

George ZOGRAFOS[1]

Introduction

Before the advent of the "social media era," consumers had limited sources of information to help them make decisions about tourism products and services. Most information about accommodations came from advertising brochures and travel agents (Hu and Wei, 2013). In recent years, however, traditional sources have been largely replaced by social media. Travelers now rely on user-generated content on social platforms to plan and decide on their hotel stays (Browning and Sparks, 2013). In the tourism industry, social media serves various purposes for consumers, such as sharing travel experiences, interacting with others, connecting with people from different destinations, and purchasing travel-related products and services (Munar and Jacobsen, 2014).

While advancements in information and communication technologies have significantly impacted consumer behavior in tourism, it was the rise of Web 2.0 that fundamentally changed how tourists select and buy tourism products and services online (Gohil, 2015). This shift is especially evident in the hotel sector, where decision-making has undergone a dramatic transformation. Social media now plays a role in every stage of the consumer decision-making process, influencing prospective tourists by shaping, guiding, and sometimes altering their initial choices (Fotis et al., 2012).

Nevertheless, the specific ways in which consumer-tourists use social media and how it ultimately influences their purchasing decisions, particularly in the hospitality industry, remains an area that requires further investigation (Cox et al., 2009; Ayeh et al., 2012; Gohil, 2015).

This research aims to explore how social media affects consumer behavior in the process of hotel selection, thus contributing to a deeper understanding

[1] Cons. Proff., Hellenic Open University, School of Social Sciences, Department of Tourism Management, zografos.georgios@ac.eap.gr

of its role and impact on consumer behavior in the tourism and hospitality sectors. To achieve this, a literature review on hotels, social media, electronic marketing, and their influence on tourists' decisions will be conducted. Additionally, a quantitative survey using a questionnaire will be administered, providing valuable insights on the topic.

The Impact of the Internet on Tourism

Tourism services and products are now available to customers and guests around the clock, throughout the entire year. In contrast, traditional travel agencies operate only during specific hours. The continuous availability of organized travel products is made possible by the shift towards digital platforms, particularly through online tourism portals.

The transformation in tourism communication, driven by the widespread adoption of information and communication technologies (ICT), has significantly altered the industry. In the 1970s, computerized reservation systems (CRS) emerged, followed by global distribution systems (GDS) in the 1980s, and finally the widespread use of the Internet in the 1990s—all of which revolutionized business practices within the tourism sector (Cooper et al., 2008). These developments not only reshaped the structure of the tourism industry but also enhanced how tourists perceive and interact with service providers. ICT now plays a central role in facilitating the exchange of information, knowledge, products, and services, while also supporting essential functions and driving innovation in tourism (Buhalis and Low, 2008).

The first computerized reservation systems were introduced in the airline industry and became known as Computer Reservation Systems (CRS). These systems are used to store, manage, and process travel-related data and transactions (Neuhofer & Buhalis, 2012). Originally developed for airlines, CRS technology later expanded to other sectors such as hotels, car rentals, and additional tourism services.Large companies that utilize CRS for booking and ticketing have also integrated into the Global Distribution System (GDS).

The Internet of Things (IOT)

The term Internet of Things (IoT) was first introduced by Kevin Ashton in 1999. IoT refers to a network capable of connecting anything, at any time and from anywhere, using technologies such as RFID (Radio Frequency Identification), WSN (Wireless Sensor Networks), and 3G/4G/5G mobile communications. These connections operate under a shared protocol to identify, locate, monitor, and manage smart objects (Mingjun et al., 2012). IoT presents new opportunities to bridge the physical and digital realms, facilitating more interactive and personalized connections between tourists, hotels, tourism products, and destinations.

Smart tourism represents a modern evolution of the tourism industry, incorporating IoT, cloud computing, Big Data, Artificial Intelligence (AI), and other cutting-edge technologies to enhance tourist experiences, drive industrial development, and improve administrative oversight (Wu, 2017). While the tourism sector is sometimes perceived as slow to adopt new technologies, it continues to evolve by embracing innovations such as IoT, which supports greater automation, personalization, and improved customer experiences. Additionally, it helps streamline daily operations within hotels and travel agencies.

Smart tourism has emerged through the integration of various technological advancements, including smartphones, mobile platforms, cloud services, and the Internet of Things. Today, multiple technologies work in tandem—such as web and mobile apps, cloud computing, IoT, and Big Data analytics—to deliver smart tourism services. Among these, IoT plays a particularly vital role, offering innovative solutions and enhancing existing systems. By leveraging IoT, the tourism industry can boost operational efficiency and provide more customized, responsive experiences for travelers.

Electronic Word of Mouth

Word of mouth (WOM) is recognized as a powerful element in the business process, involving informal communication about a brand, product, or service shared between friends, family, and members of social groups. In the digital age, this concept has evolved into electronic word of mouth (eWOM), which refers to any positive or negative statements made by potential, current, or former customers about a product or company. These messages are shared online and can reach a wide audience of individuals and organizations.

eWOM is closely linked to the rise of user-generated content (UGC) in digital marketing, and its definition often overlaps with concepts like viral advertising.

Viral marketing strategies frequently utilize formats such as blogs, social media, and consumer-generated content to spread messages quickly and organically (Akar & Topcu, 2011).

Marketers have developed a variety of strategies to intentionally spark eWOM. These efforts aim to integrate brands into consumers' conversations by leveraging UGC to support promotional objectives. The impact of eWOM on consumer purchasing behavior is significant, which is why these practices—often referred to as eWOM advertising—are widely used. A common example includes brands encouraging customers to post product reviews, regardless of whether they are positive or negative, to help influence other consumers' buying decisions.

Traditional WOM, originally defined as oral, non-commercial, and interpersonal communication between acquaintances, is perhaps the oldest known method of promoting products and services (Cheung et al., 2010). For instance, when a consumer encounters something new, unusual, or interesting, they are likely to share the experience with friends or acquaintances. While traditional WOM relies on face-to-face contact, in the online space, such messages can quickly become viral if they resonate with a broader audience.

This traditional form of communication has evolved into digital eWOM, which differs in its permanence and reach. Unlike spoken messages that disappear after being shared, eWOM messages remain accessible for extended periods and can be found across a variety of online platforms—such as blogs, product review sites, social media channels, and online forums or groups.

Factors that Shape the Hotel Selection Process

According to Kotler et al. (2009), consumers typically follow a five-step process when making a purchase decision: recognizing a need, searching for information, evaluating alternatives, making the purchase, and deciding whether or not to repurchase. Among these stages, the information search phase is particularly significant for today's consumers. In recent years, a growing number of consumers have taken control over how much information they seek and what kind, increasingly turning to social media as a primary source for reviews, recommendations, and comments.

The significance of the Internet as an information source in the tourism industry has been well-documented across numerous studies. Social media, in particular, has emerged as a key area of interest for understanding tourist behavior, purchasing decisions, and corporate communication strategies.

A number of researchers have explored the influence of social media on consumer purchasing behavior. Kaplan and Haenlein (2010) examined the opportunities and challenges of using social media in marketing, emphasizing its value in understanding consumer decision-making. Milano et al. (2011) analyzed the relationship between online social media and tourism websites in Italy, concluding that websites integrated with social media were more effective and efficient in tourism marketing than those relying solely on traditional promotional tools.

Sigala et al. (2012) investigated corporate use of social media in the tourism and hospitality sectors, highlighting various ways organizations can leverage social platforms to synthesize and share information. Zivkovic et al. (2014) further emphasized the strong influence of social media on tourist behavior, noting that many travelers rely heavily on online information during the decision-making process. Arat (2016) focused on the role of social media in shaping tourism demand through content analysis, exploring how hotels utilize Facebook accounts and the features they employ in their communication strategies.

Overall, these studies reveal that social media is widely accepted by tourists as a tool for booking, trip planning, gathering information, and sharing travel experiences. The findings consistently show that social media serves as a powerful marketing tool within the tourism industry, significantly influencing consumers' purchase decisions. In fact, the process of purchasing tourism products online is closely linked to social media marketing efforts (Arat, 2016).

Social Media in Tourism

Beyond its significant role in people's daily lives and social interactions, social media has become an essential component of the tourism industry (Gohil, 2015). The concept of "Travel 2.0" reflects the evolution of the travel sector into a new era of online platforms and interactive communication, fundamentally transforming the ways travelers search for, evaluate, purchase, and experience tourism-related products and services (Hudson & Thal, 2013). Among the numerous social media platforms available today, those most commonly used in the tourism and hospitality sectors include TripAdvisor, Facebook, YouTube, Instagram, Twitter, MySpace, and Flickr (Sigala et al., 2012).

Social media plays a critical role throughout all stages of the travel experience—before, during, and after a trip (Fotis et al., 2012). Researchers widely agree that travelers rely heavily on social media at the pre-trip stage for planning, information gathering, and decision-making. During this stage, social platforms help shape consumer behaviors related to searching, discovering, evaluating, trusting, and sharing travel-related information. User-Generated Content (UGC), such as reviews and personal experiences, provides valuable insights into destinations, accommodations, and restaurants, guiding potential tourists in their choices (Sigala et al., 2012).

Most travelers consult online reviews of hotels and travel services, with social media serving as a vital decision-support tool. Consumers often place greater trust in peer-generated content than in information from official tourism websites, travel agents, or traditional advertising channels (Fotis et al., 2012). However, despite the perceived authenticity of UGC, concerns persist over its credibility and accuracy—an important factor that significantly influences consumer decisions.

From a marketing perspective, social media has become a powerful tool for tourism businesses in fostering engagement, enhancing communication, and managing customer relationships. Although initially created for entertainment purposes, social platforms have evolved into crucial communication channels that shape consumer behavior, particularly in hospitality and tourism. Platforms like Facebook allow global interaction, content sharing, direct communication with businesses, and peer-to-peer engagement (Draboshova & Balko, 2017).

Consequently, social media now forms a vital part of digital marketing strategies, enabling companies to create consumer-centric messages. Platforms such as LinkedIn, Facebook, Twitter, and YouTube allow tourism and hospitality businesses to connect with wide audiences, boost brand awareness, and collect immediate customer feedback. One of the most valuable functions of social media is its ability to influence consumer behavior through interactions with past travelers who share their experiences.

Social media offers a cost-effective marketing solution, promoting two-way communication between businesses and consumers. It allows customers to publicly share their opinions, engage with companies, and receive timely responses. Therefore, social media is more than a promotional tool—it serves as a platform for interaction, customer service, and brand credibility enhancement. In an era of shrinking marketing budgets, many businesses are turning away from traditional methods and investing more in social media, given its high return on investment.

Unlike conventional marketing, social media provides real-time analytics, such as follower growth, engagement rates, and customer sentiment. Platforms like Facebook, Twitter, and YouTube allow businesses to monitor conversations, respond to reviews, and analyze customer feedback, which helps refine service offerings and branding strategies.

Moreover, social networking platforms enable tourism and hospitality businesses to better understand customer preferences, expectations, and levels of satisfaction. These insights help organizations evaluate staff performance, identify service gaps, and promote positive word-of-mouth marketing—ultimately attracting new customers at a lower cost and enhancing brand reputation.

Lastly, integrated marketing communication, supported by technological advancements, plays a pivotal role in brand development. In this context, a tourism company's brand is shaped by its public perception, service quality, and the strength of its customer relationships.

Tourist purchasing behavior has undergone a significant transformation with the advent of new technologies and the proliferation of online booking platforms. In an effort to capture the interest and loyalty of hotel guests, these websites often feature reviews to help consumers find services that align with their preferences. Online reviews have become a critical factor in consumer purchasing decisions, providing customers with the ability to compare businesses and access more detailed information about service offerings.

The increasing use of social media—particularly in the form of online comments, reviews, and ratings shared by hotel guests on platforms such as TripAdvisor, Expedia, and Yelp—has significantly influenced the decision-making process of potential travelers (Stringam & Gerdes, 2010). Research consistently shows that prospective guests tend to place greater trust in peer-generated reviews than in information from official marketing sources, such as hotel websites or company advertisements (Sparks, Perkins & Buckley, 2013).

This shift has also emphasized the importance of customer engagement in shaping and enhancing service quality. Guests' ability to assess and evaluate the quality of hotel services through reviews contributes meaningfully to ongoing discussions about service improvement. Consumer feedback not only influences other potential guests but also plays a vital role in the continuous development of quality assurance standards. When these standards are informed by both customer experiences and industry practices, they become more relevant and effective in meeting fundamental guest needs.

Review platforms are thus becoming an increasingly important source of information, and their influence on travel behavior is growing. Specifically, the content of user-generated reviews is changing how consumers compare products, with vacation planning now largely driven by online resources—especially customer review websites.

Furthermore, Liu, Schuckert, and Law (2015) explored the motivational aspects of review writing on platforms like TripAdvisor. Their study found that users are incentivized to post reviews through a ranking system that rewards active participation with higher user status. However, the findings also revealed that as a reviewer's rank increases, the quality of their reviews tends to decline. Additionally, high-ranking reviewers are less likely to post extreme ratings, suggesting that status influences both the frequency and tone of user contributions.

Yoo and Gretzel (2011) further argue that personality traits play a crucial role in determining whether individuals post reviews online. According to their research, those who engage in online reviewing often exhibit more egocentric tendencies, which may influence both the content and motivation behind their feedback.

The Research

This study highlights the significant role of social media in the hotel selection process. It explores modern electronic marketing strategies, the evolving characteristics of tourism in the digital era, and the transformative communication methods introduced by social media as a theoretical framework to assess the effectiveness of hotel e-marketing.

Social media directly shapes consumers' perceptions of hotel brands, ultimately affecting their purchasing intentions and decisions. Additionally, consumer experiences on social media and interactions with the brand are expected to influence attitudes toward hotel websites and buying behavior. Moreover, the study incorporates the intention to engage in electronic word-of-mouth (eWOM) as a crucial marketing tool emerging from social media, which plays a key role in shaping consumer decisions.

The research proposal that is formulated is as follows:

SOCIAL MEDIA PLAYS AN IMPORTANT ROLE IN THE CHOICE OF HOTELS BY POTENTIAL TOURISTS/TRAVELERS

To confirm it, three more specialized research proposals were investigated. In the end, the investigation of these three proposals would verify (to what extent) the main research proposal.

Hypothesis 1: The better the reviews and ratings of a hotel accommodation on social media, the greater the chances of bookings from its users (purchase decisions).

Hypothesis 2: A more favorable customer attitude toward the social media platform where a hotel maintains an active presence leads to a more positive perception of the hotel brand.

Hypothesis 3: The more positive a customer's attitude toward a hotel brand, the higher the likelihood of sharing positive electronic word-of-mouth (eWOM) about the brand through social media posts.

The research took place in 2022, involving a questionnaire which aimed finding out perceptions of potential customers who were about to select hotels in the next month. For the research purposes, a questionnaire was created and was sent to 100 randomly chosen potential hotels guests.

Conclusions

Based on the extensive body of research cited in the previous sections, it is evident that social media plays a crucial role in the hotel selection process in today's digital age. The partial confirmation of the third research hypothesis, alongside the full confirmation of the first and second, reinforces the argument that a strong online presence—both on mainstream social media platforms and on specialized tourism websites with integrated social functions—is a key factor in attracting bookings. These bookings are significantly influenced by electronic word-of-mouth (eWOM).

Potential travelers are increasingly guided by social media content and peer-generated reviews, which can strongly influence both initial consideration and final decision-making. The growing amount of time that people spend on social media, the convenience of mobile access, and the reliance on platforms such as Facebook for information have effectively transferred much of everyday interaction and information-gathering to the digital sphere. This shift has inevitably affected the way consumers make choices regarding hotel accommodations, whether for leisure or business purposes.

Tourism, being a highly visual industry, lends itself well to digital marketing. The internet—flooded with images—provides an ideal environment for promoting hotels, both through paid advertising and organic content such as guest-generated posts. Within this digital ecosystem, influencers have emerged as powerful players. Derived from traditional word-of-mouth marketing, influencer marketing has evolved into a highly profitable digital strategy aimed at increasing brand awareness, customer engagement, and direct bookings in the hospitality sector. eWOM, with its ability to spread rapidly and reach vast audiences, has amplified the impact of influencers, YouTubers, bloggers, and other content creators.

These individuals often hold substantial influence, credibility, and trust among today's consumers. However, their authority and objectivity are frequently questioned. As many influencers are compensated for their promotions, their endorsements are effectively "paid recommendations," which may not always reflect genuine user experience or expertise. Despite this, people tend to trust recommendations from social media contacts or familiar influencers more than traditional advertising methods, such as TV commercials.

Consumers increasingly seek "authentic" advice and have become desensitized to conventional advertising, which they often ignore. However, the broad definition of an "influencer" poses risks for both hotels and consumers. Fake followers, artificial engagement, and the commodification of influence—such as purchasing bots to boost metrics—undermine the authenticity and reliability of influencer marketing. Moreover, the entrepreneurial nature of many influencers means their content is often financially motivated, a fact that consumers may not fully consider.

Despite these challenges, the digital economy and internet technologies continue to profoundly reshape the tourism sector. They have revolutionized how tourism services are marketed, how businesses communicate with customers, and how tourist experiences are delivered. While tourism services are inherently local, their reach has become global due to digital platforms. The information-rich nature of these services makes them especially conducive to digital transformation.

Looking ahead, Generation Z, Millennials, and future generations—digital natives who are deeply familiar with and dependent on technology—will make up the largest portion of global tourists by 2040. Their preferences and behaviors will continue to shape how tourism services are designed, marketed, and distributed. Therefore, it is imperative that tourism businesses adapt to these emerging expectations. Failure to invest in digitalization could result in reduced competitiveness and even market exit.

To support this transition, governments have a role to play in creating favorable conditions for tourism businesses to thrive in the digital age. This includes not only maximizing business opportunities but also ensuring consumer protection and addressing the unique challenges faced by traditional tourism enterprises navigating digital transformation.

References

Akar, E., and B. Topc¸u. (2011). An examination of the factors influencing consumers' attitudes toward social media marketing. *Journal of Internet Commerce* 10 (1): 35–67.

Arat, T. (2016). The Role of Social Media in Tourism. *Proceedings of the 25th International Academic Conferences, International Institute of Social and Economic Sciences.* Paris, France, 0609 September, pp. 3752.

Ayeh, J., Leung, D., Au, N. and Law, R. (2012), Perceptions and strategies of hospitality and tourism practitioners on social media: An exploratory study. In Information and Communication Technologies in Tourism, (pp. 1–12) Vienna, Springer.

Browning, V., S., Sparks, B., A. (2013). The Influence of Online Reviews on Consumers' Attributions of Service Quality and Control for Service Standards in Hotels. *Journal of Travel and Tourism Marketing*, Vol. 30 No. 1–2, pp. 23–40.

Buhalis, D. & Licata, M. C. (2002). The Future eTourism Intermediaries, *Tourism Management*, 23, 207–220.
Buhalis, D., & Law, R. (2008). Progress in information technology and tourism management: 20 years on and 10 years after the Internet The state of eTourism research. *Tourism Management*, 29(4), 609–623.
Cheung, C., Chiu P.Y., Lee M., (2010). Online social networks: Why do students use facebook? *Computers in Human Behavior,* 27 (2011), 1337–1343.
Cooper, C., Fletcher, J., Fyall, A., Gilbert, D. & Wanhill S., (2008). *Tourism principle and practice*. 4th ed. Harlow, England: Prentice Hall.
Cox, C., Burgess, S., Sellitto, C. and Buultjens, J. (2009). The Role of User-Generated Content in Tourists' Travel Planning Behavior. *Journal of Hospitality Marketing and Management*, Vol. 18 No. 8, pp. 743–764.
Drahosova, M., Balco, P., (2017). The analysis of advantages and disadvantages of use of social media in European Union. *The 7th International Symposium on Frontiers in Ambient and Mobile Systems (FAMS 2017)*
Fotis, J., Buhalis, D. and Rossides, N. (2012), Social media use and impact during the holiday travel planning process. In Information and Communication Technologies in Tourism (pp. 13–24). Vienna, Springer.
Gohil, N. (2015). Role and Impact of Social Media in Tourism: A Case Study on the Initiatives of Madhya Pradesh State Tourism. *International Journal of Research in Economics and Social Sciences*, Vol. 5 No. 4, pp. 8–15.
Hudson, S. and Thal, K. (2013). The Impact of Social Media on the Consumer Decision Process: Implications for Tourism Marketing. *Journal of Travel & Tourism Marketing,* Vol. 30 No. 1–2, pp. 156–160.
Hu, F. and Wei, G. (2013). The impact of the knowledge sharing in social media on consumer behavior. *The Thirteen International Conference on Electronic Business*. Taiwan, Vol. 1 No. 1, pp. 71–102.
Kaplan, A.M., Haenlein, M. (2010). Users of the World, Unite! The Challenges and Opportunities of Social Media. *Business Horizons*, No 53, pp. 59–68: http://dx.doi.org/10.1016/j.bushor.2009.09.003.
Kotler P., Keller L. K., Mairead, B., Goodman, M., Torben, H., (2009). *Marketing management*. Essex: Pearson Education Limited.
Liu, X., Schuckert, M., & Law, R. (2015). Online Incentive Hierarchies, Review Extremity, and Review Quality: Empirical Evidence from the Hotel Sector. *Journal of Travel & Tourism Marketing.* Advance online publication. doi:10.1080/10548408.2015.1008669

Middleton, T.C., Alan, F., Morgan M. Ronchhod. (2009). *Marketing in travel and tourism*. Oxford: Elsevier Ltd.
Milano, R., Biaggio, R., Piatelli, R., (2011). The effects of online social media on tourism websites. *18th International Conference on Information Technology and Travel & Tourism*. Innsbruck, Austria, 26–28 January.
Miorandi, D., De Pellegrini, F., Sikari, S., (2012). Internet of Things: Vision, Applications and Research Challenges. *Ad Hoc Networks* 10(7): https://DOI:10.1016/j.adhoc.2012.02.016.
Munar, A. M. and Jacobsen, J. K. S. (2014). Motivations for sharing tourism experiences through social media. *Tourism Managem*ent Vol. 43, pp. 46–54.
Neuhofer, B., & Buhalis, D. (2012). Understanding and Managing Technology Enhanced Tourist experiences. *2nd Advances in Hospitality and Tourism Marketing & Management Conference Proceedings*, Corfu Greece
O'Connor, P. (2003). Online intermediaries — Revolutionising travel distribution. *Travel and Tourism Analyst*, February.
Sethi, P., Sarangi, S., (2017). Internet of Things: Architecture, Protocols and Applications. Journal of Electrical and Computer Engineering: https://doi.org/10.1155/2017/9324035
Sigala, M., Gretzel, U. and Christou, E. (2012), *Social Media In Travel, Tourism And Hospitality: Theory, Practice And Cases*, VT: Ashgate, Farnham, Surrey, Burlington.
Sparks, B. A., Perkins, H. E., & Buckley, R. (2013). Online travel reviews as persuasive communication: The effects of content type, source, and certification logos on consumer behavior. *Tourism Management* 39, 1–9.
Stringam, B. B., & Gerdes Jr, J. (2010). An analysis of word-of-mouse ratings and guest comments of online hotel distribution sites. *Journal of Hospitality Marketing & Management* 19(7), 773–796.
Yoo, K. H., & Gretzel, U. (2011). Influence of personality on travel related consumer-generated media creation. *Computers in Human Behavior* 27(2), 609–621
Zivkovic, R., Gajic, J. & Brdar, I. (2014). The Impact of Social Media on Tourism, *Singidunum Journal of Applied Sciences Supplement*. 758761
Zografos, G., (2004) Innovative hotel products: The role of architecture. *Greek Technical Chronics*, September, 36–38 (In Greek)

The Role of Ethnic Culinary Culture in Destination Attractiveness

Yener OĞAN[1]

Introduction

Culture is the total of elements such as knowledge, art, belief, tradition, habit, behaviour and morality that belong to a society. Ethnicity is a group of people who identify with each other such as language, religion, history, culture, tradition-custom, lineage-ancestry (Yildirim & Ogan, 2024). Therefore, each ethnic community has its own cultural characteristics which form the identities of societies. One of the most important elements that reflect the cultural characteristics of a destination is culinary practices. These practices can be listed as dishes, food and beverage products, cooking techniques, service styles, tools and equipment, and rituals. Ethnic cuisines not only contain the culinary practices of a destination but also differentiate destinations from each other. Therefore, ethnic cuisines have a special place for destinations in terms of culture. Ethnic cuisines best represent the cultural heritage and geographical characteristics of destinations. In recent years, there has been increasing interest among tourists in food and beverages. One of the main reasons for this interest is that food and beverages go beyond being a physiological need as food and beverages offer opportunities such as getting to know, interacting, discovering and experiencing the cultural characteristics of a destination. Therefore, food and beverages have become an important element of destination attractiveness (Mankan, 2017). In this context, the current study explains ethnic cuisine culture by including the concepts of destination and culture and evaluates the role of ethnic cuisine culture in destination attractiveness.

Destination Attractiveness

Destination is a word of French origin, and it is used in the sense of "touristic place" in terms of tourism. In the literature, the word destination has

[1] Associate Professor, Kırşehir Ahi Evran University, Kaman School of Applied Sciences, Department of Gastronomy and Culinary, oganyener@gmail.com, https://orcid.org/0000-0002-1523-8498

meanings such as attraction place, tourism region, touristic place, touristic station, and tourist destination (Cakici & Aksu, 2007; Kozak, 2008; Davras & Uslu, 2019; Ogan, 2021). Ulker (2010) defines the word destination as a place or region that has tourist attractions. According to Golgeli and Ilhan (2024), it is a place where tourists stay temporarily to participate in tourism activities. Therefore, the destination can be a different geographical area such as a country, island or city that offers a touristic experience and contains touristic products. In other words, it can be stated as the places offered to the use and service of tourists having certain characteristics and attractions (Akay, 2019). In this context, some features which destinations should have are called destination attractiveness. According to Morrison (2013), the basic characteristics of a destination are listed as follows:

- There should be a place where tourists can stay.
- There should be transportation, infrastructure, accommodation, food and beverage services for tourists.
- There should be a tourism marketing effort.
- A coordinated organizational structure should be created.
- There should be an image of the destination in the minds of tourists.
- Laws and regulations are required to be introduced by the authority.
- There should be cooperation and coordination among tourism stakeholders.

According to Buhalis (2000), characteristics of the destination consist of attractions, accessibility, facilities, planned tours, events and other ancillary services. According to Mill and Morrison (2012), they are listed as tourist attractions, facilities, accessibility, package tours, events and support services. In this context, the basic elements of destinations are listed in Table 1 (Mill & Morrison (2012; Akyol, 2020).

Table 1. Destination elements

Attractions	Transpor-tation	Businesses	Events	Support Services	Infrastructure
Natural	Highway	Commercial	Festivals	Shopping	Energy
Cultural	Railway	Comple-	Concerts	Banking	Water Road
Artificial	Airway	mentary	Competitions	Healthcare	Other
	Seaway	Private Sector		Local Businesses	

Source: created by the author.

The attractiveness of a destination is formed by various elements such as natural tourism resources (mountain, waterfall, cave, canyon, and lake), cultural tourism resources (mosque, castle, bridge, food and folklore) or artificial tourism resources (Ogan, 2021). In today's conditions, these attractions alone are not sufficient for destinations. Therefore, they must be supported by other destination elements as destination attractiveness is shaped within a wide framework such as transportation, infrastructure, businesses, support services, events, and gastronomy. In this context, the elements that affect tourists' preference for a destination over another constitute the attractiveness of the destination.

Culture and Cuisine

Culture is one of the elements that best reflects the lifestyle of societies in a way culture is the total of material and spiritual values created throughout the historical process and have reached the present day (Ozturk, 2021). Regardless of the cultural characteristics of a society, it has basic components such as language, religion, beliefs, norms, rules, values, attitudes, traditions, behaviour, customs and traditions, symbols, morals, laws, and technology (Akay, 2018). Therefore, all elements belonging to humanity are evaluated under the culture. In this context, the basic characteristics of the concept of culture can be listed as follows (Ogan & Aksoy, 2021).

- It belongs to a certain group or society,
- It has the distinguishing features of a certain group or society,
- It consists of learned behaviours,
- It can be learned, developed and changed later,
- It is passed on from generation to generation,
- It includes values,
- It is shared by individuals,
- It is one of the most basic elements of solidarity,
- It provides information about social structure,
- It is the determinant of social identity,
- It limits individuals by placing them within some rules,
- It consists of elements that complement each other.

According to Ozturk (2021), there are differences between the food habits of societies because of the culture. Elmas (2021) stated that food is one of the most characteristic products that show the difference of cultures. In order to have an idea about the culinary culture of a society, the history of that society, eating habits, the products they produce, and the practices such as processing and storing the products are references (Celik et al. 2021). Throughout history, cultural interactions have enabled each society to create its own culinary culture between societies because of reasons like migration, war, trade and religion. (Sariisik, 2019). The word kitchen literally means a physical space as "the place where food is cooked." In addition, it is an important element that encompasses a culture as "the entire food culture" (Ogan (2021). Therefore, the culinary practices of a society are one of the most characteristic elements that distinguish that society from other societies and reflect its identity. In this context, culinary culture describes a cultural structure shaped by the processes of preparing, cooking, storing and consuming food and beverages, as well as space, tools, traditions, beliefs and practices (Kadioglu, 1997). In other words, cuisine is a part of a community's cultural identity and provides important clues about the history and values of that society.

Ethnic Culinary Culture

The concept of ethnicity can be described as a group of people with language, religion, race, nation, tribe or cultural origins. Akinci (2019) stated that for a group to be an ethnic group, it must be different in all its parts, including language, religion, literature, art, customs and traditions, and social life. The concept of ethnicity is a cultural aspect of a person's identity. When cultural values, beliefs, traditions, practices, languages, religion, food and heritage are shared they can be identified with an ethnic origin. Therefore, ethnicity distinguishes a group from others. In other words, ethnicity can be expressed as a group of people who identify with each other based on common characteristics. These characteristics may include tradition, lineage, language, history, society, culture, nation, religion, physical differences or social characteristics. In addition, thanks to ethnicity, it becomes easier for people living together to recognize their own identities and establish social relationships (Yildirim & Ogan, 2024). In this context, the common characteristics that create an ethnic group are shown in Figure 1.

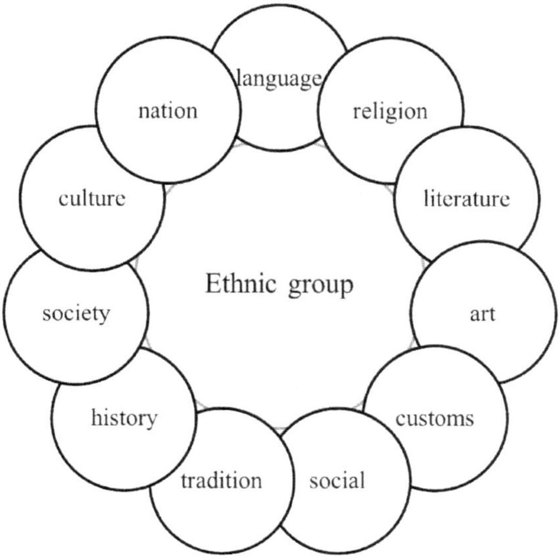

Figure 1. Characteristics of ethnic groups

Source: created by the author

There are similar structures in the cuisines of the world. However, each cuisine has its own unique characteristics. These features are the main elements that distinguish a cuisine from others. For example, some of the ethnic cuisines of the world can be listed as Turkish cuisine (molasses, bulgur, yoghurt, tarhana), Italian cuisine (pasta, pizza), Laz cuisine (kale, anchovy, corn), Indian cuisine (street food, spices), Japanese cuisine (fish, rice), Mexican cuisine (corn, beans, fruit), French cuisine (cheese, sauces). Therefore, each cuisine has its own distinctive elements such as product, dish, food, technique, service, etc. Boch et al. (2021) stated that ethnic cuisines are specific to a community, region, culture or social unit. In this context, ethnic cuisines include foods and beverages belonging to a culture and reflecting the traditions and customs, religions, heritage and origins of the culture to which it belongs (Gudek, 2024). Lee et al. (2014) stated that ethnic food is invaluable cultural treasure as ethnic foods constitute the characteristic products of the cuisine of a region that is socially and culturally accepted by the people (Karizaki, 2017). Therefore, ethnic culinary culture can be expressed as the cultural heritage, foods, dishes and the culinary practices of a society related to the geography to which it belongs. In other words, ethnic culinary culture includes all the dishes, utensils, ingredients, cooking techniques, rituals and serving methods specific to a particular region or culture (Ogan, 2025).

Ethnic Culinary Culture as a Destination Attractiveness

In its simplest definition, tourism can be defined as the travels of people for various purposes such as resting, having fun, seeing, and getting to know each other. According to the World Tourism Organization (2025), tourism is a social, cultural, and economic phenomenon that requires people to travel to countries or places outside their ordinary environment for personal or commercial/professional purposes. People who participate in tourism activities are called tourists. Regardless of the purpose for which tourists travel, eating and drinking behaviours, one of their most basic needs, form an inseparable part of their experiences. In recent years, eating and drinking behaviours have gone beyond being a basic requirement in a way that they have become the main determining factor in choosing a destination (Ogan, 2021). Dogdubay et al. (2011) stated that visiting a destination is the desire to discover culinary cultures and experience new tastes among the expectations of tourists. Therefore, eating and drinking behaviours contain important cues specific to a culture and are an indispensable part of that culture.

Ethnic cuisine refers to the culinary traditions adopted by a society. Each ethnic cuisine has its own culinary practices such as food, ingredients, cooking techniques, presentation and utensils. The most characteristic practices of ethnic cuisines include local dishes and local products. Several studies have shown that local dishes are effective tools for destination discovery and marketing (Cohen & Avieli, 2004; Yuncu, 2010). In this context, the contributions of local elements to a tourism destination are shown in Figure 2 (Ogan, 2021).

When Figure 2 is examined, it can be seen that the local elements make significant contributions to destinations, including "agriculture, originality, identity, advertisement, image, brand, and development." To experience the local flavours of a destination, the differences of the destinations should be revealed (Aydogdu & Duman, 2017) because this situation plays an important role in the destination choice of tourists. Cicek and Pala (2017) stated that destinations tend to create brand value to attract the attention of tourists and influence their purchasing behaviours. Therefore, ethnic cuisines can

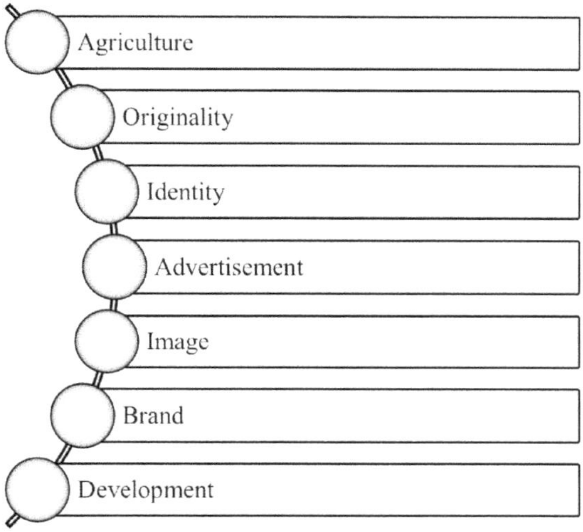

Figure 2. Contribution of local elements to the destination

Source: created by the author

be considered as an opportunity for destinations in this respect. In addition, ethnic cuisines contribute to the development of destinations by creating an advantage in tourism marketing (Aksoy & Sezgi, 2015).

Another aspect that differentiates a destination is that food and beverages specific to the destination are produced and presented in accordance with the ethnic culinary culture. In this way, it is possible to increase tourist demand as well as to ensure awareness of the destination-specific culinary culture. Sevim and Ogan (2021) stated that the most significant contributions of tourism to national economies are expanding trade volume and vitalizing the economy. Therefore, increasing tourist demand provides economic benefits to destinations in many ways such as employment and foreign exchange inflows.

Destinations can combine local culture and tradition thanks to ethnic culinary culture. However, the traditional structure must be preserved. Batmaz and Bicici (2021) drew attention to the fact that the cultural heritage that constitutes all the accumulations of the society it belongs to should be protected and observed with good strategies. Undoubtedly, ethnic culinary culture also has the feature of being a cultural heritage. Therefore, for the sustainability of destinations, ethnic culinary culture, which has the feature of a cultural heritage, should be preserved and passed on to the next generations. Karaduman (2023) stated in his study that foods are to be evaluated in terms of sociological and social dimensions as ethnic dishes stand out as an element that increases social interaction within the social structure. With the social role of ethnic dishes, different cultures come together and a common ground is achieved. Therefore, ethnic dishes also serve as a bridge among the culture, traditions and identities of the destination.

Educational activities for stakeholders, including local people, employees, businesses, and non-governmental organizations, are the elements that increase the quality of destinations. It should be essential to ensure that all stakeholders adopt behaviours which are in line with educational goals (Ogan et al., 2016). In their study, Acuner et al. (2021) stated that educational activities contribute to the development of tourism supporting this situation. Cekic et al. (2023) stated that geographically indicated products play an important role in the competitiveness of destinations and the development of tourism. A large portion of geographically indicated products consists of food and beverages. A significant portion of these products also reflect ethnic cuisines. Therefore, geographically indicated products should be evaluated in the development of the destination. Akkoc (2024) stated that touristic activities contribute to the destination in many ways, such as branding, image, and development. Therefore, organizing activities such as events, festivals, and competitions that promote ethnic cuisines are important practices in terms of increasing the attractiveness of destinations.

Conclusion

Tourists visiting a destination have desires to discover culinary cultures and experience new tastes. Therefore, ethnic cuisines can be an effective tool in discovering and marketing a destination. In this context, the role of ethnic cuisine culture in destination attractiveness was evaluated in the study. Ethnic cuisine refers to culinary traditions adopted by a society. Each ethnic cuisine has its own culinary practices such as food, ingredients, cooking techniques, presentations, and utensils. Ethnic cuisines not only contain culinary practices specific to a destination but also differentiate destinations from each other. In this context, ethnic cuisines are the best representatives of the cultural heritage and geographical characteristics of destinations.

Ethnic cuisines are specific to a community, region, culture or social unit. In this respect, ethnic cuisines have a unique structure and destinations must definitely evaluate this situation since ethnic cuisines can put destinations in an advantageous position in tourism marketing. They have great potential to enhance the competitiveness of destinations and contribute to their branding. Therefore, destinations that can achieve this will increase tourism demand and provide both social and economic benefits. Ethnic cuisines are a cultural heritage. This heritage is to be protected and its sustainability must be ensured. In particular, geographical indication studies can contribute to the protection of this cultural heritage. Educational activities for all stakeholders in destinations can likely provide a sustainable tourism understanding. In addition, destinations should be supported with activities such as events, festivals, and competitions. In this context, ethnic cuisines differentiate destinations from each other and can be accepted as both a sole and supporting element of attractiveness in tourists' destination preferences.

References

Acuner, E., Kucuk, K., Keskin, C., & Ergin, E. (2021). Research on the local residents' attitudes towards tourism: The case of Huser Plateau. *Gümüşhane Üniversitesi Sosyal Bilimler Dergisi, 12*(2), 266–283.

Akay E.S. (2018). *Kültürel Coğrafya: İnsan-Kültür-Mekân.* İstanbul Üniversitesi Açık ve Uzaktan Eğitim Fakültesi Ders Notları

Akay, S.N. (2019). *Investigation of gastronomic tourism as destination attractiveness: an example of Sirince village* (master thesis). Nevşehir Hacı Bektaş Veli University, Institute of Social Sciences, Nevşehir.

Akinci, A. (2019). The origin of nationalism: ethnicity, nation and nation relations. *İnsan ve İnsan, 6*(21), 413–430.

Akkoc, T. (2024). The effect of socio-economic consequences of festivals on residents' tourism support. *GSI Journals Serie A: Advancements in Tourism Recreation and Sports Sciences, 7*(1), 260–279.

Aksoy, M., & Sezgi, G. (2015). Gastronomy tourism and southeastern Anatolia region gastronomic elements. *Journal of Tourism and Gastronomy Studies, 3*(3), 79–89.

Akyol, C. (2020). *A research on the development of tourism in the eastern black sea region: Artvin destination example* (doctoral thesis).

Sakarya Uygulamalı Bilimler University, Institute of Postgraduate Education, Sakarya.
Batmaz, N.Y., & Bicici, G. (2021). A field study on the protection of tangible cultural heritage in Turkey: Kırıkkale-Delice example. *Uluslararası Yönetim Akademisi Dergisi, 4*(1), 97–110.
Boch, A., Jiménez, T., & Roesler, K. (2021). Mainstream flavor: ethnic cuisine and assimilation in the united states. *Social Currents, 8*(1), 64–85.
Buhalis, D. (2000). Marketing the competitive destination of the future. *Tourism Management*, 21, 97–116.
Cohen, E., & Avieli, N. (2004). Food in tourism: Attraction and impediment. *Annals of Tourism Research, 31*(4), 755–778.
Cultural heritage, (2025). *What is ethnic cuisine?* https://yemekulturleri.com/mutfak-tarihi-ve-kulturu/kulturel-miras-ve-etnik-mutfaklarin-onemi/144538 date of access: 28.01.2025
Cakici, A.C. & Aksu, M. (2007). Çekim yeri seçiminde grup etkisi: Yerli turistler üzerine bir araştirma. *Anatolia Turizm Araştırmaları Dergisi, 18*(2), 183–194.
Celik, S., Curoğlu, Ö.B., & Sandikci, M. (2021). Determining the perceptions of the living in the region about the regional cuisine culture: The case of Sakarya. *Journal of Gastronomy, Hospitality, and Travel, 4*(2), 626–636.
Cekic, I., Bulut, Z., & Ogan, Y. (2023). Investigation on domestic tourist attitudes regarding geographically indicated products. *Artvin Çoruh Üniversitesi Uluslararası Sosyal Bilimler Dergisi, 9*(1), 242–255.
Cicek, E., & Pala, U. (2017). A Study on destination branding: Case of Tarsus. *Çağ Üniversitesi Sosyal Bilimler Dergisi, 14*(2), 25–47.
Davras, O. ve Uslu, A. (2019). Effect on destination satisfaction of the destination selection factors: A research on english tourists in Fethiye. *Manas Sosyal Araştırmalar Dergisi, 8*(1), 679–696.
Dogdubay M., Sarioglan M., Saatci G., & Su Eroz S. (2011). *Destinasyon pazarlamasında yerel yönetimlerin gastronomik öğeleri kullanma eğilimlerinin ölçülmesine yönelik bir araştırma.* V. Ulusal Gastronomi Sempozyumu, 26–33.
Elmas, B. (2021). Nutrition and culinary culture of Hakkâri province. *Journal of Social, Humanities and Administrative Sciences, 4*(1), 49–59.
Golgeli, K., & Ilhan, H. (2024). Perceptions of local people of Cappadocia region towards destination brand image. *Gümüşhane Üniversitesi İletişim Fakültesi Elektronik Dergisi, 12* (1), 544–575.

Gudek, M. (2024). Ethnic food consumption: mediating role of food involvement and food neophilia. *Aydın Gastronomy, 8*(2), 265–279.
Kadioğlu, N. (1997). *Türk mutfağının, Akdeniz mutfak kültürünün genel özellikleri yönünden değerlendirilmesinin önemi.* Ankara: Kültür Bakanlığı Hagem Yayınları.
Karaduman, A.U. (2023). Food as a sociological phenomenon; Food and social meaning. *Asia Minor Studies, 11*(1), 107–115.
Karizaki, V.M. (2017). Iranian dates and ethnic date-based products. *Journal of Ethnic Foods, 3*(4), 204–209.
Kozak, N. (2008). *Turizm Pazarlaması.* Ankara: Detay Yayıncılık.
Lee, J.H., Hwang, J., & Mustapha, A. (2014). Popular ethnic foods in the United States: A historical and safety perspective. *Comprehensive Reviews in Food Science and Food Safety, 13*(1), 2–17.
Mankan, E. (2017). The example of gastronomy museums in Turkey as an attractive factor in destination marketing. *Turkish Studies (Elektronik), 12*(4), 641–654.
Mill, R.C., & Morrison, A.M., (2012). *The Tourism System,* (7th ed.) Dubuque, Iowa: Kendall/Hunt Publishing.
Morrison, A.M. (2013). *Marketing and Managing Tourism Destinations.* New York: Routledge.
Ogan, E., Ogan, Y., & Kamaci, A. (2016). Meslek yüksekokulu öğrencilerinin iş hayatına hazır bulunuşluklarının belirlenmesi. *Kastamonu Üniversitesi İktisadi Ve İdari Bilimler Fakültesi Dergisi, 14*(4), 118–143.
Ogan, Y. (2021). Destinasyon Çekicilik Unsuru Olarak Yöresel Mutfak. Karasakal, S. & Doğan, O. (Ed.), *Destinasyon Yönetiminde Yeni Yaklaşımlar* (187–208). Ankara: Nobel Yayıncılık.
Ogan, Y., & Aksoy, M. (2021). Gastronomide Kültür ve Lezzet. Oğan, Y. (Ed.) *Gastronomi Araştırmaları* (233–238). Konya: Çizgi Kitapevi.
Ogan, Y. (2025). An ethnographic insight into the Laz cuisine culture in Artvin (Turkey). International Journal of Gastronomy and Food Science, 101132.
Ozturk, Y. (2021). Evaluation of Turkish cuisine culture from the past to the present. *Journal of Social and Humanities Sciences Research, 8*(67), 582–590.
Sevim, U., & Ogan, E. (2020). Evaluation of causality between tourism sector with the real exchange: Sample of Turkey. *Gümüşhane Üniversitesi Sosyal Bilimler Dergisi, 11*(3), 858–869.

World Tourism Organization, (2025). *Definition of Tourism.* https://www.unwto.org/glossary-tourism-terms. date of access: 28.01.2025
Yildirim, B.E., & Ogan, Y. (2024). A study on the culinary cultures of the societies of the North Caucasus Region. *Artvin Çoruh University International Journal of Social Sciences, 10* (Kafkasya Araştirmalari Özel Sayısı), 36–54.
Yuncu, H. (2010). *Sürdürülebilir turizm açısından gastronomi turizmi ve Perşembe Yaylası.* X. Aybasti-Kabataş Kurultayı, Ankara, 28–34.

Benchmarking in Gastronomy Tourism Management

Gülser YAVUZ[1]

Introduction

The global business world faces the challenge of responding swiftly and with high-quality products to the changing and diversifying needs of customers in a highly competitive environment, while also managing increasing costs. In the global world created by developments such as inter-country and inter-region collaborations, advancements in transportation and communication technologies and liberal policies, there have been many changes in the roles, structures, innovations and business practices of companies (Jain & Yadav, 2006). This environment has forced decision-makers to explore and implement various creative management methods. One of these management models is benchmarking (Purmala & Debora, 2021).

Benchmarking emerged in the 1940s as a component of modern quality management principles and has since undergone many developments (Moriarty, 2011). In the early 1980s, its application by Xerox, which yielded successful results, led to its widespread adoption, making it a popular management approach in the business world. The company defines benchmarking as "the search for industry best practices that lead to superior performance" (Wireman, 2004). Today, with the development of innovative applications (Jack, 2009), benchmarking is used in many fields and has become increasingly popular in both the hospitality and tourism industries (Wöber & Fesenmaier, 2004).

The Concept of Benchmarking

Benchmarking is defined by Camp (1995), a benchmarking expert at Xerox, as a process that allows an organization to improve its performance by comparing its products, services and processes with recognized references.

1 Ass. Prof., Mersin University, Tourism Faculty, Department of Gastronomy and Culinary Arts, gulser@mersin.edu.tr

This method can also be described using Deming's continuous improvement model, the Plan-Do-Check-Act (PDCA) cycle (Tuominen, Kitaygorodskaya & Helo, 2009). The stages of the benchmarking process include: identifying the benchmarking subject and partners, determining data collection methods and competitive gaps, predicting performance, adopting findings, setting goals, developing action plans, implementing and monitoring (Camp, 1995). As a systematic and organized mechanism used to seek successful practices, benchmarking involves observing activities, exchanging information, discovering new methods and adopting and applying them within the context of the organization's conditions (Purmala & Debora, 2021).

Benchmarking can be applied in four ways: "Internal Benchmarking," which compares practices within the organization; "Competitive Benchmarking," which evaluates practices against direct competitors; "Functional Benchmarking," which involves comparing with leading organizations outside the sector; and "General Benchmarking," which refers to examining the best practices recognized nationally and internationally (Sitnikov & Giurca Vasilescu, 2008). Internal benchmarking can help organizations identify their strengths, weaknesses, opportunities and threats (SWOT), thus enhancing their economic efficiency. As the first learning step before other types of benchmarking, internal benchmarking is the most commonly applied form (Wöber, 2002).

Benchmarking in Tourism Industry

The tourism industry, which has become a giant sector due to millions of people traveling, has been deeply shaken, particularly due to the COVID-19 pandemic. Entrepreneurs in the tourism industry are using this critical situation as an opportunity to examine themselves and explore opportunities to adopt new models (Csapody, Ásványi & Jászberényi, 2021). Today, tourism businesses need to continuously research and analyze both internal and external environments, effectively collect and process information using information technologies and follow "best new practices" such as flexibility, creativity, innovation and environmental friendliness (Xiang, 2017).

When examining studies conducted in the tourism industry, it can be said that the benchmarking method has been successfully implemented (Martin & Tomas, 2012). However, tourism covers a vast area consisting of different sectors and industries. Therefore, performance metrics alone, with specific indicators, cannot comprehensively reflect the status of this economic and social activity (Zins, 2014). Terziyska (2017) emphasizes that given the broad scope of activities covered by the tourism sector, there are various types of benchmarking, each with its unique characteristics. Benchmarking in tourism can include comparisons of tourism businesses, tourist destinations, non-profit organizations (tourism associations, destination management organizations, national tourism agencies, etc.), cultural institutions (museums, galleries, etc.) and events.

Wöber (2002) examines these applications under three headings: comparison of for-profit tourism businesses (e.g., accommodation providers, restaurants, tour operators, airlines), comparison of non-profit tourism businesses (e.g., tourism organizations, etc.) and destination comparison (national, regional or local comparisons). For instance, Zins (2014) compared the activities of regional tourism organizations related to destination management. Another example could be a program launched by the UK government around 20 years ago. This program, which aimed to improve the performance of small and medium-sized enterprises in various sectors, including food production and tourism and to explore potential future challenges, is quite striking. Under this program, known as the Industry Forum Adaptation Program, funded by the government, eighty-nine "best practice" businesses from the tourism industry (hotels, bars, restaurants, entertainment activities, etc.) were examined (Kyriakidou & Gore, 2006). The OECD conducts comparative studies of national tourism policies in countries as part of best practices monitoring and learning and publishes reports on these studies for many countries (Zins, 2014).

Kyriakidou & Gore (2006) highlighted the importance of organizational culture and found that the best-performing SMEs in the tourism industry share cultural elements that consist of supportive values such as building the future together, collaboratively determining missions and strategies, teamwork and organizational learning.

Benchmarking in Gastronomy Tourism

Gastronomy is a broad discipline related to enjoying the best in food and beverages, encompassing everything that involves what is eaten and drunk

(Scarpato, 2002). The food and beverage market is not a single market, but rather a collection of various markets, each with different products, processes and needs (Jain & Yadav, 2006). According to the World Tourism Organization (2019), gastronomy tourism is a type of tourism characterized by experiences related to food and related products and activities that visitors engage in while traveling. It includes authentic, traditional and/or innovative culinary experiences, as well as activities such as visiting local producers and participating in food events.

In gastronomy tourism, although competitive benchmarking is often preferred, broader comparisons beyond just sector evaluations can unlock innovative practices (Sitnikov & Giurca Vasilescu, 2008). For example, marketing strategies from different industries may incorporate the presence of well-regarded individuals in advertisements and such strategies can be used as a model to evaluate the promotion of a destination or gastronomy event. Today, influencers who participate in promotions in the cosmetics industry can also be encouraged to advertise in restaurants, hotels or festivals.

When examining scientific research on benchmarking, it is evident that studies related to gastronomy tourism have increased in recent years. This study explores the benchmarking method in the field of gastronomy tourism by reviewing international literature, focusing on categories such as destinations, agriculture, food industry, restaurants, accommodation businesses and events.

Destination Benchmarking

In recent years, destination management has significantly changed towards more professionalism and goal orientation, largely due to the contributions of regional tourism organizations (Zins, 2014). Benchmarking is also applied in destination management studies within tourism. Wöber & Fesenmaier (2004) used the benchmarking method to examine the efficiency of state tourism advertising programs in the United States. The results of the study, which employed Data Envelopment Analysis, showed significant efficiency differences between state tourism advertising programs. It was found that while some states' tourism advertising efforts were highly efficient, the majority were inefficient.

Kozak (2002) conducted a benchmarking study comparing the destinations of Turkey and Mallorca. In the research, which was carried out through interviews, observations and surveys, the satisfaction levels of British tourists were compared and the performance of these destinations was analyzed. The results showed that the biggest differences were in local transportation services, hospitality, customer service and price levels. Terziyska (2017) conducted a benchmarking study on wine tourism destinations in Bulgaria. In a study conducted in several stages, an indicator system adapted to current conditions was developed, surveys were conducted with wine producers and data were collected and analyzed using winery-specific indicators. A wide range of metrics were established, from wine quality and price to production area, tourist numbers and profiles. Similarly, in a study by Alebaki, Liontakis & Koutsouris (2020), wine tourism destinations were examined. The research, which was conducted with wine producers on the island of Crete, aimed to identify factors that would increase resilience in wine tourism through benchmarking.

Gastronomy tourism is seen as the type of tourism that is most compatible with sustainable destination development due to its reliance on local resources and the possession of authentic food and beverage cultures (Williams & Williams, 2022). In this context, it is undoubtedly important to emphasize the vital importance of sustainability in the field of gastronomy. Today, there is increasing effort to preserve local products, support producers and protect, maintain and promote intangible cultural heritage elements. The slow food movement, the certification of geographically indicated products and practices such as UNESCO Creative Cities of Gastronomy are indicators of this approach.

Benchmarking studies are also being conducted in the field of sustainability. For example, Nair, Hussain & Ragavan (2015) in their research sought answers to how innovations and new practices in Asia could contribute to the development of more sustainable and responsible rural tourism through benchmarking. They also presented some of the best innovative approaches that could be adopted from countries outside of Asia. The authors noted that despite the global focus on rural tourism, there remains a perception and concern about "greenwashing."

Benchmarking in Agriculture and the Food Industry

The plant-based and animal-based foods that form the foundation of gastronomy are obtained through agricultural activities. These food and beverage products are either directly presented to consumers as fresh items or processed in the industry to be transformed into packaged foods. Although agricultural activities have developed and production has increased with technological advancements, the availability and affordability of food have become increasingly difficult due to recent global developments (FAO, 2023). Despite this negative situation, it has been emphasized that scientific studies in the field of benchmarking in agriculture have been delayed and insufficient (Jack, 2012).

De Salvo et al. (2013) used the benchmarking method to compare the extra virgin olive oil valuation and promotion experiences in Italy and Spain, two world leaders in high-quality olive oil production. Olive farming is an inseparable part of local culture and it is understood that in both countries, olive oil has developed a new awareness, being associated with socio-economic benefits as a strong identity element of national culture. It was highlighted that developing and promoting olive oil strategies have great potential for the development of gastronomy tourism and could be useful for increasing the income of agricultural businesses.

The food industry consists of many interconnected elements, including supply chains, food producers, processors, wholesalers, storage facilities, retailing, logistics and customers (Tuominen, Kitaygorodskaya & Helo, 2009). Due to its industry scope and economic contribution, it is considered highly significant. Demographic changes in society, changes in consumer behavior, women's participation in the workforce and the increase in dining outside the home have significantly affected and developed the food and beverage sector (Jain & Yadav, 2006).

Mann (1998) states that the food sector in the United Kingdom is the largest industry, employing a large number of people and accounting for 9% of the Gross Domestic Product. However, due to the long-standing trade deficit, the industry required a mechanism to be competitive on a global scale. In 1996, the "Food and Beverage Industry Comparative Assessment and Self-Assessment Initiative" was launched in the country, creating an opportunity for businesses to develop practices such as learning and catching up with world standards. This initiative includes business excellence conferences, comparative assessment events, publications to spread best practices in the sector and self-assessment programs according to the European Business Excellence Model. The researcher examined the best practices in food companies with high scores according to the European Business Excellence Model criteria through interviews and observations. These criteria are presented under the headings: leadership, policy and strategy, human resource management, resource management, process management, customer satisfaction, environmental and social impact and result evaluations.

The supply chain is defined as a system of interconnected materials and information flows involving material suppliers, production facilities, distribution services and customers (Tuominen, Kitaygorodskaya & Helo, 2009). The food supply chain has significant impacts on sustainability, including fulfilling human needs, providing employment and economic growth and its effects on the natural environment. Yakovieva, Sarkis & Sloanetkin (2009) note that benchmarking focuses on data analysis and uses many tools. However, applications that fully integrate sustainability elements into supply chain benchmarking are limited. The authors conducted interviews with supply chain experts in the United Kingdom and created an index consisting of economic, social and environmental sustainability dimensions in agriculture, food processing, wholesale, retailing and catering. The analysis revealed that the chicken supply chain performed better than the potato supply chain in terms of sustainability performance.

Tuominen, Kitaygorodskaya & Helo (2009) examined the differences between Finnish and Russian supply chains to identify the causes of low productivity in the Russian food industry. In the field of supply chain performance measurement, the Supply Chain Operations Reference model is a commonly used benchmarking tool. In the comparison, interviews were conducted with food industry company managers. The results showed that the productivity gap between Finnish and Russian food industry companies was largely due to differences in the business environment and the technology used. It was determined that the logistics costs for Russian businesses were approximately twice as high as those of Finnish companies.

Benchmarking in Hospitality Businesses and Restaurants

The Ritz-Carlton Hotel is one of the most well-known examples of benchmarking in the hospitality industry. In its efforts to meet quality goals in 1996, the hotel identified 19 critical processes that were vital to its customers. Among these, demands such as clean, spacious, fully-equipped rooms and services that are better, faster, cheaper and more reliable were highlighted. Standards were developed based on these criteria (Camp, 1995).

Jessica Hwang & Lockwood (2006) examined the barriers to small and medium-sized enterprises (SMEs) in the UK tourism sector adopting best practices. Through interviews conducted with award-winning businesses, eight key areas of best practice were identified and analyzed within SMEs. The findings indicated that customer-focused goals, planning and control, partnerships and networking, internal and external communication, consistent standards, strategic workforce management, cash flow and performance management were factors that facilitated the adoption of best practices. Barriers to implementation included changes in demand, lack of skilled labor, absence of competitive benchmarking and location-related challenges.

In gastronomy, high-quality restaurant service is also an important factor that influences tourist routes and motivates travel (Nezdoyminov Iaromenko & Bedradina, 2023). The indicators of high-quality service can be more clearly observed in restaurants. For instance, guides such as Gault & Millau and the Michelin Guide, which rate restaurants, provide clear criteria. Restaurant managers can both strive to achieve these ratings and use these indicators as a reference for self-assessment.

Csapody, Ásványi & Jászberényi (2022) used the benchmarking method to investigate the sustainable practices of high-quality restaurants in Europe. The research focused on restaurants listed in the Michelin Guide and those that have received the Michelin Green Star for Gastronomy and Sustainability. The performance criteria included the procurement of raw materials and menu items, restaurant operations and design, local community involvement and social responsibility.

In their research, Nezdoyminov Iaromenko & Bedradina (2023) examined the quality and safety of local restaurant services in Odessa, one of Ukraine's most popular tourist destinations, which was severely impacted by the COVID-19 pandemic, using the benchmarking method. Reviews from TripAdvisor, filtered for COVID-19 safety, were recorded and analyzed according to various indicators. Benchmarking indicators included the "Protocols of the Safe Travels Program: Global Protocols and Stamp for the New Normal" and six indicators compliant with quarantine restrictions for food businesses in EU countries. These indicators were: the maximum number of people per table, table spacing, the wearing of masks and gloves by employees and guests, temperature checks, use of disinfectants, order completion time, service culture quality, guest satisfaction with employees and the alignment of menu items with consumer expectations. The authors developed guidelines and suggestions aimed at improving service quality and customer satisfaction.

Benchmarking in Events

Events are considered an important motivational factor in tourism and a key element in the development of many destinations (Getz, 2008). In gastronomy tourism, various events such as festivals, fairs, competitions, scientific events, workshops and gastronomy routes can be organized either independently or in combination with other events.

Millions of tourists are attracted to destinations through gastronomy events held around the world. Examples of prestigious international events include culinary competitions such as Bocuse d'Or Europe, Culinary Olympics, San Pellegrino Young Chef, festivals like Taste of Amsterdam, Oktoberfest, Bordeaux Wine Festival, Taste Florence Food Festival and food fairs like Anuga, SIAL Paris, Gulfood and WorldFood Moscow. What distinguishes these events from the many others held, most of which remain local, is a critical question in the context of benchmarking.

Benchmarking can be applied in gastronomy events through a comprehensive review to identify best practices. By systematically examining the activities of event organizers from the planning stage to the execution and analyzing every step in the process, ideas for standardization and sustainability can emerge. Csapody, Ásványi, & Jászberényi (2021) examined the state of gastronomy events following the COVID-19 pandemic. In their research, which employed the benchmarking method, they identified high-quality gastronomy festivals in Europe and investigated which sustainable practices were being implemented. The findings were grouped into four main dimensions: ingredients and dishes, organization, communication and programs and local communities and culture. The study showed that popular gastronomic trends, such as the slow food movement, zero-kilometer food and farm-to-table concepts, had an impact on festivals focused on sustainability, with local products receiving more emphasis in the themes of gastronomy festivals. The results highlighted that the concept of "thinking locally" was exemplified across all dimensions of the events.

Results

The adoption of benchmarking methods in the field of gastronomy can enable business managers, event organizers and destination managers in the industry to gain a sustainable competitive advantage and remain competitive. Based on the findings of various studies in the literature, there are some key considerations that should be taken into account. First and foremost, it is important to determine which comparison indicators and criteria should be included in a comparative analysis. Increasing and diversifying research in this area could prove to be beneficial.

Today, industrially produced food products, which offer ready, fast and convenient options for consumers, are gradually replacing traditional dishes made from raw ingredients (Jain & Yadav, 2006). While this trend fosters industry development, it also poses a threat to gastronomy tourism by moving away from authenticity and locality. Local products can be replaced by fast, ready-to-eat food items. Indeed, the motivation of gastronomy tourists is to experience the destination's unique local cuisine culture and authentic elements (Williams & Williams, 2022). Therefore, examples of successful gastronomy cities that preserve and promote local gastronomic elements should be followed. In this regard, data collection methods such as reports from organizations like the WTO and primary insights from tourists visiting the destination can be utilized. It is recommended to examine destinations that have successful sustainability practices, such as preserving agricultural areas, promoting slow food and green initiatives and supporting organic farming, within the scope of benchmarking.

References

Alebaki, M., Liontakis, A., & Koutsouris, A. (2020). Benchmark analysis of wine tourism destinations: Integrating a resilience system perspective into the comparative framework. In Proceedings of the 2nd International Research Workshop in Wine Tourism, Santorini, Greece.

Camp, R.C. (1995), Benchmarking –The Search for Industry Best Practices that Lead to Superior Performance, ASQC QUality Press, Milwaukee.
Csapody, B., Ásványi, K. & Jászberényi, M. (2021). Sustainable best practices or european culinary festivals. *ToSEE*, 6, 155–165.
Csapody, B., Ásványi, K., & Jászberényi, M. (2022). Green paths for foodies after COVID-19: Sustainable practices of European high-end restaurants. In Current issues in tourism, gastronomy and tourist destination research (pp. 378–385). Routledge.
De Salvo, P., Hernández Mogollón, J. M., Di Clemente, E. y Calzati, V. (2013). Territory, tourismand local products. The extra virgin oil's enhancement and promotion: a benchmarking Italy-Spain. *Tourism and Hospitality Management*, 19(1), 23–34.
The State of Food and Agriculture (FAO) (2023). Revealing the true cost of food to transform agrifood systems. https://literatur.thuenen.de/digbib_extern/dn059499.pdf
Getz, D. (2008). Event tourism; definition, evolution and research. *Tourism Management*, 29(3),pp. 403–428
Jack, L. (2009). Benchmarking in food and farming, Routledge Publishing
Jain, R., & Yadav, O. P. (2006). Driving improvement opportunities in the food processing industry through benchmarking. *International Journal of Process Management and Benchmarking*, *1*(4), 352–369.
Jessica Hwang, L. & Lockwood, A. (2006), Understanding the challenges of implementing best practices in hospitality and tourism SMEs," *Benchmarking: An International Journal*, 13 (3), pp. 337–354.
Kyriakidou, O., & Gore, J. (2005). Learning by example: Benchmarking organizational culture in hospitality, tourism and leisure SMEs. *Benchmarking: An International Journal*, 12(3), 192–206.
Kozak, M. (2002). Destination benchmarking. *Annals of Tourism Research, 29*(2), 497–519.
Mann, R., Adebanjo, O., & Kehoe, D. (1998). Best practices in the food and drinks industry. *Benchmarking for Quality Management & Technology, 5*(3), 184.
Martin, L., & Tomáš, K. (2012). Tourism destination benchmarking: Evaluation and selection of the benchmarking partners. *Journal of Competitiveness*, *4*(1), 99–116.
Moriarty, J. P. (2011). A Theory of Benchmarking. *Benchmarking: an International Journal*, 18(4), 588–611

Nair, V., Hussain, K., Lo, M. C., & Ragavan, N. A. (2015). Benchmarking innovations and new practices in rural tourism development. *Worldwide Hospitality and Tourism Themes, 7(5), 530–534.*
Nezdoyminov, S., Iaromenko, S., & Bedradina, G. (2023). Safety and quality of restaurant service as factor of restoring tourist mobility in the gastronomic tourism destination of Ukraine. *Almatourism-Journal of Tourism, Culture and Territorial Development*, *13*(24), 117–137.
Purmala, Y. A., & Debora, F. (2021). A systematic literature review of benchmarking implementation in various industries. *Indonesian Journal of Industrial Engineering & Management*, *2*(1), 35–51
Sangwan, K. S., & Choudhary, K. (2018). Benchmarking manufacturing industries based on green practices. *Benchmarking*, *25*(6), 1746–1761.
Scarpato, R. (2002). Gastronomy as a tourist product, The perspective of gastronomy studies. In A.M. Hjalager ve G. Richards (Eds.), Tourism and gastronomy (pp. 51–70). Routledge.
Sitnikov C, Giurca Vasilescu L (2008). A global vision over benchmarking process: benchmarking based enterprises, MPRA paper, https://EconPapers.repec.org/RePEc:pra:mprapa:9482
Terziyska, I. (2017). *Benchmarking wine tourism destinations – the case of Bulgaria. International Journal of Wine Business Research, 29(4), 384–400.*
Tuominen, T., Kitaygorodskaya, N., & Helo, P. (2009). Benchmarking Russian and Finnish food industry supply chains. *Benchmarking: An International Journal, 16(3), 415–431.*
UNWTO (2019). Guidelines for the Development of Gastronomy Tourism. https://www.e-unwto.org/doi/pdf/10.18111/9789284420957.
Williams, H.A. & Williams Jr, R.L. (2022). A Process Map for Creating a 6+ Gastro-cluster Network: Facilitating Sustainable Gastronomic Destination Development in Africa and the World. *African Journal of Hospitality, Tourism and Leisure,* 11(SE2):1765–1786.
Wireman, T. (2004). *Benchmarking best practices in maintenance management.* Industrial Press Inc.
Wöber, K. W. (2002). Benchmarking Studies in the Tourism and Hospitality Industries. CABI Publishing
Wöber, K. W., & Fesenmaier, D. R. (2004). A multi-criteria approach to destination benchmarking. *Journal of Travel & Tourism Marketing, 16*(2–3), 1–18.
Yakovieva, N., Sarkis, J., & Sloan, T. W. (2009). Sustainable benchmarking of food supply chains. *George Perkins Marsh Institute Working Paper,2.*

Xiang, Z., Kothari, T., Hu, C., & Fesenmaier, D. R. (2007). Benchmarking as a strategic tool for destination management organizations: a proposed framework. *Journal of Travel & Tourism Marketing*, *22*(1), 81–93.
Zins, A. H. (2014). Internal Benchmarking for regional tourism organizations: a case example. *Tourism Analysis*, *19*(4), 413–424.

Spatial Perception and Social Interaction in Café Spaces: Customer Experiences

Fatma KOÇ[1]

Introduction

Sensory perceptions shape experiences; therefore, the concept of 'place', which helps us internalize the world, plays a crucial role in architectural spaces. When a space becomes individualized through our senses and the meanings we attribute to it, it gains the identity of a 'place'. Every place has its own unique identity. Yi-Fu Tuan (1977) defines place as a space endowed with human characteristics. At this point, it is essential to discuss the concept of '*genius loci*' or the 'spirit of place', which grants a space its distinctive character and atmosphere.

This concept, which denotes architecturally and experientially memorable spaces, is explained by Christian Norberg-Schulz (1980) as follows: 'A person can truly live in a place only when they can define themselves in relation to their surroundings, adapt to them, or, in short, meaningfully experience their environment'. The places that hold significance for us and become unforgettable are not merely physical spaces but lived environments intertwined with memories, senses, and experiences.

Spatial perception enables individuals to comprehend their surroundings through sensory experiences and interact with their environment. In this context, the milieu, or setting, refers to the space where individuals coexist and engage in interactions. The concept of milieu is not merely confined to a physical structure; rather, it encompasses social interactions among individuals and should be considered alongside processes of socialization. The dynamic relationship between space and individuals is strikingly encapsulated in Winston Churchill's statement: 'We shape our buildings; thereafter, they shape us' (Edgü, 2010).

Both space and individuals undergo transformation over time, making the way individuals define, perceive, and behave within a given space critically

1 Asst. Prof. Siirt University, School of Social Sciences, Department of Hotel, Restaurant and Catering Services, fatma.koc@siirt.edu.tr

important. Spatial experience is not limited to visual perception alone; auditory, tactile, olfactory, and kinesthetic inputs also play a significant role in constructing a meaningful understanding of the environment. As individuals navigate a space, they learn about their surroundings through sensory experiences. The sounds of traffic or birds chirping, the sensation of a cool breeze, the texture of cobblestone pavement beneath their feet, the scent of the sea, or the aroma of freshly baked bread from a nearby bakery all contribute to spatial awareness (Edgü, 2021: 219).

The Concept of Perception and Spatial Perception

Perception is the process through which an individual transmits impressions acquired from the environment to the brain via sensory organs, encompassing both physical and psychological dimensions. According to Ramachandran and Blakeslee (1999), perception involves not merely the transmission of sensory input but also the formation of mental representations. Three fundamental factors influence perception: the characteristics of the perceiver, the perceived, and the environment. For instance, while physical and social conditions determine the quality of perception, past experiences, motivations, and cultural biases also shape it.

On the other hand, spatial perception differs from object perception in its structure. This distinction arises from factors such as the scale's magnitude and complexity, the observer's role as a dynamic part of the environmental system, and their goal-directed movement (Bar and Neta, 2006; Ramachandran and Hirstein, 1999). According to Friedenberg and Bertamini's (2015) perceptual uncertainty hypothesis, fragile forms such as triangles are perceived as less attractive.

Perception is the process of making sensory information meaningful, allowing us to interpret the external world through mental images. Perception is not simply the transmission of sensory input to the brain; it is a process that is shaped by both constant and changing visual stimuli. Our brain assumes that the environment is orderly and stable, which allows us to recognize objects under different conditions. The human body serves as a reference point for perceiving the size of surrounding objects. For example, children may control the space under a table based on their own size. Similarly, tall buildings or crowded environments evoke feelings of power, authority, or constriction. These perceptions are shaped by the structure of the space and the way we interpret that space (Edgü, 2021: 224).

Café Spaces

Cafés are spaces where individuals choose to relax, read books, listen to music, engage in conversation, and consume food and beverages while socializing in their daily lives. Although the café industry has significantly evolved, many establishments still face various issues related to physical comfort. To enhance customer satisfaction, café operators must consider multiple factors, including the quality of food and beverages, staff performance, hygiene standards in the kitchen, spatial organization, thermal comfort, visual comfort, and acoustic conditions. However, structural comfort is often overlooked, and user experience is not sufficiently prioritized. Given their significant role in social life, cafés, like any other built environment, must ensure optimal comfort conditions to enhance customer satisfaction (Güven and Şenkal Sezer, 2019: 184).

Today, cafés have evolved into spaces where people can both meet their dining needs and socialize, influenced by changes in modern lifestyles. Over time, these establishments have undergone transformations and adopted diverse concepts. This variety allows a café to cater to different expectations simultaneously. Contemporary cafés serve as multifunctional environments where individuals can spend time with friends while also providing a setting for business meetings. Additionally, they have become social venues for celebrating special occasions and hosting small events. (Dedeoğlu & Doğan, 2021: 133).

Method

The aim of this study is to understand the relationship between spatial perception and social interaction in café environments. In this context, the study was conducted with 18 participants from a globally operating retail coffee chain offering premium coffee and café experiences. The study focuses on the impact of spatial factors (such as the taste of coffee, crowding, flavors offered in the space, volume of music, etc.) on customer experience and spatial perception.

Throughout the research process, field notes were continuously analyzed, and qualitative data were examined using an inductive content analysis approach.

Since codes and sub-codes were derived through interaction with the data, an inductive approach was adopted. During the data coding process, the data were summarized, and main codes were created. These main codes accurately represented the research questions. Based on the information obtained from the participants, sub-codes were generated considering word frequency. The resulting codes were then connected, and a hierarchical coding model was established. Visual representations were created using the code-sub-code model.

Findings

The table and visuals below were created from the data obtained from the participants.

Table 1. Demographic data of the participants

	Age	Job	Gender	Frequency of choosing this place
P1.	24	Student	Male	3–4 times in a week
P2.	32	Sales consultant	Female	1–2 times in a week
P3.	40	Tradesman	Male	Rarely
P4.	21	Student	Male	Almost every day
P5.	48	Lawyer	Male	3–4 times in a week
P6.	34	Accountant	Male	Once every 2 weeks
P7.	28	Graphic designer	Female	Almost every day
P8.	37	Teacher	Female	3–4 times in a week
P9.	29	Engineer	Male	3–4 times in a week
P10.	35	Unemployed	Female	1–2 times in a week
P11.	22	Student	Female	3–4 times in a week
P12.	41	Psychologist	Female	Once every 2 weeks
P13.	27	Barista	Male	3–4 times in a week
P14.	30	Teacher	Male	Almost every day
P15.	19	Student	Female	3–4 times in a week
P16.	36	Housewife	Female	1–2 times in a week
P17.	32	Ad. designer	Female	3–4 times in a week
P18.	24	Student	Female	3–4 times in a week

The sample group consists of 8 male and 10 female participants, aged between 19 and 48, with diverse professional backgrounds. The participants were also asked about the frequency with which they choose this café.

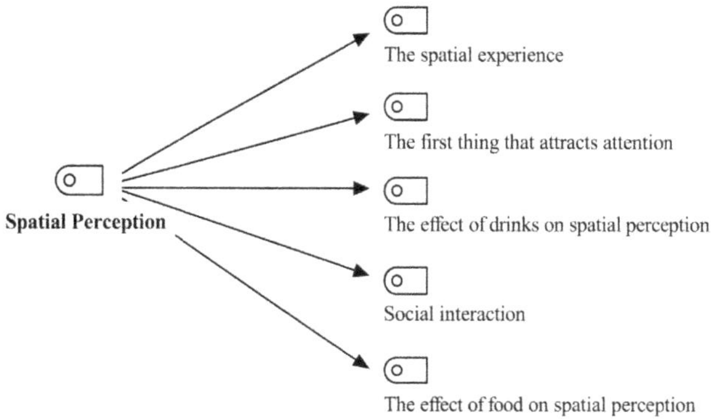

Figure 1. Hierarchical code model of the study

The hierarchical code model of the study is given in Visual 1. Based on the responses provided by the participants to the study questions, the codes include the spatial perception they associate with the café, the first noticeable element in the space, the food offered in the space, the beverages offered, and the social interaction within the space.

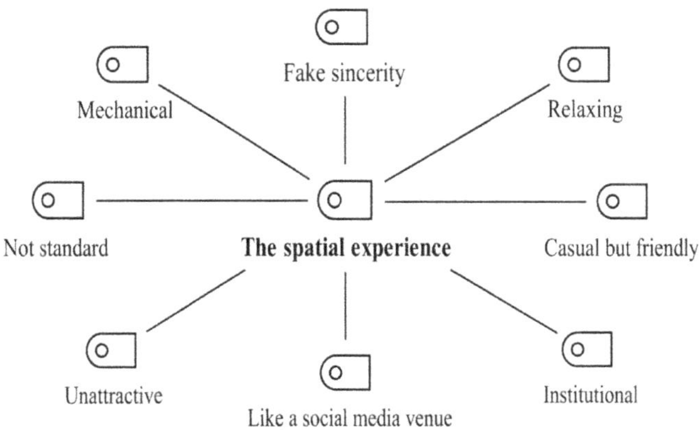

Figure 2. Code-subcode model for spatial experience

The participants' responses to the question 'What is your perception of this space?' are provided in Figure 2. Some participants described the café as providing a relaxing atmosphere, attributed to the scent of coffee, the lighting, and the soft music (P1). In contrast, others characterized it as a place that conveys a mechanical, planned, and corporate feel (P3, P8). A few participants also described the space as 'the real-life version of social media', noting a sense of fake sincerity and politeness (P10, P15).

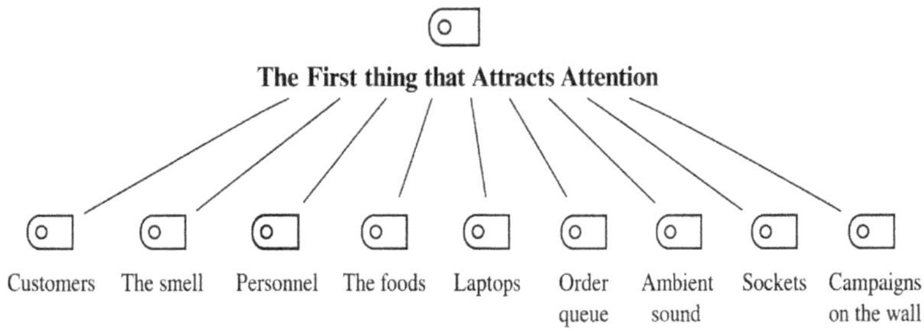

Figure 3. The first thing that attracts attention in the place

The participants' responses to the question 'What is the first thing that captures your attention here?' are presented in Visual 3.

Some participants mentioned the smell of freshly roasted coffee beans (P1, P2, P18), background music (P3) and the long queue (P4) when the café is crowded. The campaigns on the wall (P5, P6, P10) were also frequently highlighted. The display of desserts and sandwiches in the display case (P7, P8) was prominent. Participants also mentioned other elements such as the customer profile (P9, P17), the prevalence of laptops on the tables (P12). Some participants focused on the fast movement of employees (P11,P14, P15) and the availability of sockets (P16).

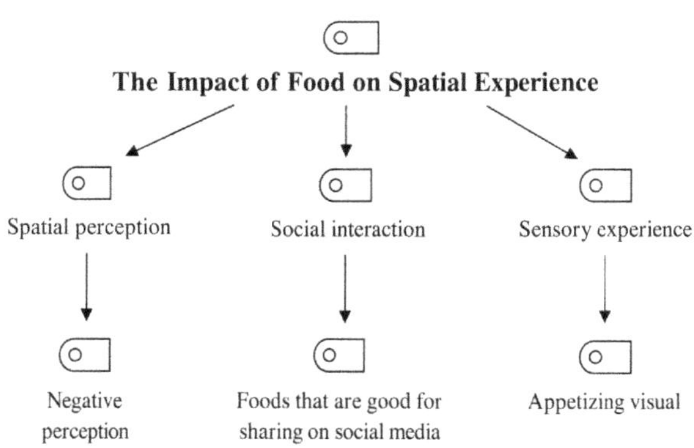

Figure 4. The impact of food on spatial experience

The responses of participants to the question, 'What is the effect of the food offered at this space on spatial perception?' are presented in Visual 4.

It was found that the main reason for the negative perceptions of food among participants (P2, P3, P5, P7, P9, P16) stemmed from an imbalance between price and performance. Despite the high prices of the food, participants expressed that the expected taste was not achieved. On the other hand, some participants (P6, P10, P12, P14, P17) stated that the food appeared appetizing. However, rather than being appetizing, some participants (P1, P4, P8, P11, P13, P15, P18) noted that the food was more suitable for sharing on social media platforms.

This perception suggests that, for some participants, the food served in the café is valued more for its visual appeal and potential to enhance one's social media presence rather than its actual taste or quality. These participants seem to prioritize the aesthetic qualities of the food—such as its presentation, color, and styling, over its practical or sensory attributes. This trend could reflect the growing influence of social media in shaping consumer behavior, where visual content and the act of sharing experiences online have become a significant part of dining culture.

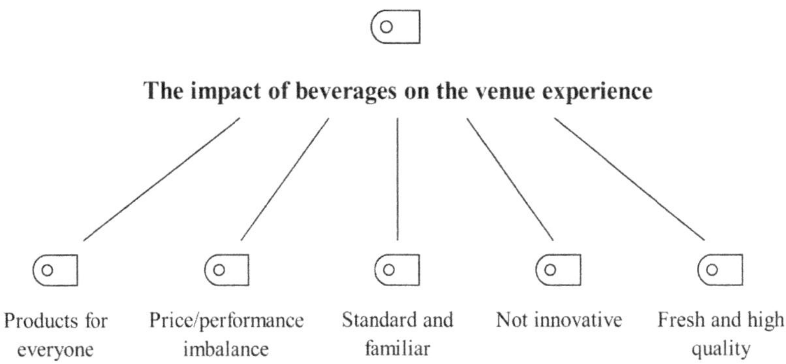

Figure 5. The impact of beverages on the venue experience

The responses of participants to the question, 'What is the effect of the beverages offered at this space on spatial perception?' are presented in Visual 5.

Many participants expressed satisfaction with the wide variety of beverages and their ability to cater to different tastes. This group (P2, P12, P13) generally emphasized alternatives beyond coffee and showed an interest in exploring different flavors. However, some participants (P5, P16) complained about the price/performance imbalance despite the high quality of the beverages. Although the beverages were of high quality, it was noted that the prices were high and that beverages of the same quality could be offered at more affordable prices. While some participants (P9, P11) mentioned that certain aspects of the beverages, such as their standard and familiarity, influenced their perception of the venue, others found the lack of innovation in the beverage options and the excessive similarity to older versions of updates (P8, P15, P17) to make the venue feel monotonous.

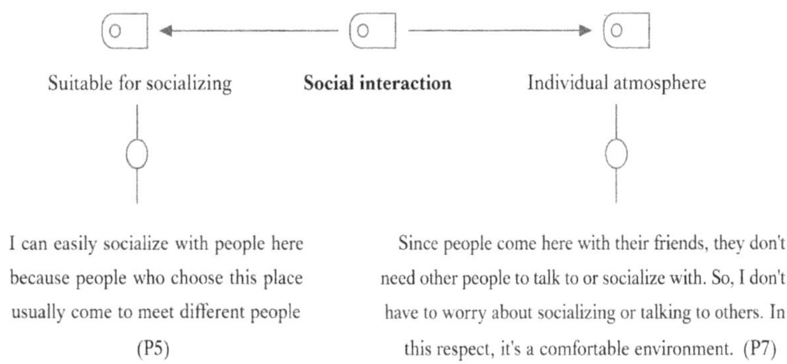

Figure 6. Single case model of social interaction in space

In Visual 6, a single case code model is given in line with the participants' answers to the question 'How would you evaluate the social interaction in this place?'.

The responses from P5 and P7 reveal two contrasting perspectives on social interaction in the venue, reflecting the diversity of social preferences. P5's response, categorized under 'Suitable for socializing', indicates that the venue is conducive to meeting and interacting with different people. This participant finds it easy to socialize here because they perceive the environment as one where others are also looking to meet new people, which facilitates social interaction. On the other hand, P7's response, categorized under 'Individual atmosphere', reflects a different approach to socializing. This participant suggests that the venue's atmosphere is more focused on personal interactions, where people typically come with their friends and do not feel the need to engage with others. For P7, this creates a comfortable environment where socializing or talking to strangers is not a concern, and the space feels more individualistic rather than collective.

These responses suggest that the venue caters to different social needs: some patrons are looking for opportunities to interact with strangers, while others prefer a more private and familiar social setting. Both types of experiences contribute to the venue's overall social dynamic, offering a range of environments for different social preferences.

Conclusion

In recent years, changes in the labor market and developments in transportation and communication infrastructures have significantly transformed social interaction patterns. Professional activities and leisure activities are now mostly carried out outside the home. This means that interactions between acquaintances and strangers are not limited to shared spaces (Steigemann, 2017: 45).

One of these shared spaces is the café. While this situation reshapes the role of spaces in social interaction, the unique characteristics of each space have also affected how people perceive and give meaning to that space. By giving spaces an identity (Apaydın & Eren, 2013), it has differentiated the ways people perceive and give meaning to café spaces.

Café spaces in the food and beverage sector play an important role in individuals' daily lives as areas where they prefer to rest, chat with friends, meet new people, and social activities such as studying, working on their computers, and enjoying snacks and drinks. In the context of cafe businesses, deeply analyzing the mechanisms of spatial perception and how individuals interpret this space constitutes a research area in revealing the complex dynamics of interaction between people and space. This study aims to address elements such as food and beverages offered in the space, striking elements in the space, and social interaction with a holistic approach.

In the perception of space, it is observed that the café ambiance is interpreted in various ways. While one group evaluates the space as an environment that provides calmness and peace, another group defines the area as an overly organized, impersonal and mechanical atmosphere. In addition, some participants describe the space as a reflection of social media in real life and draw attention to the artificiality of the politeness displayed. This diversity reveals that the spatial experience in a café environment is completely subjective.

When examining the impact of the food offered in the café on spatial perception, participants predominantly emphasized its suitability for social media sharing rather than its appetizing qualities. This suggests that, for some individuals, the visual appeal and social media shareability of the food take precedence over its taste or quality. The aesthetic aspects of the food—such as its presentation, color, and decorative details—are perceived as more significant than its sensory experience. This trend reflects the growing influence of social media in shaping consumer behavior, highlighting the increasing importance of visual content in contemporary dining culture.

Moreover, the café experience appears to be centered not on the quality of service provided but on whether it is deemed share-worthy. Instead of engaging in social interactions with people physically present in the space, participants tend to prioritize online engagement, leading to a diminished emphasis on the direct, in-person experience of the café environment.

The spatial perception of café environments is often associated with fostering social interaction, as they are generally seen as spaces that encourage people to connect. However, perspectives on social engagement within these venues can vary. While some individuals perceive cafés as ideal places for socializing, where meeting new people is natural and encouraged, others experience them as spaces that allow for personal interactions without external social pressures.

In such cases, the environment provides a balance—offering opportunities for those who wish to engage with others while also respecting the need for personal space. This dual perception suggests that café spaces can simultaneously serve as social hubs and as places where individuals can maintain a sense of privacy without feeling obligated to interact. Given the evolving nature of café spaces and their role in social interaction, future studies could explore additional dimensions of spatial perception and user experience in these settings. Additionally, investigating the psychological effects of café environments on well-being, productivity, and social belonging could provide valuable insights into how these spaces influence human behavior.

Further studies could also examine how cultural and regional differences shape café experiences, particularly in relation to consumer preferences and the role of social media in defining the aesthetic and functional aspects of these spaces. Finally, longitudinal research tracking shifts in café culture over time—especially in response to societal changes, economic fluctuations, and technological advancements—could contribute to a deeper understanding of the dynamic relationship between people and café environments.

References

Apaydın, B., & Eren, C. (2013). Yapıların otel olarak yeniden kullanım bağlamında mekânsal dönüşümlerinin analizi: Four Seasons oteli-Sultanahmet Cezaevi örneği. *Tasarı Kuram Dergisi, 9(15),* 73.

Bar, M., & Neta, M. (2006). Humans prefer curved visual objects. *Psychological Science, 17,* 645–648.

Dedeoğlu, M., & Doğan, R. K. (2021). An analysis on coffee culture and coffee-cafe spaces Konya example. *IDA: International Design and Art Journal, 3(1),* 132–145.

Edgü, E. (2010). Evin Direği, Sokağın Bekçisi. Kadın ve Mekân Tutsaklık mı? Sultanlık mı? içinde (72–79), Ayşen Akpınar, Gönül Bakay, Handan Dedehayır (ed), İstanbul: Turkuvaz Kitapçılık ve Yayıncılık.

Edgü, E. (2021). Hayatta kalma güdüsü: Bir mekânsal algı süreci. *Coğrafi Bilimler Dergisi, 19(1),* 217–241.

Friedenberg, J., & Bertamini, M. (2015). Aesthetic preference for polygon shape. *Empirical Studies of the Arts, 33(2),* 144–160.

Güven, E. (2019). Kafelerde kullanıcı memnuniyetinin konfor koşulları açısından değerlendirilmesi: Görükle/Bursa örneği. Mimarlık ve Yaşam, 4(1), 183–196.

Norberg-Schulz, C. (1980). Genius loci: Towards a phenomenology of architecture. New York: Rizzoli.

Ramachandran, V. S., & Blakeslee, S. (1999). Phantoms in the brain: Probing the mysteries of the human Mind. William Morrow Paperbacks.

Ramachandran, V. S., & Hirstein, W. (1999). The science of art: A neurological theory of aesthetic experience. *Journal of Consciousness Studies, 6*(6–7), 15–51.

Steigemann, A. M. (2017). Social practices in a café: Community through consumption?. *Geographica Helvetica, 72(1),* 45–54.

Tuan, Y. F. (1977, 1997). Space and Place: The Perspective of Experience. Minneapolis: University of Minnesota Press.

Compensation Practices in Restaurants and Tipping

Fatih AKYOL[1]

Introduction

Organizations must employ all their resources in an effective and efficient manner. Restaurants that produce service have many inputs like any other business. However, the human capital in service business is distinct from other production inputs. Because human capital has much more importance in the service sector. The satisfaction of customers in tourism enterprises is mostly directly proportionate to the performance of employees. It is therefore recommended that organizations consider their employees as internal customers. Because the satisfaction of guests, who are external customers, depends on the satisfaction of internal customers by the organization.

Compensation is a significant factor to employees' job satisfaction and overall motivation. Therefore, it is thought to be important to attract and retain qualified personnel in restaurants through appropriate compensation practices. Because if employees perceive a discrepancy between their expectations and the reality of their remuneration, they may become dissatisfied and seek alternative employment opportunities. Employees experiencing job dissatisfaction who choose to remain within the organization may exhibit signs of discontent and provide poor service. This phenomenon is especially damaging for restaurants which require a high level of guest interaction. Restaurateurs should have an interest in compensation policies for a variety of reasons, including retaining qualified employees, reducing turnover, meeting organizational goals and improving service quality.

Retaining qualified employee is a particular difficulty in the hospitality sector. The turnover rate for hospitality employees was 62.2% in 2013, which is relatively high compared to the 42.2% turnover rate for the private sector as a whole. The consequences of turnover for restaurants include reduced productivity, loss of institutional knowledge and increased

1 Dr., Kırşehir Ahi Evran University, Vocational School of Social Sciences, Department of Tourism and Hotel Management, fatih.akyol@ahievran.edu.tr

training costs (Mascho & Mao, 2017). Therefore, considering the weight of the hospitality sector in overall employment, restaurateurs need to be well-informed and familiar with compensation practices to reduce turnover of service staff.

It is evident that compensation policies have a significant impact on organizational performance. Research to date suggests that human resource management practices, including compensation practices, are an important instrument for achieving organizational goals (Becker & Gerhart, 1996). In this study, primarily, the compensation practices in restaurants will be explained, thereafter the advantages and disadvantages of these practices will be mentioned. A general framework will be provided by evaluating the aspects of the remuneration methods.

Compensation Practices in Restaurants and Tipping

The generation of goods and services by tourism enterprises involves the utilization of various resources and inputs. In the context of a service business, human capital is the most valuable one of those resources. In return for their labour, companies are required to provide some income to their employees.

Research has demonstrated that establishments within the restaurant industry utilize a variety of approaches to remunerate their service personnel, with different menu pricing methods being employed across different geographical regions and cultures. In practice, these methods are (a) service-inclusive pricing method, (b) service charge method added to the bill and (c) voluntary tipping method paid by the guests (Lynn, 2006).

Service-inclusive pricing method is combining the cost of food and labor into one price. While service-inclusive pricing offers guests a single price, voluntary tipping and service charge methods offer split prices. In other words, they price separately for food and service (Wang & Lynn, 2007).

A service charge can be defined as a service fee added to the guest's bill (Lin & Namasivayam, 2011). In the US, this fee is around 15–20% of the total account (before taxes) (Brown & Rolle, 1991; Mealey, 2019). The distinction between service charge and voluntary tipping is that the former is mandatory and predetermined by restaurant management, whereas the latter are discretionary and given directly to the server by the guest (www.restaurant365.com). The voluntary tipping method differs from the other two methods in that the majority of the service staff's wages are provided by the guests, not the employer (Becker, Bradley & Zantow, 2012; Lynn, Zinkhan & Harris, 1993).

The term 'tip' refers to the monetary payment made by a guest to an employee, in exchange for satisfactory service. This payment is made voluntarily by the guest and is not a result of any obligation or legal requirement (Zeytinoğlu, 2014). Steingass (2005), in his Persian-English dictionary, defines the tip as a gift, a present. Collins Cobuild English Dictionary (1995) defines the tip as an amount of money given to an employee, such as a waiter, in order to thank them for his services.

There are different opinions about the first appearance of the tipping phenomenon. Brenner (2001) argues that tipping first appeared in England in the 16th century, first in coffee shops and later in pubs. Similarly, Frankel (1990) states that in London coffee shops, guests who expect a little more special service leave change in boxes marked "T.I.P." (To Insure Promptness), and this practice can be considered the genesis of the modern tipping tradition (Azar, 2004; Conley, 2003; Lynn, Zinkhan & Harris, 1993; McCarty et al., 1990). Christensen (2001) states that tipping first emerged in medieval Europe. It is known that until the American Civil War (1865), tipping behavior was very rare in America, but it was an established tradition in Britain during this period (Mentzer, 2013). When Americans who traveled to Europe and had to tip there returned home countries, they continued to do so to show that they had traveled abroad and learned European customs (Azar, 2004; Christensen, 2001; Cook, 2000).

At the present time, in the global context of the food and beverage industry, the practice of tipping is most prevalent in the US (Bershidsky, 2015; Lynn & Withiam, 2008; Ogbonna & Harris, 2002). According to a survey of over a thousand US waiters, 95% of respondents worked in restaurants where tipping is customary (Wang & Lynn, 2017). However, it should be noted that the practice of tipping in this country is purely a form of remuneration (Akyol, 2018).

In the US, service workers operating within a tipped system typically receive a modest hourly wage, amounting to $2.13 (Norris, Taylor Jr. & Taylor, 2024; Robinson, 2001). This amount falls short of the minimum wage. On top of this, waiters supplement their income with tips from guests in addition to the meal bill (Lin & Namasivayam, 2011). In other words, tipping is an important source of income (Namasivayam & Upneja, 2007). Azar & Tobol (2008) and Holland (2009) report that many restaurant workers in the US earn more than half of their income from tips. Tipping as a method of remuneration (Akyol, 2019) has a significant economic size in the US. According to Azar (2011), the annual value of tipping in the food and beverage industry in the US is 47 billion dollars.

Who and how much guests tip is largely determined by traditions (Lynn & Withiam, 2008; Wessels, 1997). Remuneration of service personnel through tipping is a social phenomenon rather than an economic necessity (Lynn, Zinkhan & Harris, 1993). According to the findings of etiquette books, the tipping rate in America in the early 1900s was approximately 10%. However, a survey of 30,911 individuals residing in New York City revealed that the tipping rate has increased from 15% to 18.7% (Berman, 2007). It has been documented that this figure can exceed 20% in some luxury restaurants (Margalioth, 2006).

Economists argue that the practice of tipping derives from its role as an incentive for service providers to deliver optimal service. However, studies to date have failed to demonstrate that tipping has a significant effect on service quality (Lynn, 2006). As a result of the meta-analysis of these studies, the average correlation between the variables was 0.11 (Lynn & McCall, 2000). The conclusion is that tipping is not a strong motivator for service quality, contrary to what many people think (Lynn & Withiam, 2008). However, there are some studies (Bodvarsson, Luksetich & McDermott, 2003; Conlin, Lynn & O'Donoghue, 2003; Liraz, 2012; Lynn & Grassman, 1990; Lynn & Graves, 1996) that have revealed a positive relationship between tip amount and service quality evaluation.

Tipping is a distinctive economic phenomenon because people tip even though they are not obliged to (Azar, 2004). A considerable amount of research has been devoted to investigate the motivations behind the act of tipping. One hypothesis that has been suggested to explain this behavior is the intention to receive better service in the future. However, this theory assumes that the guest intends to revisit the restaurant. But in contrast to this idea, research shows that most people will tip even at a restaurant they will never visit again (Azar, 2003). Consequently, the underlying rationales for tipping have been the focus of extensive research. Various psychological factors have been identified, such as conforming to social norms, influencing others, fear of judgment, gratitude for good service, and empathy with the staff member (Azar, 2003; Azar, 2004; Azar, 2007).

Based on behavioral science theories and tipping literature, Lynn (2015) identified five reasons for tipping and two reasons for not tipping, which are generally referred to as the "Tipping Motives Framework (TMF)." The TMF is shown in Figure 1.

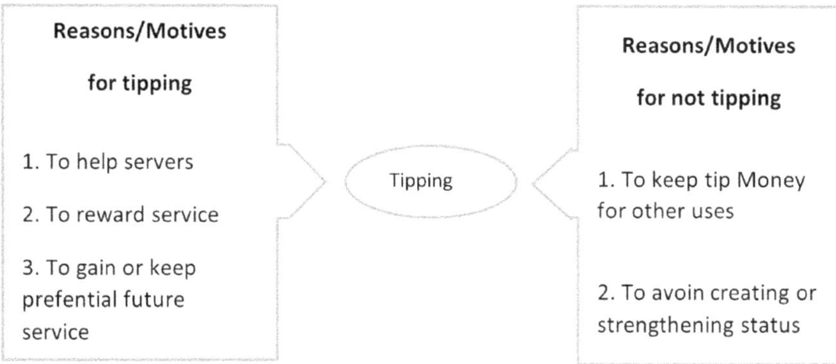

Figure 1. The Tipping Motives Framework (TMF)
(Lynn, 2015)

Tipping is also known to cause undesirable behaviors in service personnel. First, tipping may lead service personnel to behave more carelessly towards group members who tip less (Brewster, 2015; Lynn, Kwortnik & Sturman, 2011). For example, in the US, black people tip less than white people (Lynn & Haysbert, 2003). Other group members who are likely to receive poorer quality service under a tipping system are young people, women, the elderly and foreigners, as they tip less (Lynn, 2006).

The tipping system may cause waiters to focus exclusively on their own tables and thus not pay much attention to the needs and wants of other guests. If the pooling system is not applied in tip distribution, it would be unreasonable to expect waiters to assist their colleagues. Because helping other waiters will mean that they will reduce the time they spend on their own tables (Lynn & Withiam, 2008).

Tipping system can also lead to service staff giving food or drinks to customers for free (Lynn & Wang, 2013). For example, a bartender may pour more drinks for a customer than they should in order to get more tips (Lynn, Kwortnik & Sturman, 2011). A study has shown that when service personnel give free products that their employer would normally charge for, customers are more likely to (Strohmetz et al., 2002).

In addition, the tipping system is known to allow for income tax evasion (Margalioth, 2006). In contrast to fixed wages, tips can be easily hidden from the employer and the government (Macnaughton & Veall, 2001). The US Internal Revenue Service (IRS) estimates that half of all tip income is not reported (Lynn, 2006).

Some problems with voluntary tipping practice have led managers to consider the alternative pricing methods (Lynn & Wang, 2007). In order to reach a decision regarding the most appropriate compensation method, it is necessary for managers to consider a number of factors. The following discussion will address some of the most important of these factors.

Guest Preferences: Guest perspectives on the compensation of service personnel may vary. Despite the ongoing controversy surrounding the practice, tipping remains an important element of the hospitality industry (Lynn, 2006). A survey of the extant literature reveals a dominance of research results from the US. This phenomenon can be attributed to the fact that a significant proportion of research on tipping is conducted in the US. According to the results of research on guest preferences in the US over the last 50 years, most people prefer guaranteed waiter salaries to a tipping system. But for the same people, a tipping system is much more preferable than a service charge added to the bill (Lynn & Withiam, 2008).

Nominal Prices: It is important to emphasize the issue of nominal price and actual price when discussing pricing systems (service-inclusive pricing, service charges and voluntary tipping). The nominal price forms only part of the guest check. In other words, the final amount (actual price) includes any tax, service charge or gratuity added on top of the nominal price (Lynn, 2006). As tipping is a common practice throughout the United States (Miller, 2005), restaurant menus typically list the nominal price. The actual price is the total cost of goods and services to the guest, including all unspecified charges. Therefore, the nominal price is lower under the tipping and service charge methods than under the service-inclusive pricing method.

Research shows that presenting the price of a product or service separately, rather than as a consolidated whole, reduces the guest's perception of total cost and thus increases demand (Xia & Monroe, 2004). As a result, since food and service are priced separately in the tipping and automatic service charge methods, they will be perceived as cheaper than the service-inclusive pricing method. This will provide a competitive advantage to those restaurants that choose to implement these two systems (Wang & Lynn, 2007).

Price Discrimination: In some cases, charging different prices for the same service to different guests can increase revenue and profits. In instances where guests exhibit divergent price sensitivity, and the prices charged match with each guest's propensity to pay and demand from price-insensitive guests is less than the firm's production capacity, the business has the potential to optimize its revenue and profit by earning more from non-price-sensitive guests without losing price-sensitive guests. The tipping system is a price discriminating method because it allows some guests to pay more than others for the same service. This is because although all guests have to pay the same nominal price for the service, the amount they tip will vary (Lynn, 2006).

Price-sensitive guests will tip less than non-price-sensitive guests in order to reduce their expenditure. However, if the demand of non-price sensitive guests is less than the production capacity, the chairs that would otherwise remain empty will be filled by price sensitive guests. Therefore, the company will profit from this situation. On the other hand, if the demand of price-insensitive guests is high, there will be no need for price-sensitive guests, and the business will choose one of the service charge or service-inclusive pricing methods instead of the tipping system in order to appeal to a higher-level market group (Lynn & Withiam, 2008).

Server Incentives: It is a commonly held view that the practice of tipping has a positive impact on the encouragement of upselling and the delivery of good service. In contrast, the service charge method is considered to encourage upselling, but has no effect on providing good service. The service-inclusive pricing method is assumed to have no incentive for either case (Lynn, 2006).

Employee Recruitment and Retention: These three remuneration methods, which operate differently from each other, will have different effects on employee recruitment and retention. There are three different effects of the tipping system that do not apply to alternative methods (Lynn & Withiam, 2008). First, this system improves the pool of potential staff and therefore the quality of the existing service staff. The rationale behind this phenomenon can be explained by the observation that service staff who provide low-quality services tend to receive low tips. Consequently, as their remuneration is lower, they will seek alternative employment opportunities. Second, tipping provides a competitive advantage for the employer (Lynn & Withiam, 2008; Mkono, 2011) as it attracts and retains more talented employees (Lynn, Kwortnik & Sturman, 2011) than other compensation methods. Talented people earn more in the tipping system than in other systems. Finally, tipping attracts employees who are less professional than alternative methods of remuneration. This is due to the fact that the uncertain amount of earnings and the daily hot cash are more suitable for young people, part-timers or temporary workers. However, the opposite is true for people who are professionals and want to work permanently and for the long term (Lynn, 2006).

Exhibit 1 summarizes the known advantages and disadvantages of service personnel remuneration methods. All issues are framed in terms of the extent to which each issue is a benefit to the restaurant that favors tipping, service charge or service-inclusive pricing (shown with thumbs up or thumbs down). It also identifies the conditions under which each benefit is particularly strong or important.

Table 1. Summary of the benefits offered by service-inclusive pricing, service charges and voluntary tipping.

Benefit	Tipping	Service Charge	Service-inclusive Pricing	Benefit is Especially Strong or important When ...
Match guest preferences	👍	👎	👍	Rarely-preferences are weak and guests will accept all options
Lower distribution costs (through lower nominal prices)	👍	👍	👎	Distributor commissions are high
Increase demand (through lower nominal prices)	👍	👍	👎	Guests are sensitive to price
Higher profits (through price discrimination)	👍	👎	👎	Guests differ in price sensitivity and price sensitive guests are needed to fill seats

Benefit	**Tipping**	**Service Charge**	**Service-inclusive Pricing**	**Benefit is Especially Strong or important When …**
Decrease demand from lower socio-economic consumers	👎	👍	👍	Demand is strong and more exclusive, upscale guests are desired
Lower psychological costs to guests (through payment bundling)	👎	👍	👍	Payments of tips and bill are separated in time or place
Inspire up-selling	👍	👍	👎	Servers are trained how to sell and feel competent to do so
Inspire good service	👍	👎	👎	Rarely-tips and tip policies are only weakly related to service
Inspire equal service for all guests	👎	👍	👍	Large proportion of customer base made up of women, foreigners, ethnic minorities, teenagers and the elderly
Redistribute income from tipped staff to non-tipped staff	👎	👍	👍	Tipped employees earn significantly more than non-tipped employees
Attract qualified employees (through high income potential)	👍	👎	👎	Tips are high compared to wages for other jobs
Attract more professional employees (through guaranteed income)	👎	👍	👍	High and stable wages
Reduced tax payments	👍	👎	👎	Servers do not declare large amounts of income from tips
Reduced risks of adverse impact lawsuits	👍	👍	👍	Tips will not be pooled and protected classes will receive a lower tip

(Lynn, 2006; Lynn & Withiam, 2008)

When the exhibit is analyzed, it is seen that all three systems have positive and negative aspects. Therefore, restaurateurs should evaluate the advantages and disadvantages of each system according to the conditions of their own business and decide on the system to be implemented accordingly (Lynn & Withiam, 2008).

Results

In the contemporary business environment, where labour-intensive restaurants face fierce competition, retaining qualified employees has become imperative for gaining a competitive advantage. Restaurateurs should use compensation method as a tool to attract and retain qualified employees. Because employees rely on their compensation to maintain stability in their lives, while restaurateurs rely on their employees' abilities and performance to remain competitive in the marketplace. Earlier theories of motivation suggested that in this social exchange between employees and employers, compensation acts as an important stimulus that enhances individual performance. In service-inclusive pricing and service charge systems, management has complete control over employee compensation, whereas in the tipping system, the guests are in control.

The implementation of service-inclusive pricing, service charges and voluntary tipping offer different sets of costs and benefits for each party involved. But no one policy is the optimal solution in all cases for all organizations. Reducing nominal prices, increasing profits through price discrimination, encouraging up-selling, attracting talented employees and reducing tax payments are the main benefits of voluntary tipping. Tipping is mostly common in the US, but it should be noted that the practice of tipping in this country deviates from tipping in service-inclusive pricing method (in which employee wages are incorporated into the menu prices) or in service charge method (in which a service fee is added to the bill). Unlike the other two methods, in the US, tipping is purely a form of remuneration. Although it is said that it is not obligatory in the US, the fact that it is an established social norm actually makes this practice obligatory. Because people tend to act in accordance with social norms. But in recent years, there has been a growing debate in the US about whether restaurants should give up tipping.

Although tipping is a common phenomenon in various service sectors around the world, it is important to note that these practices vary significantly from one geographical location to another. Consequently, a distinct research approach is necessary for each sector, country or culture.

Instead of randomly choosing any of these methods or following the practices in their regions, restaurateurs should review the advantages and disadvantages of each method and choose the one that best suits their own business. It would be appropriate to adopt and implement a system ensuring fair income distribution among employees, encouraging them to provide high quality service and rewarding them.

Cross-cultural research should be conducted on the remuneration of restaurant staff in order to have a better understanding of the differences between the three compensation practices. Restaurateurs that employ the most appropriate compensation method for their culture and their unique conditions will achieve their organizational goals with greater ease.

References

Akyol, F. (2018). Örgütsel Adalet Algısının Örgütsel Bağlılığa Etkisi: Bahşiş Havuzu ile Bireysel Bahşiş Sisteminin Karşılaştırılması. (Doctoral Dissertation), Nevşehir Hacı Bektaş Veli University, Institute of Social Sciences, 2018.

Akyol, F. (2019). Örgütsel Adalet Algısının Örgütsel Bağlılığa Etkisi: Bahşiş Havuzu ile Bireysel Bahşiş Sisteminin Karşılaştırılması. *Journal of Tourism and Gastronomy Studies*, 7 (3), 1933–1957. DOI: 10.21325/jotags.2019.456

Azar, O. H. (2003). The Implications of Tipping for Economics and Management. *International Journal of Social Economics*, 30, 9/10. DOI:10.1108/03068290310492878

Azar, O. H. (2004). The history of tipping—from sixteenth-century England to United States in the 1910s. The Journal of Socio-Economics, 33, 745–764. DOI:10.1016/j.socec.2004.09.043

Azar, O. H. (2007). The Social Norm of Tipping: A Review. *Journal of Applied Social Psychology*, 37(2), 380–402. DOI:10.2139/ssrn.370081

Azar, O. H. & Tobol, Y. (2008). Tipping as a Strategic Investment in Service Quality: An Optimal-Control Analysis of Repeated Interactions in the Service Industry. *Southern Economic Journal*, 75(1), 246–260. DOI:10.2307/20112038

Azar, O. H. (2011). Business strategy and the social norm of tipping. *Journal of Economic Psychology*, 32, 515–525. DOI:10.1016/j.joep.2011.03.018

Becker, B., & Gerhart, B. (1996). The impact of human resource management on organizational performance: Progress and prospects. *Academy of Management Journal*, 39(4), 779–801. https://doi.org/10.2307/256712

Becker, C. Bradley, G. & Zantow, K. (2012). The underlying dimensions of tipping behavior: An exploration, confirmation, and predictive model. *International Journal of Hospitality Management*, 31, 247–256. DOI:10.1016/j.ijhm.2011.05.006

Berman, T. (2007). Zagat Survey's 2006 Top Restaurant Survey. http://www.alternet.org/story/34988/tipping_in_america (15.03.2025).

Bershidsky, L. (2015). Hold the tip. No gratuities could mean better food. https://www.bloomberg.com/news/articles/2015-10-22/the-end-of-tipping-could-mean-better-restaurant-food?leadSource=uverify%20wall. (20.03.2025).

Bodvarsson, O. B., Luksetich, W.A. & McDermott, S. (2003). Why do diners tip: rule-of-thumb or valuation of service? *Applied Economics*, 2003, 35, 1659–1665.

Brewster, Z. W. (2015). Perceptions of intergroup tipping differences, discriminatory service, and tip earnings among restaurant servers. *International Journal of Hospitality Management*, 46 (2015), 15–25. DOI:10.1016/j.ijhm.2015.01.001

Brown, N. & Rolle, S. A. (1991). Tip versus service charges: The Iowa scene. *The Cornell Quarterly*, 32, 75–81. 10.1016/0010-8804(91)90020-r

Christensen, J. N. (2001). Analysis of Casino Table Game Tipping By Comdex Conventioneers. (Master's thesis), University of Nevada, Las Vegas.

Collins Cobulild English Dictionary. (1995). HarperCollins Publishers, London.

Conley, D. (2003). Tip Jars and the New Economy. *Chronicle of Higher Education*, 1/3/2003, Vol. 49 Issue 17.

Conlin M., Lynn, M. & O'Donoghue, T. (2003). The Norm of Restaurant Tipping. *Journal of Economic Behavior & Organization*, Vol. 52 (2003), 297–321. https://doi.org/10.1016/S0167-2681(03)00030-1

Cook, S. (2000). A History of 'Handing It Over'. Christian Science Monitor, 10/23/2000, 92 (232).

https://www.restaurant365.com/blog/understanding-service-fees-vs-tips/

Holland, S. J. (2009). Tipping as Risk Sharing. *The Journal of Socio-Economics*, 38 (2009), 641–647.

Lin, I.Y. & Namasivayam, K. (2011). Understanding restaurant tipping systems: a human resources perspective. *International Journal of Contemporary Hospitality Management*, Vol. 23, 923–940. DOI:10.1108/09596111111167533

Liraz, H. M. (2012). Service quality and tipping: The moderating role of the quality of food. *International Journal of Hospitality Management*, 31 (2012), 1327–1329. DOI:10.1016/j.ijhm.2012.02.009

Lynn, M., & Grassman, A. (1990). Restaurant Tipping: An Examination of Three 'Rational' Explanations. Cornell University, School of Hospitality Administration.

Lynn, M., Zinkhan, G.M. & Harris, J. (1993). Consumer Tipping: A Cross-Country Study. *Journal Of Consumer Research*, Vol. 20, December, 1993. DOI:10.1086/209363

Lynn, M., & Graves, J. (1996). Tipping: An Incentive/Reward for Service? Cornell University, School of Hospitality Administration. https://doi.org/10.1177/109634809602000102

Lynn, M., & McCall, M. (2000). Gratitude and gratuity: a meta-analysis of research on the service-tipping relationship. *Journal of Socio-Economics*, 29 (2000), 203–214. DOI:10.1016/S1053-5357(00)00062-7

Lynn, M., & Haysbert, C. T. (2003). Ethnic Differences in Tipping: Evidence, Explanations, and Implications. *Journal of Applied Social Psychology*, 33(8), 1747–1772. DOI:10.1111/j.1559-1816.2003.tb01973.x

Lynn, M. (2006). Tipping and its alternatives: A comparison of tipping, service charges, and service-inclusive pricing. Cornell Hospitality Reports, 6(5), 6–16.

Lynn, M. & Wang, S. (2007). The Effect of Tipping and Other Ways of Compensating Employees on Perceived Restaurant Expensiveness. Cornell University, School of Hospitality Administration.

Lynn, M. & Withiam, G. (2008). Tipping and its alternatives: business considerations and directions for research. *Journal of Services Marketing*, 22/4, 328–336. DOI:10.1108/08876040810881722

Lynn, M., Kwortnik, R. J. & Sturman, M. C. (2011). Voluntary tipping and the selective attraction and retention of service workers in the USA: an application of the ASA model. *The International Journal of Human Resource Management*, 22(9), 1887–1901. DOI:10.1080/09585192.2011.573968

Lynn, M. & Wang, S. (2013). The indirect effects of tipping policies on patronage intentions through perceived expensiveness, fairness, and quality. *Journal of Economic Psychology*, 39, 62–71. DOI:10.1016/j.joep.2013.07.003

Lynn, M. (2015). Service gratuities and tipping: A motivational framework. *Journal of Economic Psychology*, 46, 74–88. https://doi.org/10.1016/j.joep.2014.12.002

Macnaughton, A. & Veall, M. R. (2001). Tipping and the McQuatters Formula. *Public Finance Review*, March 2001, 29, 99–107. DOI:10.1177/109114210102900201

Margalioth Y. (2006). The case against tipping. *Journal of Labor and Employment Law*, Vol.9(1), 117–145.

McCarty, J. A., Shrum, J. L., Conrad-Katz, T. E. & Kanne, Z. (1990). Tipping as a Consumer Behavior: A Qualitative Investigation. *Advances in Consumer Research*, 17, 723–728.

Mascho, E. & Mao, Z. (2017). An exploratory study of managerial compensation in the restaurant industry. *Journal of Human Resources in Hospitality & Tourism*, Vol. 16, No. 3, 270–287. DOI:10.1080/15332845.2017.1253432

Mealey, L. (2019). Restaurant Tipping and Dining Gratuity Guide. https://www.liveabout.com/the-basics-of-restaurant-tipping-2888420. (25.03.2025).

Mentzer, M. S. (2013). The Payment of Gratuities by Customers in the United States: An Historical Analysis. *International Journal of Management*, 30(3), 108–120.

Miller, B. (2005). Tipping as a compensation method for restaurant employees: An exploration of its effect on employee organizational commitment. International Hospitality and Convention Summit Proceedings, June 5–7, Las Vegas, Nevada.

Mkono, M. (2011). Tipping Practices and Policies in Zimbabwe's Hotel Industry: Impacts on Restaurant Waiters' Service Delivery and Work Relationships. *Journal of Foodservice Business Research*, 14, 414–425. DOI:10.1080/15378020.2011.624054

Namasivayam, K. & Upneja A. (2007). Employee Preferences for Tipping Systems. *Journal of Foodservice Business Research*, 10:2, 93–107. DOI:10.1300/J369v10n02_07

Norris, C., Taylor Jr. S. & Taylor, D. C. (2024). Just the tip: exploring the tipped restaurant employee perspective. *International Hospitality Review*, Vol. 38 No. 2. pp. 221–238. DOI:10.1108/IHR-02-2023-0011

Ogbonna, E. & Harris, L. C. (2002). Institutionalization of Tipping as a Source of Managerial Control. *British Journal of Industrial Relations*, 40(4), 725–752. DOI:10.1111/1467-8543.00254

Robinson, T. B. (2001). A Comprehensive Look at the Tip Wage. *Compensation & Benefits Review*, 43(4), 214–219. DOI:10.1177/0886368711407995
Steingass, F. (2005). A Comprehensive Persian-English Dictionary. Çağrı Yayınları, İstanbul.
Strohmetz, D., Rind, B., Fisher, R. & Lynn, M. (2002). Sweetening the til: the use of candy to increase restaurant tipping. *Journal of Applied Social Psychology*, 32(2), 300–309. https://doi.org/10.1111/j.1559-1816.2002.tb00216.x
Wang, S. & Lynn, M. (2007). The Effects on Perceived Restaurant Expensiveness of
Tipping and Its Alternatives. *Cornell Hospitality Reports*, 7(3), 6–17.
Wang, S., & Lynn, M. (2017). The effects of service charges versus service-included pricing on deal perception. Retrieved (18.03.2025), from Cornell University, School of Hotel Administration site: http://scholarship.sha.cornell.edu/articles/1030.
Wessels, W. J. (1997). Minimum Wages and Tipped Servers. *Economic Inquiry*, Volume 35, April 1997, 334–349. DOI:10.1111/j.1465-7295.1997.tb01914.x
Xia, L. & Monroe, K. B. (2004). Price partitioning on the internet. *Journal of Interactive Marketing*, 18 (4), 63–73. https://doi.org/10.1002/dir.20017
Zeytinoğlu, E. (2014). Bahşişin Ücret Olarak Kabul Edilip Edilmeyeceğinin İrdelenmesi. *İstanbul Ticaret Üniversitesi Sosyal Bilimler Dergisi*, 13 (26), Güz 2014/2, 57–71.

Sustained Mathematical Thinking in Traditional Gastronomy

Çağnur ÇÖREKLİ[1]

Introduction

Mathematics and mathematical thinking, as one of the most ancient and concrete products of human cognitive evolution, shed light on the functioning of the mind and the understanding of human cognitive development (Wynn, 2002; Van der Waerden, 2012). Although some mathematical principles, as universal outputs of this cognitive evolution, manifest in different forms across cultures due to factors such as geography, climate, beliefs, and traditions, they have persisted unchanged across various cultures and historical periods (Noble & Davidson, 1996; Robson & Stedall, 2009). These mathematical principles and patterns, commonly observed in fields such as science, art, and astronomy, are also evident in traditional culinary practices. These principles are primarily related to proportions, symmetry, fractal structures, and optimization strategies. For instance, the golden ratio was employed in the construction of the Egyptian pyramids and Greek temples (Bartlett, 2014; Foutakis, 2014), specific ratios were also applied in traditional bread-making and beer and in their trading (Halawa, 2023; Robson & Stedall, 2009), and food shapes, textures, and even geometrical properties and harmony were metaphorically linked to ethical and spiritual states in Pythagorean dietary systems (Kingsley, 1995), as well. Similarly, the fractal patterns found in cauliflower and broccoli in nature are mirrored in decorative culinary arts and more (Barrett & Peleg, 1995; Taylor, 2021). These examples demonstrate that while culinary traditions may change over time, fundamental mathematical principles remain constant. Focusing on mathematical principles in gastronomy presents a rich area of study, offering insights into both aesthetic principles and functional efficiency. Therefore, this chapter will explore the intersection of mathematics, cognitive science and gastronomy and highlight

1 PhD., Recep Tayyip Erdogan University, Faculty of Art and Science, Department of Mathematics, cagnur.corekli@erdogan.edu.tr

the enduring role of mathematical principles in shaping culinary traditions worldwide, by drawing upon existing academic research.

Roughly Mathematical Comparison of Bread and Beer Production in Ancient Egypt and Modern Times

Ancient Egypt is considered the birthplace of leavened bread, possibly discovered by accident when natural yeast or ale caused dough to rise (Civitello, 2011). The interaction of gluten with fermenting agents caused the dough to rise, marking a significant development in baking, then bread has earned sustained a 'divine' meaning, like the gift of God, and sustained it as the same with the ancient and modern world. Bread and beer were fundamental components of daily nutrition in Ancient Egypt, where their production followed mathematical ratios (Halawa, 2023; Robson & Stedall, 2009), and were reflected its centrality in the daily life, economy, and ritual practices of the civilization. The Ancient Egyptians used grain to produce both bread and beer, and the way grain was allocated played a crucial role in determining product quality (Nicholson & Shaw, 2000), even the using as the currency of the payment of the construction of the Great Pyramids of Giza (Halawa,2023; Robson & Stedall, 2009) placed in Rhind and Moscow papyri and in many texts (Allen, 2007). In modern times, food production is optimized through scientific methods and technological advancements. However, the Ancient Egyptian concept of ***pefsu*** or ***pesu***, 'cooking ratio' as a measure of bread and beer quality based on grain dilution (Gillings, 1965; Robson & Stedall, 2009), can be mathematically compared to modern efficiency and quality metrics in food science. Thus, it has been explored that the historical continuity and transformation of quality assessment in bread and beer production by comparing the mathematical ratios used in Ancient Egypt to those applied in contemporary food production. The Ancient Egyptians used grain to produce both bread and beer, and the way grain was allocated played a crucial role in determining product quality. The concept of *pefsu* or *pesu*, the mathematical quality measure, was a unit used to express the ratio of one *heqat*, a standard unit of volume used for measuring grain, flour, and beer, and approximately equaled to 4.8 liters, of grain to the number of loaves or jars of beer, produced as the following (Robson & Stedall, 2009):

$$\text{Pefsu/Pesu} = X/Y$$

Where X=the number of loaves of bread/jars of beer, Y=the number of heqat of grain,

- *Lower pefsu* refers to higher quality and more nutritious product
- *Higher pefsu* refers to lower quality and more diluted product.

Also,, it was normed that the quality *pefsu ratios* ranged between 5–45 and 2–6 for bread and beer, respectively. For example; pefsu 5 for bread means that 1 heqat of grain used to produce 5 loaves of bread i.e. denser and more nutritious bread while pefsu 45 for bread means that 1 heqat of grain used to produce 45 loaves of bread i.e. lighter and lower quality bread, and pefsu 5–20 can be normed for high quality bread offering to elite, temples, and skilled workers (Robson & Stedall, 2009; Helck, 1975). Likewise, pefsu 2 for beer means that 1 heqat of grain used to produce 2 jars of beer i.e. strong and nutritious beer while pefsu 6 means 1 heqat of grain used to produce 6 jars of beer i.e. more diluted beer, and pefsu 2–4 can be normed for high quality strong beer offering to elite, temples, and skilled workers, (Robson & Stedall, 2009; Helck, 1975). The mathematical problems in the Rhind and Moscow papyri indicate that the Ancient Egyptians used these ratios to optimize food production and trade value (Robson & Stedall, 2009), for example, one of problems in Rhind papyrus; *"If 100 bread loaves "of pefsu 10" are to be evenly exchanged for x loaves "of pefsu 45." Find x,"* that is solved as 450 after finding Y, the number heqats of grain, is needed for 100 bread loaves of pefsu 10, 10=100/Y, and then X, the number of bread, of pefsu 45 from 10 heqats of grain, 45=X/10. In modern and contemporary bread and beer production has been scientifically measured using modern techniques, such as protein content, fermentation time, and specific gravity measurements. Modern high-quality bread has been determined by factors such as protein content, hydration ratio, and fermentation time. While whole wheat flour like emmer wheat (12–15% protein) means denser and more nutritious bread, white flour like modern wheat (8–10% protein) means lighter and lower nutritional value. In roughly mathematical comparison, for bread, 1 heqat of grain has been approximately equivalent to average 3.5 kg of emmer wheat, 3.36–3.60 kg from 0.70–0.75 kg/lt density of emmer wheat (Markowski, Żuk-Gołaszewska & Kwiatkowski, 2013), and emmer wheat gives a flour extraction rate 65% to obtain 2.275 kg of flour with modern milling. High quality bread typically uses 70% hydration for artisan bread, so water has been needed 1.59 kg for 2.275 kg of flour, then total dough yields 3.865 kg of dough and it has been produced about 6–9 of baked loaves, Ancient Egypt estimated ceremonial offering 400–600 gr high quality bread (Robson & Stedall, 2009; Nicholson & Shaw, 2000). after nearly 12% baking loss with modern technique, which exactly corresponding pefsu 5–10. Thus, a pefsu 5–20 bread in Ancient Egypt may be comparable to a modern whole wheat or sourdough bread in terms of nutritional density.

Likewise, for beer, 1 heqat of grain has been approximately equivalent to average 3.2 kg of barley, 3.02–3.36 kg from 0.63–0.70 kg/lt density of barley (Markowski et al., 2010). The modern beer quality has been measured using specific gravity (SG), fermentation time, and malt concentration, such as SG 1.060 - 1.080 referring high-malt and premium-quality beer and representing 60–80 grams of sugar dissolved per liter of beer wort, averaging 70 gr. With modern malting assumption, total malt extract with 70% brewhouse efficiency yield 672 grams of sugar from 300 grams sugar per kg of malted barley (total extract= 3.2 kg x 300 g/kg x 0.70). Then, the volume of beer at the average SG 1.070 has been divided total extract by sugar per liter i.e. 672/70= 9.6 liters, so from 3.2 kg of barley, using modern brewing with 70% efficiency and targeting a high-quality beer SG of 1.070, it can be produced approximately 9.6 liters of strong, full-bodied beer that means to equal to pefsu 2–4 of 4–5 liters of standard jars of Ancient Egypt and of sometimes figured *des-jar which held 2.5 to 3 liters* (Geller, 1993). Thus, modern brewing typically yields high malt craft beers exactly corresponding a pefsu 2–4 beer in Ancient Egypt. Also, this consistency in Mesopotamian accounting, whereby a fixed quantity of grain corresponds to a predetermined volume of beer of uniform quality, reflects a strict adherence to standardized brewing ratios. In contrast, Egyptian brewers or scribes appear to have employed a more flexible approach, adjusting the dilution of the wort to produce beers of varying strength, while maintaining the ability to calculate and document the resulting pefsu values with precision. By this direct mathematical comparison of bread and beer production ratios between Ancient Egypt and modern times, the highest-quality bread and beer in Ancient Egypt correspond closely to modern premium products in both nutritional and production in terms of grain-to-product ratios. Mathematically, the highest-quality Ancient Egyptian bread and beer exhibit striking similarities to modern high-quality food production in terms of nutritional value and production efficiency although the beer production in Ancient Egypt involved some loss of yield from the old milling.

The Relationship Between Shape and Taste Perception in Ancient and Modern Contexts

The association between shape and taste perception has long been embedded in human cognition. Research in experimental psychology has demonstrated that round shapes are often associated with sweetness and softness, while angular shapes correspond to bitterness and hardness (Juravle, Olari, & Spence, 2022). This phenomenon is not only observed in contemporary cognitive science but also deeply rooted in historical food practices.

Ancient civilizations such as Egypt, Greece, and Rome strategically utilized food geometry to enhance taste perception, establish cultural symbolism, and create hierarchical food distributions. Thus, it has been explored how Ancient Egyptian, Greek, and Roman civilizations utilized food geometry to shape taste expectations, comparing these practices with modern cognitive science research. In Ancient Egypt, bread and beer were staples of daily life, and bread was produced in a variety of forms such as round, oval, triangular and conical forms pointed pyramidal sign and Horus and animal figured shape reflecting both practical baking techniques and cultural preferences. Also, sometimes fruit, spices, seeds, or honey were added for flavor (Mehdawy & Hussein, 2010; Nicholson & Shaw, 2000). The bread crust was often decorated with cuts, small holes, or strips of dough on top, so their shapes influenced their perceived quality and taste such as

- Round breads like lotus blossom bread were associated with soft textures and were often enriched with honey or dates, aligning with modern studies linking roundness to sweetness (Mehdawy & Hussein, 2010; Nicholson & Shaw, 2000).
- Angular, dense breads were less refined, coarser, and perceived as harsher, often used for rations in tombs and military supplies (Mehdawy & Hussein, 2010).
- Beer was fermented and stored in round vessels like wide circular shapes and compressed spherical shapes, reinforcing warmth and nourishment

associations, akin to modern cognitive associations between roundness and pleasantness (Mehdawy & Hussein, 2010; Wang, Friedman & Baba, 2021; Nicholson & Shaw, 2000).

In Ancient Greek thought, the connection between natural elements and geometric shapes was most notably articulated by Plato in *Timaeus*, where he associated the four classical elements (earth, air, fire, water) with the five regular solids, later termed Platonic solids (Zeyl, 2000; Vlastos, 2006). This conceptual framework, wherein each element's essence was reflected in geometric form, extended beyond metaphysics and into cultural practices, influencing food presentation and consumption, such as

- Round foods (cheese wheels, fruits like figs and grapes) were considered harmonious and pleasant such as grapes and grape vines associated with *Dionysus*, the god of vine, and the symbol of ressurection due to be reborn in springtime (Civitello, 2011). After wine production in Ancient Greece, the Greeks drank wine sweetened with honey (Civitello, 2011) and drinking wine usually was compliance with the meat of sacrified animals in the *Symposium* ceremony (Civitello, 2011), aligning with modern findings that associate rounded forms with sweetness and umami flavors.
- In Pythagorean-Platonic systems, shapes had moral/ontological meaning (Horky, 2013). Angularity could metaphorically signify harshness, rigidity, difficulty, while roundness suggested harmony and ease. Also, Galen, a prominent Greek physician, and Theophrastus, a student of Aristotle, explored how various food and the characteristics of plants influenced the body's balance and effected on humans (Galenus, & Powell, 2003; Scarborough, 1978), so *the humoral theory* was common in Ancient Greek medicine associated certain food qualities with physiological effects. Thus, Angular foods, such as hard cheeses and certain types of bread, were regarded as harsh and intense, so this gives symbolic weight to angular foods being viewed negatively, mirroring modern associations of angularity with bitterness.

In Ancient Rome, Romans were highly aware of the psychological impact of food presentation (Civitello, 2011), using decoration and shape as a means to enhance taste perception (Beerden, 2018), such as

- Desserts and confections were rounded or curved, reinforcing their association with sweetness such as *globi/globus*, honey-dipped dough balls, *libum*, sacrificial cake, made from cheese, egg, flour and honey, baked as small and roundedloaves and used in rituals and religious ceremonies, and other sweet cakes are *spira*, a spriral version of sacrificed cakes, *spaerita*, formed into large spheres, *encytum*, shaped with similar a ring doughnut (Déry, 2000), *mustaceum cake*, shaped like a rhombus with rounded corners, served at weddings to wish sweetness and harmony to couples (Wilson, 2005).
- Flat and more angular breads, such as *Roman panis quadratus* (Cardenas, 2023), were typically associated with functionality rather than indulgence. This aligns with modern findings that angular shapes are often perceived as harder and less palatable (Spence & Ngo, 2012), reinforcing the symbolic function of these breads in Roman culture as emblems of social cohesion and frugality rather than luxury (Waring, 2025)
- In ancient Rome, tableware was meticulously designed to enhance the dining experience, with various shapes and materials employed to serve different types of dishes such as *Terra sigillata*, a type of fine red-gloss pottery widely used in Roman dining settings featured various shapes, including bowls and plates and indicating their use in serving diverse food items (Oyen, 2016). Such round bowls and plates were preferred for serving savory dishes, a trend still reflected in modern plating psychology where functionality and aesthetic appeal were both valued (Hally, 1986).

These practices suggest that Antiquity intuitively associated shape with taste perception, a notion still supported by modern experimental psychology. These examples highlight that shape was intentionally employed to influence taste perception in antiquity, much as it is today which recent experimental psychology research confirms that humans systematically associate certain shapes with specific taste properties (Ogata et al., 2023).

Results

This chapter has demonstrated that the core components of mathematical cognition, such as proportion, symmetry, geometric structuring, and optimization. These are not abstract skills developed solely through formal education but are instead grounded in embodied, multisensory, and culturally transmitted experiences. The analysis of pefsu ratios in Ancient Egyptian bread and beer production illustrates how early civilizations applied mathematical reasoning to ensure efficiency, quality, and resource management, practices that mirror modern scientific approaches to food production.

Moreover, the symbolic and sensory connections between shape and taste across Ancient Egyptian, Greek, and Roman cuisines reveal a longstanding cognitive framework wherein geometry shaped perception, social rituals, and moral symbolism. These associations, now validated by experimental psychology, highlight the continuity of cross-modal correspondences in human cognition. These patterns suggest that mathematical thinking is not solely a mental abstraction but is also an embodied and multisensory cognitive experience.

Each of these examples illustrates how mathematical cognition has co-evolved with human cultural and physiological development, with gastronomic practices acting as key cultural carriers of this cognitive continuity. Culinary traditions have preserved not only knowledge of taste and nutrition but also intuitive models of proportion, classification, and structural reasoning, transmitted across generations through shared rituals and practices. Recognizing this allows us to reframe mathematics not just as a formal discipline, but as a dynamic, culturally embedded mode of human thought.

In conclusion, the multilayered intersection of mathematics, cognitive science, and gastronomy makes visible the profound link between the human mind's capacity for abstraction and its cultural memory. Reframing mathematical thought through the lens of taste, rhythm, structure, and communal ritual allows us to recognize that mathematics lives not only in symbols and numbers, but also in sensory, bodily, and social experience. Even after thousands of years, we have still shared surprising emotional and cognitive roots with the ancient world. This holistic perspective offers a deeper understanding of the human dimension of mathematics, enriching both ethnomathematics and cognitive science.

References

Allen, J. P. (2007). The ancient Egyptian pyramid texts. Society of Biblical Lit.

Barrett, A. H., & Peleg, M. (1995). Applications of fractal analysis to food structure. LWT-Food Science and Technology, 28(6), 553–563.
Bartlett, C. (2014). The design of the Great Pyramid of Khufu. Nexus Network Journal, 16(2), 299–311.
Beerden, K. (2018). Moderation, refined luxury, or extravagance? Fattened animals and ancient Roman norms and values. Food, Culture & Society, 21(4), 505–520. https://doi.org/10.1080/15528014.2018.1480641
Cardenas, M., Yarza, I., Matterne, V., & Arranz-Otaegui, A. (2023). Disentangling the production of the panis quadratus from Pompeii: A new interdisciplinary perspective. International Journal of Gastronomy and Food Science, 32, 100729.
Civitello, L. (2011). Cuisine and culture: A history of food and people. John Wiley & Sons.
Déry, C. A. (2000). Milk and Dairy Products in the Roman Period. In Milk: Beyond the Dairy: Proceedings of the Oxford Symposium on Food and Cookery 1999 (pp. 117–125). Totnes, Devon, UK: Prospect Books.
Foutakis, P. (2014). Did the Greeks build according to the golden ratio?. Cambridge Archaeological Journal, 24(1), 71–86.
Galenus, & Powell, O. (2003). On the Properties of Foodstuffs (De alimentorum facultatibus) (p. xxvi). Cambridge University Press.
Geller, J. (1993). Bread and beer in fourth-millennium Egypt. Food and Foodways, 5(3), 255–267.
Gillings, R. J. (1965). The Addition of Egyptian Unit Fractions. The Journal of Egyptian Archaeology, 51(1), 95–106.
Halawa, A. (2023). Influence of the traditional food culture of Ancient Egypt on the transition of cuisine and food culture of contemporary Egypt. Journal of Ethnic Foods, 10(1), 11.
Hally, D. J. (1986). The identification of vessel function: A case study from northwest Georgia. American Antiquity, 51(2), 267–295.
Helck, W. (1975). Bier. Lexikon der Ägyptologie, 1, 789–792.
Horky, P. S. (2013). Theophrastus on Platonic and 'Pythagorean'Imitation. The Classical Quarterly, 63(2), 686–712
Juravle, G., Olari, E. L., & Spence, C. (2022). A taste for beauty: On the expected taste, hardness, texture, and temperature of geometric shapes. i-Perception, 13(5), 20416695221120948.
Kingsley, P. (1995). Ancient philosophy, mystery, and magic: Empedocles and Pythagorean tradition. Oxford University Press

Markowski, M., Majewska, K., Kwiatkowski, D., Malkowski, M., & Burdylo, G. (2010). Selected geometric and mechanical properties of barley (Hordeum vulgare L.) grain. International Journal of Food Properties, 13(4), 890–903.
Markowski, M., Żuk-Gołaszewska, K., & Kwiatkowski, D. (2013). Influence of variety on selected physical and mechanical properties of wheat. Industrial Crops and Products, 47, 113–117.
Mehdawy, M., & Hussein, A. (2010). The Pharaoh's kitchen: Recipes from ancient Egypt's enduring food traditions. American Univ in Cairo Press.
Nicholson, P. T., & Shaw, I. (Eds.). (2000). Ancient Egyptian materials and technology. Cambridge university press.
Noble, W. & Davidson, I. (1996) Human evolution, language, and mind: A psychological and archaeological inquiry. Cambridge University Press.
Ogata, K., Gakumi, R., Hashimoto, A., Ushiku, Y., & Yoshida, S. (2023). The influence of Bouba-and Kiki-like shape on perceived taste of chocolate pieces. Frontiers in Psychology, 14, 1170674.
Oyen, A. V. (2016). How things make history: the Roman Empire and its terra sigillata pottery (p. 174). Amsterdam University Press.
Robson, E., & Stedall, J. (Eds.). (2009). The Oxford handbook of the history of mathematics. Oxford University Press. ISBN 978-0-19-921312-2.
Scarborough, J. (1978). Theophrastus on herbals and herbal remedies. Journal of the History of Biology, 353–385.
Spence, C., & Ngo, M. K. (2012). Assessing the shape symbolism of the taste, flavour, and texture of foods and beverages. Flavour, 1, 1–13.
Taylor, R. P. (2021). The potential of biophilic fractal designs to promote health and performance: A review of experiments and applications. Sustainability, 13(2), 823.
Van der Waerden, B. L. (2012). Geometry and algebra in ancient civilizations. Springer Science & Business Media.
Vlastos, G. (2006). Plato's universe. Parmenides Publishing.
Wang, J., Friedman, R., & Baba, M. (2021). Predynastic beer production, distribution, and consumption at Hierakonpolis, Egypt. Journal of Anthropological Archaeology, 64, 101347.
Waring, B. (2025). Sung, drawn and quartered: the roman ideogram of bread (part 2). DIAITA: Food-Heritage, (2), e0203-e0203.

Wilson, C. (2005). Wedding cake: A slice of history. Gastronomica, 5(2), 69–72.
Wynn, T. (2002). Archaeology and cognitive evolution. Behavioral and brain sciences, 25(3), 389–402.
Zeyl, D. J. (2000). Timaeus (Vol. 9). Hackett Publishing.

The Effect of Information Sources on Food Image

Ayşe SÜNNETÇİOĞLU[1], *Betül TABAN*[2]

Introduction

Cuisine and food have an important place among the factors that attract tourists. It has been observed that local cuisine ranks first among the reasons why tourists visit a destination (Lopez-Guzman and Canizares-Sanchez, 2012; Okumuş, Okumuş and McKercher, 2007). Tourists' perception of culinary image has become one of the main factors affecting their travel decisions (Okumuş, Okumuş and McKercher, 2007). From a holistic perspective, culinary image can be defined as "the totality of cognitive and emotional perceptions arising from the characteristics of a destination's food and beverages, culinary products, culinary culture and related activities, environmental factors, food and beverage establishments, and social interaction around food and beverages during a visit to the destination" (Yılmaz, 2021: 62). Culinary image is handled with two approaches: reflected and perceived image (Görür, 2019: 37). In the projected image, information sources are used to create a culinary image (Chuang, 2009; Fisher et al., 2012; Frochot, 2003; Horng and Tsai, 2010). In their study, Horng and Tsai (2010) analyzed the content of culinary and gastronomy websites of countries such as Hong Kong, Japan, Korea, Singapore, Taiwan and Thailand to promote East Asian culinary tourism. The study examined the capacity of government websites to promote traditional foods, restaurants, gastronomy tours, recipes and culinary culture. In this context, they concluded that the quality of written and visual materials used on websites can directly affect tourists' perception of the destination. The second approach refers to tourist perceptions of destination food (Ab Karim and Chi, 2010; Ab Karim et al., 2011; Seo, Kim, Oh, and Yun, 2013; Seo and Yun 2015; Seo, Yun, and Kim, 2017).

1 Çanakkale Onsekiz Mart University, Gelibolu Piri Reis Vocational School, Travel, Tourism and Leisure Services Programme, aysesuunetcioglu@comu.edu.tr, 0000-0002-4295-9776.

2 Çanakkale Onsekiz Mart University, Institute of Postgraduate Education, betuultaban@gmail.com, 0009-0000-0498-8261.

Food image plays an important role in the construction of the overall perception of a destination (Okumuş, Okumuş and McKercher, 2007). Food image also has an impact on the intention to visit and revisit the destination (Ab Karim, Lia and Salleh, 2010; Ab Karim and Chi, 2010, Ab Karim et all, 2011; Nelson 2016, Lai et al. 2017; Demircan et al, 2019; Polat, 2020). Its effect on tourists' satisfaction has also been revealed in various studies (Ab Karim, Lia, and Salleh, 2010; Ab Karim et all, 2011; Aktaş, 2021). It can be stated that information sources are effective on image (Hanlan and Kelly 2004; Phau, Shanka and Dhayan, 2009). Yıldırım and Albayrak (2019) obtained information about street food in Istanbul can be evaluated as two different groups. The first one is the information that tourists receive from reference groups such as friends, relatives and acquaintances. The other group consists of information obtained through information technologies such as social media channels, YouTube, and the internet. Tourists stated that the information they received from reference groups led them to eat certain street foods. Boz et al. (2014) finds out that which sources people heard about the festival, it was revealed that friends and family recommendations, the internet and social media were the most common sources of information for ho participated in the Local Flavors Festival in Bozcaada.

When it was investigated from The variety of information sources used also significantly affects the cognitive evaluations made by the individual (Baloğlu, McCleary, 1999; Frías, Rodríguez and Castañeda, 2008). In this context, it is predicted that the information sources used are effective on food image. Accordingly, the hypothesis of the study was formed as follows:

H_1: *Information sources have a positive effect on food image.*

Method

The empirical research model was adopted in the study. Relational survey model was applied as the research model in the study (Bursal, 2014: 155–157; Karasar, 2016). The model was realized with a cross-sectional approach (Karasar, 2016: 112). People aged 18 and over who had visited Bozcaada before. While determining the sample, convenience sampling technique was applied. Bell's (2016) information sources scale with 11 items and Likert's (1932) 5-point Likert scale was used. In order to measure food image, the food image scale of Promsivapallop and Kannaovakun (2019), whose reliability was tested by

Nakilcioğlu (2022), was used to measure food image with 26 statements and 5-point Likert type. Data were collected online between January 2–29, 2024. Correlation Analysis was applied to reveal the relationship between information sources and culinary image. Pearson correlation coefficient was used among the correlation coefficients.It was applied to examine the relationship between variables before regression analysis. Regression Analysis was applied to test the hypothesis of the research.

Reliability Analysis

In reliability analysis, Cronbach's Alpha value should be greater than 0.70, Corrected Item-Total Correlation should exceed +0.200 and Squared Multiple Correlation should exceed +0.300. The Cronbach's Alpha value of the information sources scale is 0.873. At the same time, all 11 statements are below this value. When the Corrected Item-Total Correlation and Squared Multiple Correlation values are examined, it is seen that each statement exceeds the desired values. In other words, the information source scale is reliable and none of the statements violate reliability. The Cronbach's Alpha value of the food image scale is 0.951. At the same time, the Corrected Item-Total Correlation and Squared Multiple Correlation values of all statements are within the desired range. In other words, the food image scale is a reliable scale and none of the scale statements violate reliability.

Validity Analysis

The results of the exploratory factor analysis applied to the information sources scale used in the study are reported in Table 1. When the results of the exploratory factor analysis of the information source scale were examined, Bartlett's Sphericity value (Coşkun, Bayraktaroğlu and Yıldırım, 2012: 272) was found to be 1079.182 and Keiser-Meyer-Olkin Measure of Sampling Adequacy (KMO) was found to be .855. Thus, it was revealed that factor analysis was feasible.

While the co-occurrence values obtained as a result of principal component analysis were between ,586-,730, the factor loadings were between ,632-,818. The values obtained show that the items have a good relationship with the relevant factor. When the eigenvalue ($\lambda 1$) is taken into account in deciding the number of factors (Alpar, 2016: 609; Altunışık, Coşkun, Bayraktaroğlu and Yıldırım, 2012: 274; Karagöz, 2017: 404), the resulting scale has 3 dimensions. The total variance explained is 65.614%.

Table 1. Results of Exploratory Factor Analysis of Sources of Information Scale

	Statement	Co-Originality	Factor Load	Values
Intermediary Websites	High street travel agent.	,619	,764	λ:4,889 Variance: %44,447
	Local tourist board's web site internet	,696	,758	
	Direct communication by phone or email (e.g. local tourism board, airline, hotel, tour operator, attraction)	,586	,740	
	High street travel agent's web site	,644	,718	
Online Information Resources	Independent traveller review sites and forums (Tripadvisor, Wikitravel etc.)	,666	,769	λ:1,131 Variance: %11,940
	Online only travel agent/ price comparison site (e.g. Expedia, Laterooms etc.)	,708	,675	
	Travel guides online	,635	,647	
Social Environment	Friends and relatives	,730	,818	λ:1,015 Variance: 9,227
	Social media (Facebook, Twitter, Instagram, Blogs etc.)	,661	,741	
	News	,725	,632	

Total variance explained: 65,614%, KMO Sampling adequacy: ,855 Barlett's test of sphericity: X2 :1079,182 s.d.:55, p<0,001, Principal components, analysis with varimax rotation

The findings of the factor analysis applied to the food image scale used as the dependent variable in the study are presented in Table 2.When the results of the exploratory factor analysis were analyzed, Bartlett's Sphericity value (Altunışık, Coşkun, Bayraktaroğlu and Yıldırım, 2012: 272) was found to be 4007.286 and Keiser-Meyer-Olkin Measure of Sampling Adequacy (KMO) was found to be .947. The values obtained show the applicability of factor analysis. While the co-occurrence values obtained as a result of principal component analysis were between ,534-,760, the factor loadings were between ,575-,834. The values obtained show that the items have a good relationship with the relevant factor. When the eigenvalue ((λ1) is taken into account in deciding the number of factors (Alpar, 2016: 609; Altunışık, Coşkun, Bayraktaroğlu and Yıldırım, 2012: 274; Karagöz, 2017: 404), the resulting scale has 3 dimensions. The total variance explained is 61.282.

Table 2. Table of Findings Related to Factor Analysis of Food Image Scale

	Statement	Co-Originality	Factor Load	Values
Health and Hygiene	Bozcaada food is healthy.	,755	,834	λ :12,095 Variance: 46,520
	Bozcaada food is easily digestible.	,760	,816	
	Bozcaada food is safe to consume.	,669	,714	
	Herbs used in Bozcaada food are good for health.	,552	,691	
	Bozcaada food is aromatic.	,584	,670	
	Bozcaada food is hygienic.	,608	,658	
	Bozcaada food is tasty.	,618	,653	
	Ingredients used in Bozcaada food are fresh.	,594	,629	
	Bozcaada food gives a cultural experience.	,594	,579	
Food Tasting	Bozcaada food is a popular cuisine in the world.	,637	,754	λ:2,053 Variance: %7,895
	Bozcaada food is hot and spicy.	,492	,692	
	Bozcaada food is unique.	,596	,651	
	Bozcaada cuisine has attractive eating habits and table manners.	,546	,631	
	There are a variety of restaurant types that offer Bozcaada cuisine.	,652	,608	
	Bozcaada food is rich in flavours.	,609	,597	
	Bozcaada cuisine uses a variety of cooking methods.	,620	,590	
	Bozcaada food is exotic.	,564	,575	

(Continued)

Table 2. (Continued)

	Statement	Co-Originality	Factor Load	Values
Restaurant Selection	Services by food providers are good.	,706	,733	λ :1,785 Variance: %6,867
	Food spots are located in convenient places.	,515	,675	
	Food providers are friendly.	,569	,644	
	Bozcaada restaurants have a unique style of decoration and environment.	,554	,639	
	Bozcaada restaurants offer value for money.	,645	,632	
	There are various information guides for local food.	,643	,620	
	There are a variety of local specialties.	,642	,617	
	Bozcaada restaurants have comfortable eating surroundings.	,534	,612	

Total variance explained: 61,282%, KMO Sampling adequacy: ,947, Bartlett's test of sphericity: X2: 4007,286, s.d.:325, p<0.001, Principal components analysis with varimax rotation

Finding

Findings on Demographic Characteristics

The research sample consists of 60.2% women and 39.8% men. 32.6% of the participants are between the ages of 18–24, 17.4% between 25–34, 17.8% between 35–44, 16.1% between 45–54, 12.7% between 55–64 and 3.8% between 65 and above. 36.9% of the participants last visited Bozcaada before 2019, 8.5% in 2019, 8.9% in 2020, 9.3% in 2021, 14.4% in 2022, and 22% in 2023.

Correlation and Regression Analysis

Before the regression analysis, correlation analysis was applied to determine the amount and direction of the relationship. According to the correlation analysis results, the relationship between information sources and food image is significant at p<.05 level. Pearson Correlation Coefficient between Variables is ,508. The relationship between the variables is positive.

Simple regression analysis was applied to test the research hypothesis stating that information sources positively affect food image. The findings of the analysis are presented in Table 3

Table 3. Findings on the Effect of Information Sources on Food Image

	Unstandardized Coefficients		Standardized Coefficients		
Model	β	Std. Hata	B	t-value	p-value
Fixed	7,115	,181		39,345	,000
Social Environment	,255	,049	,337	5,172	,000
Dependent Variable: Health and Hygiene; Method: Enter, Model 1: R=.495, R^2: 24.5% (0.245), Adjusted R^2: 23.5% (0.235); F for Model: 25,050; p:.000					
Fixed	6,699	,198		33,876	,000
Social Environment	,214	,054	,261	3,970	,000
Dependent Variable: Food tasting; Method: Enter, Model 1: R=.482, R^2: 23.3% (0.233), Adjusted R^2: 22.3% (0.223); F for Model: 23,437; p:.000					
Fixed	7,113	,203		34,981	,000
Intermediary Websites	,166	,050	,250	3,304	,001
Social Environment	,235	,056	,283	4,239	,000
Dependent Variable: Restaurant choice; Method: Enter, Model 1: R=.459, R^2: 21.1% (0.211), Adjusted R^2: 20.0% (0.200); F for Model: 20.622; p:.000					

According to the regression analysis, social environment (F: 25,050, p: 0,000) has a significant effect on the health and hygiene dimension of food image. Among the information sources, social environment dimension (F: 23,437; p: ,000) has a significant effect on the food tasting dimension of food image. Social environment and intermediary websites (F:20,622; p: ,000) among the information sources have a significant effect on the restaurant selection dimension of food image. Accordingly, it can be stated that social environment, one of the information sources, is effective in the formation of food image. The change in food image is 24.5% for Model 1, 23.3% for Model 2 and 21.1% for Model 3.

These rates show that the social environment affects food image at a low level (Ferguson, 2009: 533), although it does not affect it strongly (minimum rate for strong effect: 64%).

Conclusion

This study examines how tourists' culinary image perceptions of destinations are shaped through information sources. Understanding which information sources shape the culinary image of Bozcaada will help analyze how tourists perceive the island in terms of gastronomy. As a result of the research conducted on the Bozcaada sample, it was determined that information sources have a significant and positive effect on culinary image. Correlation and regression analyses revealed that information sources are effective in the formation of food image perception. The data obtained within the scope of the research show that information obtained from Friends and relatives, Social media and news significantly affect tourists' images of Bozcaada's cuisine. In particular, recommendations from the social environment play a strong role in shaping tourists' dining experience expectations and destination preferences. This suggests that culinary image is closely related not only to the quality of the food, but also to the type and quality of information obtained about that food. When Boz et al. (2014) investigated from which sources people heard about the festival, it was revealed that friends and family recommendations, the internet and social media were the most common sources of information. In conclusion, the information sources that tourists use are powerful tools that shape the food image of a destination.

References

Ab Karim, M. F., Lia, J. J., & Salleh, M. A. (2010). Food image and the destination choice. *International Journal of Hospitality Management*, 29(2), 368–376.

Ab Karim, M. F., Chi, C. G. Q., & Salleh, M. A. (2011). The impact of food image on tourists' satisfaction and behavioral intentions. *Tourism Management*, 32(3), 614–621.

Aktaş, S. (2021). The relationship between destination food image and destination personality, satisfaction, and behavioral intention. *Journal of Hospitality and Tourism Management*, 46, 1–9.

Alpar, P. (2013). *Sosyal bilimlerde istatistik*. Detay Yayıncılık.
Alpar, P. (2016). *Sosyal bilimlerde istatistik*. 3. Baskı, Detay Yayıncılık.
Altunışık, R., Coşkun, R., Bayraktaroğlu, S., & Yıldırım, E. (2012). *Sosyal bilimlerde araştırma yöntemleri: SPSS uygulamalı*. 7. Baskı, Sakarya Yayıncılık.
Baloğlu, S., & McCleary, K. W. (1999). A model of destination image formation. *Annals of Tourism Research*, 26(4), 868–897.
Bell, E. (2016). *The role of information sources in tourism decision-making: A scale development*. Journal of Travel Research, 55(3), 389–405.
Boz, I., Çalışkan, E., & Türkay, K. (2014). Local Flavors Festival: A case study of local food perceptions. *International Journal of Gastronomy and Food Science*, 2(3), 44–50.
Bursal, B. (2014). *Araştırma yöntemleri ve istatistik*. Ankara: Detay Yayıncılık.
Chen, C. F., & Tsai, D. (2007). How destination image and evaluative factors affect behavioral intentions. *Tourism Management*, 28(4), 1115–1122.
Coşkun, R., Bayraktaroğlu, S., & Yıldırım, E. (2012). *Sosyal bilimlerde araştırma yöntemleri: SPSS uygulamalı*. Sakarya Yayıncılık.
Demircan, O., Öztürk, A., & Kocaoğlu, S. (2019). The impact of food and beverage image on revisit intention. *International Journal of Contemporary Hospitality Management*, 31(4), 1673–1689.
Fisher, R. J., Bennett, R., & Harris, R. (2012). Culinary tourism: A systematic review. *International Journal of Tourism Research*, 14(3), 209–224.
Frochot, I. (2003). Marketing French tourism destinations: The impact of food and wine imagery. *Tourism Review*, 58(1), 9–16.
Görür, E. (2019). Gastronomy tourism and the culinary image. *Tourism and Hospitality Research*, 19(1), 35–45.
Güçlü, M. (2017). *Turizmde tanıtım ve pazarlama stratejileri*. Beta Yayınları.
Hanlan, J., & Kelly, J. (2004). The role of information sources in tourism decision-making. *Tourism Management*, 25(4), 533–542.
Horng, J. S., & Tsai, C. Y. (2010). Culinary tourism in Taiwan: A study of tourism promotion for East Asian culinary tourism. *Tourism Management*, 31(4), 595–603.
Karasar, N. (2016). *Bilimsel araştırma yöntemleri*. Nobel Yayın Dağıtım.
Karagöz, S. (2017). *İstatistiksel analizler ve uygulamalar*. Detay Yayıncılık.

Lai, P. C., Wang, C. H., & Lin, M. T. (2017). The role of food image in destination branding: A case study of Taiwan. *International Journal of Tourism Research*, 19(1), 1–10.

Lopez-Guzman, T., & Canizares-Sanchez, C. (2012). Gastronomy and tourism: The case of culinary tourism in the Mediterranean region. *Tourism Review*, 67(1), 32–45.

Nakilcioğlu, İ. (2022). *Promsivapallop and Kannaovakun scale reliability*. Unpublished Master's Thesis.

Nelson, J. (2016). Houston's food image: Ethnic diversity and culinary culture. *International Journal of Hospitality & Tourism Administration*, 17(2), 125–140.

Okumuş, B., Okumuş, F., & McKercher, B. (2007). Culinary tourism as a destination attraction: An examination of the literature. *Journal of Vacation Marketing*, 13(3), 215–226.

Phau, I., Shanka, T., & Dhayan, N. (2009). Destination image and choice intention of university student travelers to Mauritius. *International Journal of Contemporary Hospitality Management*, 22(5), 758–764.

Polat, G. (2020). The effect of local food image on destination preference. *Tourism Management*, 41(1), 89–99.

Promsivapallop, P., & Kannaovakun, W. (2019). *Measuring food image in tourism destinations*. Unpublished research.

Seo, Y., Yun, S., & Kim, W. (2017). The relationship between destination food image and food intention. *Journal of Travel & Tourism Marketing*, 34(5), 601–612.

Seo, S., Kim, O. Y., Oh, S. & Yun, N. (2013).Influence of informational and experiential familiarity on image of local foods." *International Journal of Hospitality Management, 34(1),*295–308.

Seo, Y., & Yun, S. (2015). The role of food culture in destination food image. *International Journal of Tourism Research*, 16(4), 314–323.

Yıldırım, B., & Albayrak, T. (2019). *Sokak yemekleri ve turistlerin algıları: İstanbul örneği*. Unpublished Master's Thesis, Istanbul University.

Yılmaz, A. (2021). *Gastronomy tourism and the culinary image: A conceptual framework*. Istanbul: Istanbul University Press.

Zağralı, E., & Akbaba, A. (2015). Gastronomy as an attraction in destination selection. *Tourism Management Perspectives*, 15, 40–47.

A Study on the Factors Affecting the Selection of Colorful Foods

Ayşe Nur ÖZER[1] Serdar SÜNNETÇİOĞLU[2]

Introduction

Today, in addition to the beliefs and cultural values of the society we live in, there are differences in eating and drinking behaviors and preferences that vary depending on individuals' personal tastes and choices. The physiological eating and drinking needs of people to sustain their lives have created a desire to experience foods with diverse tastes, flavors, and images, a phenomenon influenced by globalization (Bozok & Yalın, 2018: 252).

The desire to experience has led to reasons such as the desire to be healthy and fit, the search for innovation, and eating better among the ever-changing gastronomy trends. In the increasingly globalized world, there is a desire to try food and beverages with different tastes, flavors and appearances with the effect of multiple reasons, and these eating and drinking activities should be included in life as an indicator of socialization for individuals (Özdemir, 2020: 18).In order to create change and differentiate, people have gone in search of differences in their eating and drinking tendencies (Güneş, et al., 2015: 4; Uluçay, 2017: 197).

The purpose of coloring food is to increase the desire to want the effect of color context on food, to enjoy and consume more (Tomita, et al., 2007: 50; Genschow, et al., 2012: 700; Piqueras-Fiszman, et al., 2012: 207; Spence, 2018: 160). In addition, color indicates the edibility and nutritional value of food (Spence, 2015: 2).

When people make food choices, they go through a decision process, taking into account many sensory and cognitive factors. Past experiences, emotions and values, current needs are evaluated consciously or unconsciously

---

* This study was presented as an oral presentation at the International Gastronomy Tourism Congress (UGTAK 2024).

1 Master student, Çanakkale Onsekiz Mart University, Postgraduate Education Institute, Gastronomy and Culinary Arts Department, ozeraysenur1@gmail.com

2 PhD., Çanakkale Onsekiz Mart University, Tourism Faculty, Gastronomy and Culinary art Department, serdarsunnetcioğlu@comu.edu.tr

(Franchi, 2012: 19). Colors can come into play at this point. For example, the color of the food can affect the choice of food by reminding the positive or negative event experienced in the past, or the color of the food can create prejudice about its taste. Colors are interpreted by attributing meanings during the perception process. Especially the color and appearance of the food is an important source of information for people to predict the taste before tasting. Color affects people's ability to accurately know the aroma in the food. (Mazlum, 2011: 126).

The aim of this study is to investigate the effects of colored foods on food choice. In the existing literature, there are studies on colored foods.Türker and Yılmaz (2023) investigated the effect of color on products and desire to eat in adult individuals. Çafa and Madenci (2020) focused on the production of colored hamburger bread and aimed to examine the natural color substances used in the coloring of these breads. Clydesdale (1993) collected the articles on the effects of color on food choice by affecting taste thresholds, sweetness perception, food preference, pleasantness and acceptability in one article. Sarıışık and Kardeş (2019) investigated the role of colors in the formation of recently trending gastronomy trends. Bozok and Yalın (2018) conducted a review on the production, types and consumption of black foods, which emerged as a new trend. Özdemir (2020) aimed to draw attention to the point of healthy eating in his study and to contribute to different studies in which purple foods are at the center.When these studies are examined, it is seen that color is mostly handled in terms of health. In addition.Within the existing literature review, it has been seen that there are not enough resources on people that address color as a factor affecting food choice, and this study aims to fill this gap.

Gastronomy and Colorful Foods

One of the important factors affecting these trends is the color factor. When we look at the context of the integrity of color with food; it is seen that it plays an important role in food selection by affecting food preference, taste, beauty of food, taste perception and preferability (Clydesdale, 1994: 85). At the same time, colors are the most indicative element as a sign for food and beverages (Cordain, et al., 2005: 345; Heseker, 2012: 48). Color has been used for many years to increase food acceptance and taste. Consumers' perception of color can be influenced by external factors. In particular, people's family, personal learning experiences, cultural background and family influence people's ideas and create lasting experiences in their memory. The presentation of a food can reduce or increase the appetite level of consumers. The colors to be used in food presentation should accurately reflect the characteristics of the food. Colors have a very important effect on taste perception. Consumers associate different colors with different foods. (Magoulas, 2009: 2; Kontukoski, et al., 2016: 3). Colors can play an indicative role about the color, appearance and taste of beverages and foods (Türker & Yılmaz, 2023: 351).

The color white represents cleanliness, righteousness, away from complexity. White color generally represents low-calorie foods and is therefore seen as healthy food. Yellow color creates a sincere and natural perception given by the feature of warm colors and is known to stimulate appetite. It also expresses transience and because of this feature, yellow color is preferred in fast food style businesses where there is an excess of consumers. Orange color creates a warm, artificial and sincere perception and is known as one of the appetizing colors. Examples of orange foods are citrus fruits with high vitamin value and pumpkin, which is widely used in Turkish cuisine. Red is an appetizing color with its vivid and bright appearance and is the first color that comes to mind in emblems and food packaging. Red is also used in packaging for jelly, jam, etc. to create an energizing and endearing effect. Purple, on the other hand, is an interesting and delicate color that determines the quality of the product. In addition, purple is among the colors that suppress appetite. The green color is one of the cold colors and brings vegetables to the memory by reminding healthy life. In addition, since it is a color that increases creativity, it is often used in the kitchens of hotels. Blue is among the colors that create an appetite-suppressant effect. The reason for this is that we do not encounter blue color in natural foods (Spence, 2015: 7; Blechert, et al., 2014: 3; Yılmaz & Erden, 2017: 267; Fişek, 2019; Sarıışık & Kardeş, 2019: 432).

Özdemir (2020) aimed to investigate whether purple foods, which are a new trend, are preferred and consumed for reasons such as some types of purple foods and the desire for healthy nutrition. As a result of the study, it is seen that purple foods are especially healthy due to their nutritional values and the anthocyanin contained in them, so it is thought to be a reason for preference. Bozok and Yalın (2018) aimed to make a compilation about the production, types and consumption of black foods, which emerged as a new trend, and to contribute to different studies on black foods by drawing attention to this trend, which is lacking in the literature. As a result of the research, no examples of this trend have been found in Turkish cuisine yet. Sarıışık and Kardeş (2019) seek the role of colors in the formation of gastronomy trends that have been trending recently. As a result of the study, gastronomy trends were examined and it was seen that the common points of the trends are health and color. In addition, it was found that consumers are more likely to buy fresh and colorful-looking foods, and colors have a great influence on social media trends. Çafa and Madenci (2020) conducted a literature review on the production of colored hamburger buns in order to bring alternative products to the consumer market and to attract consumer attention to these products. As a result, it is thought that some consumers prefer colored hamburgers only for their color and showiness. In addition, it is thought that some consumers are harmful and not suitable for consumption because they do not know the content and properties of the color substances used. Colored bread production will continue with many different colors and it is thought that these breads will remain on the agenda for a long time. Yu, Xie and Wen (2020) aimed to explain the roles of color psychology in product packaging, retail environment and online marketing. As a result of the analysis, it was found that people generally respond to brighter and more saturated colored destination images, and warm colors such as yellow and orange-yellow contribute the most to gastronomy photo shoots. Blechert et al. (2014) presented analyses of food types, individual differences, situation effects and image characteristics to explore the simplest variables related to the experimental design of food imaging studies. The results of the survey showed that using food imagery, neurocognitive measures were effective in examining the appetitive and regulatory effects of eating behavior and that red color was associated with higher degrees of arousal, while green color was associated with lower calories and lower protein, fat and carbonhydrate.

Based on the researches and literature reviews the following hypotheses were formulated.

H1. Consumers' motivation to prefer colorful foods differs with gender
H2. Consumers' motivation to prefer colorful foods differs with the age

H3. Consumers' motivation to prefer colorful foods differs with the income.
H4. Consumers' motivation to prefer colorful foods differs with the education.

Methodology

The population of this study consists of all people aged 18 and over. The sample is all people aged 18 and over residing in Turkey. As a data collection tool in the research, an online questionnaire application was realized from Google documents. The data collection period started on 20.12.2023 and the data collection phase was completed on 30.01.2024. During the data collection phase, 300 questionnaires were collected. The data were recorded electronically. The data recorded electronically were analyzed using statistical package programs as the analysis method. Firstly, reliability test and factor analysis were performed with the findings obtained. T-test and ANOVA test were conducted to examine the relationship with demographic characteristics. The questionnaire was taken from a previously published, open access and reliability tested and validated scale. It was taken from the Food Choice Questionnaire of Percott et al. (2002) and adapted to the study on the Effect of Colorful Foods on Food Choice. The scale is designed as a 5-point Likert scale and consists of 34 items. It is expressed as 1-strongly disagree, 2- disagree, 3-undecided, 4-agree, 5-strongly agree.

Cronbach's alpha value was above 80 for each statement. The cronbach alpha value of the entire scale was measured as .946 and was found to be highly reliable. It shows that the answers given to the survey statements are consistent and that the analyzes to be made can be done correctly. As a result of the factor analysis, items S6, A3 and F3 were removed because their extraction values were below .50. In the factor analysis conducted after the item was removed, K5, EK2 and A1 items were removed because they were overlapping expressions. After this process, the scale was reduced to 28 items and categorized under 5 factors: health, mood, sensory appeal, convenience and price. Then, according to the results of the reliability analysis of the dimensions, the cronbach alpha value of the health dimension was found to be ,917; the cronbach alpha value of the mood dimension was found to be ,909; the cronbach alpha value of the sensory appeal dimension was found to be ,865; the cronbach alpha value of the convenience dimension was found to be ,825 and the cronbach alpha value of the price dimension was found to be ,887

Findings

The findings on the demographic characteristics of the survey participants were analyzed under the headings of gender, educational status, age and income status. Demographic data shows that 57% of the participants were female and 43% were male. It is seen that 55.7% of the participants are between the ages of 18–29, 18% between 30–39, 16.7% between 40–49 and 9.6% between 50 and above. The educational status of the participants was determined as 21% high school, 15.3% college, 55% bachelor's degree and 8.7% master's degree, while the income status of the participants was 0-11,402 for 34.3%, 11,403-13,684 for 10.3%, 13,685-45,686 for 39.3%, and 45,687 and above for 16%.

Explanatory factor analysis was applied to the food choice scale to examine the effect of colorful foods on food choice. The Kaiser-Meyer-Olkin (KMO) coefficient and Barlett sphericity test results were significant for the food choice scale. Considering the test results of the data for the scale, it was determined that factor analysis was appropriate for the scale. As a result of the factor analysis of the food choice scale in Table 2, five dimensions with eigenvalues greater than 1 were identified. These dimensions are "health," "mood," "sensory appeal," "convenience" and "price." The factor loadings of the statements in these dimensions vary between 0.465 and 0.866, and item factor loadings greater than 0.40 indicate a well-defined structure (Kozak, 2017).As a result of the factor analysis, items S6, A3 and F3 were removed because their extraction values were below .50. In the factor analysis conducted after the item was removed, items K5, EK2 and A1 were removed because they were overlapping expressions. As a result, these five dimensions explain 67,171% of the total variance.

The "t-test" results of consumers' attitudes towards the choice of colored food in terms of gender variable are given below. There is no significant difference in the health dimension according to gender variable (p= ,305). It can be said that both male and female consumers develop a negative attitude towards the choice of colored food in the health dimension. There is no significant difference in the mood dimension according to gender variable (p= ,117). It can be said that both male and female consumers develop a negative attitude towards the choice of colored food in the mood dimension. There is no significant difference in the sensory appeal dimension according to gender variable (p= ,337). It can be said that both male and female consumers develop a positive attitude towards the choice of colored food in the sensory appeal dimension. There is no significant difference in the convenience dimension according to gender variable (p= ,375). It can be said that female consumers develop a more positive attitude towards the choice of colored food than male consumers in the convenience dimension. There is no significant difference in the price dimension according to gender variable (p= ,608). It can be said that both male and female consumers develop a negative attitude towards the choice of colored food in the price dimension. According to this result, hypothesis H1, "Consumers' motivation to prefer colorful foods differs with gender variable," is not accepted.

The results of the ANOVA test on consumers' attitudes towards choosing colorful foods and the age variable are given below. According to the food choice questionnaire, it was determined that there was no significant difference between the health sub-dimension and the age factor regarding the choice of colored food (p= ,524) According to the food selection questionnaire, it was found that there was a significant difference between the sensory appeal sub-dimension and the age factor (p= ,004). With the sensory appeal sub-dimension, the arithmetic mean of the 18–29 age range (X = 3.55) is the highest, while the arithmetic mean of the 50 and over age range (X = 3.06) is the lowest. According to the food choice questionnaire, it was found that there was a significant difference between the mood sub-dimension and the age factor regarding the choice of colorful food (p= ,000). With the mood sub-dimension, the arithmetic mean of the 18–29 age range (X=2,88) is the highest, while the arithmetic mean of the 40–49 age range (X =2,23) is the lowest. According to the food selection questionnaire, it was found that there was a significant difference between the convenience sub-dimension and the age factor (p= ,000). With the convenience sub-dimension, the arithmetic mean of the 18–29 age range (X = 3.18) is the highest, while the arithmetic mean of the 50 and over age range (X = 2.68) is the lowest. According to the food choice questionnaire, it was determined that there was no significant difference between the price sub-dimension and the age factor regarding the choice of colorful food (p= ,237). According to this result, H2 hypothesis, "Consumers' motivations to prefer colorful foods differ with the age variable," is partially accepted.

The results of the ANOVA test on consumers' attitudes towards choosing colorful foods and the income variable are given below. According to the food choice questionnaire, it was determined that there was no significant difference between the health sub-dimension and income factor regarding the choice of colored food (p= ,580). According to the food choice questionnaire, it was determined that there was no significant difference between the sensory appeal sub-dimension regarding the choice of colorful food and the income factor (p= ,456). According to the food choice questionnaire, it was found that there was a significant difference between the mood sub-dimension and income factor regarding the choice of colorful food (p= ,027). With the mood sub-dimension, the arithmetic mean of the income range of 0-11,402 (X =2,78) is the highest, while the arithmetic mean of the income range of 45,687 and above (X =2,36) is the lowest. According to the food choice questionnaire, it was found that there was no significant difference between the convenience sub-dimension of choosing colored food and the income factor (p= ,375). According to the food choice questionnaire, it was determined that there was no significant difference between the price sub-dimension and income factor regarding the choice of colored food (p= ,684). According to this result, hypothesis H3, "Consumers' motivations to prefer colorful foods differ with the income variable," is partially accepted.

The results of the ANOVA test on consumers' attitudes towards choosing colored foods and the education variable are given below. According to the food choice questionnaire, it was determined that there was no significant difference between the health sub-dimension and the education factor regarding the choice of colored food (p= ,791). According to the food selection questionnaire, it was found that there was a significant difference between the sensory appeal sub-dimension and the education factor (p= ,016). With the sensory appeal sub-dimension, the arithmetic mean of the master's degree education status (X =3,62) is the highest, while the arithmetic mean of the high school education status (X=3,31) is the lowest.

According to the food choice questionnaire, it was found that there was no significant difference between the mood sub-dimension of colorful food choice and the education factor (p= ,384). According to the food choice questionnaire, it was found that there was no significant difference between the sub-dimension of convenience regarding the choice of colored food and the education factor (p= ,074). According to the food choice questionnaire, it was determined that there was no significant difference between the price sub-dimension and the education factor regarding the choice of colored food (p= ,232). According to this result, hypothesis H4, "Consumers' motivation to prefer colorful foods differs with the education variable," is partially accepted.

Conclusions and Recommendations

It has been observed that health, sensory appeal, mood, convenience and price dimensions have a certain level of influence on the reasons why consumers prefer colorful foods. In line with the determined effects, hypotheses H2, H3, H4 were partially accepted and hypothesis H1 was not accepted. The health dimension was found to be the dimension that explained the effect of colored food on consumers' food choices at the highest level. The health dimension was evaluated positively and negatively by consumers. It is seen that consumers who choose colorful foods as healthy choose them because they are colored with natural products. Sarıışık and Kardeş (2019) state that green-colored foods are preferred because they evoke natural, healthy life and healthy nutrition. Jalil, Yunus, and Said (2011), Spance (2015), and Özcan (2018) also found that colored foods evoke a fresh and healthy feeling in consumers, especially in connection with the expressions of health and healing. However, some consumers find colored foods unhealthy. Consumers have the colors they have determined for food in their minds and it is seen that they first evaluate the food with its colors. For this reason, when the color of the food is not the expected color, it leaves a negative impression, so colored foods are considered unhealthy (Megep, 2012; Karahan, 2020). Consumers may also have a fear of trying new foods. Colorful foods may seem new, different and unhealthy to consumers and may cause eating phobia and fear of trying. The study of İflazoğlu (2019) in the literature also supports this result.

As a result of the study, it was observed that there was no effect of colored foods on food choice between the genders of consumers. The proportion of female and male consumers choosing colorful foods in food selection does not vary. It has been observed that the level of influence of female and male consumers on colored foods is generally close to each other. The effect of colored foods on food choice and the changes in different age groups were observed in the dimensions of sensory appeal, mood and convenience. In age groups, it is thought that colorful foods improve mood and contribute to positive thinking. In particular, consumers between the ages of 18–29 tend to consume colorful foods because they believe in the visuality and psychology of colors. It can be said that consumers between the ages of 18–29 prefer colorful foods for their visuality, especially because colorful foods attract attention in the social environment. In a similar study, Magoulas (2009) emphasizes that colors affect our taste, mood and inner comfort level. In this context, it was determined that colors are effective in the selection and consumption of food.The other dimension that is effective with the age factor in the choice of colorful foods is sensory appeal. Sensory appeal activates consumers' culture, social values, spiritual values and past emotions. These values affect consumers (Toker & Sulak, 2020). Consumers' tastes, values and emotions vary according to age groups. For this reason, the sensory appeal of colors to consumers also varies between age groups. As seen in the study, it can be said that colorful foods are preferred because they are beautiful to the eye according to the values of consumers and evoke memories from their old lives. On the other hand, the fact that it is foreign to the cultural and social values of consumers can be shown as a reason for not preferring colorful foods. In a similar study found in the literature, Yu, Xie and Wen (2020) state that bright and saturated-colored foods attract consumers more sensually.

It was observed that there was a difference between the age factor and the convenience dimension in consumers' choice of colorful foods. The increasing popularity of colorful foods, which is one of the gastronomy trends in recent days, has enabled businesses to give more space to colorful foods. As seen in the results of the study, consumers stated that it is easy to access colorful foods. As another result of the study, a difference was found between the income and mood dimensions that are effective in the food choice of colorful foods. It can be said that consumers with average income status resort to colorful foods to change their mood.

It is seen that colored foods have positive and negative effects on consumers. This is because colors affect the psychology of consumers. Consumers look for different options to have different experiences and show purchase intention. In this context, this study suggests that food and beverage businesses should evaluate this desire of consumers. Businesses can differentiate in the competitive environment by adding colorful versions of existing foods to their menus. It is thought that colorful foods can be useful for consumers who are hesitant to try new foods and who are prejudiced.

References

Arıker, Ç. (2012). The Relationship Between Consumers' Restaurant Selection Criteria and Demographic Characteristics. Öneri Journal, 10(38), 11–31.

Blechert J., Meule, A., Busch, N., A., Ohla, K. (2014). Food-Pics: An Image Database For Experimental Research On Eating And Appetite, Original Research Article, 5(617), 1–10.

Bozok, D., Yalın, G. (2018). New Trend in Gastronomy: Black Foods. Journal of Current Tourism Research, 2(1), 251–261.

Cordain, L., Eaton, S. B., Sebastian, A., Mann, N., Lindeberg, S., Watkins, B. A., BrandMiller, J. (2005). Origin and devolution of the Western diet: health implications for the 21st century. The American Journal of Clinical Nutrition, 81(2), 341–354.

Çafa, İ., Madenci, A., B. (2020). Latest Trend in Gastronomy Colored Breads: What Color Do You Want Your Hamburger?. Journal of Academic Social Research, 8(102), 46–54.

Franchi M. (2012). Food Choice: Beyond The Chemical Content. International Journal Of Food Sciences And Nutrition, 63(1), 17–28.

Genschow, O., Reutner, L., and Wänke, M. (2012). The color red reduces snack food and soft drink intake, Appetite, 58, 699–702.

Heseker, H. (2012). LebensmittelverzehrundNährstoffzufuhr, Ergebnisse der NationalenVerzehrsstudie II, Ernährungsbericht, 12, 40–85.

İflazoğlu, N. (2020). Gastronomy Tourism in Unesco Gastronomy Cities and Fear of Trying New Food (Food Neophobia): The Case of Hatay and Gaziantep (Published Master's Thesis). Mersin University Institute of Social Sciences, Mersin.

Jalil, N. A. Yunus, R. M., Said, N. (2011). Environmental Color Impact upon Human Behavior: A Review. Procedia - Social and Behavioral Sciences, 35, 54–62.

Karahan, S. (2020). The Effects of Color Factor on Visual and Tatsal Perceptions towards Beverages (Published Master's Thesis). T.C. Kocaeli University Institute of Social Sciences. Kocaeli.

Kontukoski, M., Paakki, M., Thureson, J., Uimonen, H., Hopia, A. (2016). Imagined salad and steak restaurants: Consumers' color, music and emotion associations with different dishes. International Journal of Gastronomy and Food Science, 4, 1–11.
Kozak M. (2017). Data analysis: Scientific research: design, writing and publication techniques. Ankara: Detay Publishing.
Magoulas, C. (2009). How Color Affects Food Choice (Published Master's Thesis). Bachelor of Arts Warner Southern University, Las Vegas.
Magoulas, C. (2009). How color affects food choices, U.N.L.V. Master of Hospitality Administration William F. Harrah College of Hotel Administration.
Mazlum, Ö. (2011). Cultural Associations of Color. Dumlupınar University Journal of Social Sciences, 31, 125–137.
MEGEP, (2012). Making Sensory Controls. Food Technology. Ankara: Ministry of National Education.
Onwezen, M., C., Reinders, M., J., Verain, M., C., D., Snoek, H., M. (2019). The Development of a single-item Food Choice Questionnaire. Food Quality and Preference, 71, 34–45.
Özdemir, B. (2020). Healthy Purple Foods in Gastronomy Trends. GSI Journals Serie B: Advancements in Business and Economics, 3(1), 16–30.
Piqueras-Fiszman, B., Alcaide, J., Roura, E., & Spence, C. (2012). Is it the plate or is it the food? Assessing the influence of the color (black or white) and shape of the plate on the perception of the food placed on it. Food Qual Prefer, 24, 205–208.
Piqueras-Fiszman, B., & Spence, C. (2012). The influence of the color of the cup on consumers' perception of a hot beverage. J. Sensory Study, 27, 324–331
Sarıışık, M., Kardeş, C., N. (2019). A Review on the Relationship between Gastronomy Trends and Colors. International Conference on Eurasian Economies, 430–438.
Spence, C. (2015). On the Psychological Impact of Food Color. Spence flavor, 4(21), 2–16.
Spence, C. (2018). Background color and its impact on food perception and behaviour, Food Qual Prefer, 68, 156–166.
Toker, A., Sulak, H. (2020). A Research on the Types of Attraction Used in Advertising Films and Their Effects on Consumers. Süleyman Demirel University Visionary Journal, 11, 117–133.

Tomita, K., Ono, M., Aiba, T., Ohtani, K. (2007). Psychological effects of tablecloth color on diners under different brightness. Jpn. Assoc. Integr. Study Dietary Habits, 18, 48–55.
Türker, A., Yılmaz, İ. (2023). Investigation of the Relationship Between Colors and Food Desire in Adult Individuals. Journal of Gastronomy, Hospitality and Travel, 6(1), 349–364.
Uluçay, D. M. (2017). A Comparative Evaluation on the Websites of Gastronomy Tourism Enterprises. Erciyes Journal of Communication, 5(2), 196–208.
Yu, C., Xie, S. Y., Wen, J. (2020). Coloring the destination: The role of color psychology on Instagram. Tourism Management, 80, 1–12.

Career and Talent Management in Food and Beverage Enterprises

Savaş BÜYÜKYILMAZ[1]

Introduction

For enterprises to succeed in a competitive environment, effective management of their existing resources is essential. One of these critical resources is human capital, which holds particular significance for food and beverage enterprises within the service sector. These enterprises determine their human resource needs based on their organizational structure, adopting an organizational career management perspective. However, this alone is insufficient; it must now be acknowledged that employees also have their own career aspirations. This individual career management refers to the goals employees aim to achieve within a specific timeframe. This situation highlights the need for a collaborative career management approach that integrates the talents required by the enterprise with employees' individual career plans.

This collaborative career planning begins with the process of identifying the competencies required based on the enterprise's job descriptions. This process extends beyond mere personnel selection and can be regarded as a talent management approach that also considers the individual's goals.

Human resources are of paramount importance in the labor-intensive food and beverage sector. Employees in these enterprises are in direct contact with guests. Therefore, their knowledge and skills must be sufficient to represent the enterprise effectively and deliver services adequately. Consequently, both employees and the enterprise need to develop a conscious career management strategy. This study examines the career management and talent management processes within the context of food and beverage enterprises.

1 Republic of Turkiye Ministry of National Education Food and Beverage Services Field Teacher, sbuyukyilmaz@gmail.com https://orcid.org/0000-0002-1162-2273

Food and Beverage Enterprises

When tourists travel away from their homes, various needs arise at their destinations. During their travel, activities addressing needs such as accommodation, food and beverage, transportation, and entertainment collectively constitute the tourism sector. Consequently, the tourism industry comprises a variety of enterprises that cater to these essential needs. These can primarily be summarized as hotels, restaurants, travel agencies, transportation enterprises such as roadways and airlines, and other related businesses (Oğan, 2021:3). Food and beverage enterprises operate institutionally in diverse settings, including schools, hospitals, and numerous other areas. Commercially operating food and beverage enterprises can be classified into categories such as restaurants, cafeterias, fast-food restaurants, ethnic restaurants, and specialty restaurants (Sökmen, 2014:9). Generally, food and beverage enterprises can be examined under three main categories. The first includes establishments like à la carte restaurants and buffets that meet a variety of dining needs. The second comprises cafeterias, where waiter service is less prevalent, offering quick meal services. The third category encompasses accommodation facilities, including four- and five-star hotels (Şahin & Ünal, 2016).

Career Management

According to the Turkish Language Institution, a career is defined as "*a stage, success, and expertise attained in a profession through time and effort.*" Employees, as individuals, have various needs, one of which is to build a career within the organization they work for. Successful organizations support their employees by addressing these needs and assisting in the planning of their career development. In today's context, human resources are among the most critical assets of organizations. The effective management of organizational climate and culture depends on human resources. Enhancing organizational efficiency and performance is tied to organizational career planning (Tunçer, 2012: 230).

For an organization to succeed, collaboration between employees and the organization must be fostered. Rather than relying solely on human resource management, which is often reduced to climbing hierarchical ranks, career management—a process-oriented approach—should be prioritized to enable organizations to compete more effectively. Within this process, the organization's goals and the aspirations of employees must be planned and directed collaboratively (İrmiş & Bayrak, 2019: 185). Accordingly, career management can be examined under two dimensions: organizational career management and individual career management.

- Organizational career management, conducted by organizations, involves aligning the needs of the individual with those of the organization. Employees with high motivation, leveraging their knowledge and skills, enable the organization to achieve success. Employees who feel valued within the organization are likely to exhibit increased organizational commitment and demonstrate organizational citizenship behavior (Yıldız, 2017: 259). Employees' loyalty to the organization extends their tenure, manifesting in three forms: affective commitment, continuance commitment, and normative commitment. Affective commitment relates to an employee's engagement with the organization (Allen & Meyer, 1990: 15). Continuance commitment is associated with the material benefits employees accrue over time, which they risk losing if they leave the organization (Karavardar, 2014: 95). Normative commitment pertains to employees viewing their decision to remain in the organization as a moral obligation, reflecting a sense of virtue rather than material gain (Taşkın & Dilek, 2010: 40).
- Individual career management: Employees must define the goals they aim to achieve in the future and plan how to reach them. This process involves managing the changes that occur during this pursuit (Yıldız, 2017: 261). Employees are aware of their own abilities and make plans based on an understanding of their strengths and weaknesses (Bedük et al., 2019: 95). Individual career management is not the organization's responsibility but rests entirely with the individual. Considering that individuals may not work for a single organization throughout their careers, they must enhance their job skills within their industry and manage their marketability (Wesarat, 2014: 5).
- In this regard, both organizations and individuals need to implement career management practices within a conscious plan. When these two career plans are successfully aligned, the likelihood of achieving the intended goals increases significantly.

Talent management

The Turkish Language Institution evaluates talent under four headings, which can be summarized as follows: "*A person's ability to understand or perform something; aptitude, capability*," "*Innate strength or capacity*," "*The boundary of a person's hereditary potential that frames their learning*," and "*The ability to absorb external influences*." After establishing their objectives, enterprises must identify the talents they require. Subsequently, finding the best talent and fostering a talent-oriented mindset become critical. This necessitates the development of an effective resource strategy (Chambers et al., 1998). Armstrong (2009) states that talent management encompasses the process of identifying, developing, recruiting, retaining, and deploying talented individuals. Accordingly, the talent management process is structured as follows:

- Attraction and retention policies and programs.
- Talent auditing.
- Role design.
- Talent relationship management.
- Performance management.
- Learning and development.
- Succession planning.
- Career management.

As the application of talent management increases, individual career management is also expected to rise (Karcıoğlu & Bakab Atasever, 2024: 14). It is essential for enterprises to utilize talent pools and succession planning to appropriately fill vacant positions and prepare managers for the future (Altınoğlu et al., 2015: 66).

After systematically identifying key positions within enterprises, the creation of a talent pool is necessary. To effectively leverage this talent pool, a differentiated human resource management approach must be established (Collings & Mellahi, 2009: 311). Talent management begins with the integrated consideration of resource provisioning and career development. A shift has occurred from a job-based payment system to an employee-based payment system. Consequently, it has been demonstrated that managing talented employees through a structured system positively impacts enterprise performance. The importance of social capital has further enhanced this process. Particularly in multinational enterprises, where competition is intensifying, the quality of leadership and management talent has gained even greater significance (Sparrow, 2019: 169). Despite economic and resource dependency challenges in small and medium-sized enterprises, attracting talented employees remains feasible. In such enterprises, establishing an organizational life cycle to identify necessary talents is crucial (Krishnan & Scullion, 2017: 439).

Career and Talent in Food and Beverage Enterprises

In hotel enterprises, emphasizing career development practices to enhance employees' career perceptions can prevent them from seeking opportunities at other organizations (Çalhan & Küçük, 2024: 41). A study conducted in five-star hotel enterprises highlighted the importance of precisely identifying the skills, knowledge, and competencies relevant to job roles, while also suggesting that promotion systems should be designed fairly based on employee performance (Kılıç, 2009: 58).

In the tourism sector, adopting an inclusive approach to talent management may prove more effective. For small and medium-sized tourism enterprises, successful human resource management can facilitate the implementation of talent management practices. Collaboration between the sector and educational institutions can enhance talent management through knowledge exchange (Johnson et al., 2019). Studies in luxury hotel enterprises indicate the adoption of talent management practices aligned with organizational culture. A robust organizational culture is essential to ensure business continuity through talented employees, necessitating the development of career plans and internal promotion opportunities (Marinakou & Giousmpasoglou, 2019). For relatively small-scale enterprises to compete with international enterprises, cultivating a performance development culture and clearly defining their talent requirements are imperative (Scott & Revis, 2008: 789).

Employees with strong communication skills are noted to perform more effectively in the food and beverage sector. Given that employees interacting directly with guests work on the front lines, high levels of communication skills and empathy are critical (Reilly, 2018: 23). Universities should increase knowledge exchange with the tourism sector, enabling enterprises to establish clear standards based on their needs. Additionally, enterprises should foster a culture of social responsibility (Yan, Jiang, & Wen, 2025: 412). Enterprises must fully define their corporate talent pools and delineate all stages of the talent identification process. Continuous and effective communication between business leaders and the talent pool is vital (Jooss, Burbach, & Ruël, 2021: 32). Acquiring and assimilating external knowledge is crucial for an enterprise. Thus, establishing a learning culture and engaging with external partners to stay current with industry trends are deemed essential. In a context requiring constant adaptation and innovation, the ability to assimilate external knowledge underscores the importance of talent. Developing innovation-oriented talents can contribute significantly to enterprise success (Agina, 2025).

Key attributes sought in employees include familiarity with the sector, specialized knowledge of operational areas, and awareness of diverse cultures. Additionally, employees must remain up-to-date across various domains. Proficiency in technical language is vital for effective communication with guests. Beyond these competencies, employees need to develop skills such as customer orientation, effective communication, entrepreneurship, strong public relations, and emotional management. Moreover, problem-solving skills are among the critical competencies to be enhanced in the tourism sector (Bagheri et al., 2020: 93).

From a human resource management perspective, employees with high performance and potential, as well as the ability to acquire and develop new skills, are highly valued (Jooss et al., 2024: 153). Practices aimed at attracting, motivating, and retaining talented individuals are also critical for crisis management (Beechler & Woodward, 2009: 283). When talent management is implemented effectively, enterprises can attract employees capable of assimilating new knowledge and improving work behaviors, thereby enhancing their potential (Algina, 2025).

Conclusion

In this study examining career and talent management in food and beverage enterprises, various insights emerge from the perspective of relevant stakeholders. Key positions within food and beverage enterprises should be systematically identified. Concurrently, the talents required for each position must be matched accordingly. A payment system tailored to employees should aim to enhance performance. Consequently, the necessary talents for the enterprise should be identified, and an organizational employee life cycle should be established.

From the perspective of food and beverage enterprises, career management should be addressed within a comprehensive framework. Collaboration with educational institutions is essential. Opportunities for internal promotion must be clearly defined, and employees' career plans should be developed.

In terms of talent pools within food and beverage enterprises, communication skills stand out as a priority. Additionally, employees' ability to effectively assimilate externally acquired knowledge is considered significant. The capacity for continuous adaptation and self-renewal among employees is also emphasized. Furthermore, a high level of problem-solving skills is desired.

These talents collectively contribute to extending employees' tenure within the enterprise. However, merely selecting these talents is insufficient; they must also be monitored and managed effectively.

References

Agina, M. F., Farrag, D. A., Khairy, H. A., Alhemimah, A., & Al-Romeedy, B. S. (2025). Talent management and innovative work behavior in tourism and hospitality businesses: Does absorptive capacity matter?. *Journal of Human Resources in Hospitality & Tourism*, *24*(2), 367–391.

Allen, Natalie J. ve Meyer, John P. (1990), The Measurement and Antecedents of Affective, Continuance and Normative Commitment to the Organization, Journal of Occupational Psychology, 63(1), 1–18.

Armstrong, M. (2009). Armstrong's handbook of human resource management practice 11th edition.

Bedük, A., Eryeşil, k. ve Öztürk, M., (2019). *Kariyer Yönetimi*. (Ed: Erdoğan Kaygın ve Yunus Zengin). Kariyer Temel Kavramlar Yönetimi Güncel Konular. Eğitim yayınevi; Konya.

Beechler, S., & Woodward, I. C. (2009). The global "war for talent." *Journal of international management*, *15*(3), 273–285.

Chambers, E. G., Foulon, M., Handfield-Jones, H., Hankin, S. M., & Michaels III, E. G. (1998). The war for talent. The McKinsey Quarterly, (3), 44.

Collings, D. G., & Mellahi, K. (2009). Strategic talent management: A review and research agenda. Human resource management review, 19(4), 304-313.
Çalhan, H., & Küçük, B. (2024). Otellerin Kariyer Yönetimi Uygulamaları ve Çalışanların Kariyer Tatmini: Kapadokya Bölgesinde Bir Araştırma. *Coğrafya, Planlama ve Turizm Stüdyoları*, *4*(1), 32-46.
İrmiş, A., & Bayrak, S. (2001). İnsan kaynakları yönetimi açısından kariyer yönetimi. *Sosyal Ekonomik Araştırmalar Dergisi*, *1*(1-2), 177-186.
Johnson, K. R., Huang, T., & Doyle, A. (2019). Mapping talent development in tourism and hospitality: a literature review. *European Journal of Training and Development*, *43*(9), 821-841.
Jooss, S., Burbach, R., & Ruël, H. (2021). Examining talent pools as a core talent management practice in multinational corporations. *The International Journal of Human Resource Management*, *32*(11), 2321-2352.
Jooss, S., Collings, D. G., McMackin, J., & Dickmann, M. (2024). A skills-matching perspective on talent management: Developing strategic agility. Human Resource Management, 63(1), 141-157.
Karavardar, G. (2014). Kariyer yönetimi ve örgütsel bağlılık ilişkisini belirlemeye yönelik görgül bir araştırma.
Karcıoğulu, F. & Bakan Atasever, S. (2024). Üniversitelerde Stratejik İnsan Kaynakları Yönetiminde Yetenek Yönetimi Uygulama Düzeyi ile Bireysel Kariyer Yönetimi İlişkisi: Atatürk Üniversitesinde Bir Uygulama, *Academic Knowledge,* 7(1), 10.5281/zenodo.12594154.
Kılıç, G., & Öztürk, Y. (2009). Kariyer yönetimi: beş yıldızlı otellerde bir uygulama. Anatolia: Turizm Araştırmaları Dergisi, 20(1), 45-60.
Krishnan, T. N., & Scullion, H. (2017). Talent management and dynamic view of talent in small and medium enterprises. Human Resource Management Review, 27(3), 431-441.
Marinakou, E., & Giousmpasoglou, C. (2019). Talent management and retention strategies in luxury hotels: evidence from four countries. *International Journal of Contemporary Hospitality Management*, *31*(10), 3855-3878.
Moslem Bagheri, Tom Baum, Abolghasem Ebrahimi & Abbas Abbasi (2020) Talent Management in the Tourism and Hospitality Industry: Evidence from Iran, Anatolia, 31:1,88-98, DOI: 10.1080/13032917.2019.1708423
Oğan, Y. (2021). *Yiyecek İçecek İşletimleri.* (Ed: Oğan, Y.) yiyecek içecek hizmetleri. Nobel yayınevi; Ankara.

Reilly, P. (2018). Building customer centricity in the hospitality sector: the role of talent management. *Worldwide Hospitality and Tourism Themes*, *10*(1), 42–56.

Scott, B., & Revis, S. (2008). Talent management in hospitality: graduate career success and strategies. *International Journal of Contemporary Hospitality Management*, *20*(7), 781–791.

Sökmen, A. (2014). *Yiyecek içecek hizmetleri yönetimi ve işletmeciliği*. Detay yayıncılık: Ankara.

Sparrow, P. (2019). A historical analysis of critiques in the talent management debate. BRQ Business Research Quarterly, 22(3), 160–170.

Şahin, S., & Ünal, A. (2016). Türkiye'de Turizm İşletmelerinin Hukuki Açıdan Sınıflandırılması. Genel Turizm. Paradigma Akademi Yayınları.

Taşkın, F., & Dilek, R. (2010). Örgütsel güven ve örgütsel bağlilik üzerine bir alan araştirmasi. Organizasyon ve Yönetim Bilimleri Dergisi, 2(1), 37–46.

Tunçer, P. (2012). Değişen insan kaynaklari yönetimi anlayişinda kariyer yönetimi. Ondokuz Mayis University Journal of Education Faculty, 31(1).

Türk Dil Kurumu (2025). Yetenek kavramı. https://sozluk.gov.tr/ 10.03.2025 tarihinde erişildi.

Türk Dil Kurumu, (2024). Kariyer teriminin tanımı. https://sozluk.gov.tr/ 01.03.2024 tarihinde erişildi.

Wesarat, P. O., Sharif, M. Y., & Majid, A. H. A. (2014). A review of organizational and individual career management: A dual perspective. International Journal of Human Resource Studies, 4(1), 101.

Yan, L., Jiang, Y., & Wen, J. (2025). Tourism talent management: empirical research among tourism academics, practitioners, and government officials. *Current Issues in Tourism*, *28*(3), 400–416.

Yıldız, D. (2017). Pozitif Örgütsel Davranış Ve Kariyer Yönetimi Arasındaki İlişki. Avrasya Sosyal ve Ekonomi Araştırmaları Dergisi, 4(11), 254–266.

Uses of Virtual (VR) and Augmented Reality (AR) Technologies in Tourism Sector

İsa UĞUR[1]

Introduction

While virtual reality and augmented reality applications, which have emerged with the rapid development of the internet and technology in recent years, have started to be used in the tourism sector as in many other fields, leading to transformation; innovative technologies such as artificial intelligence, internet of things, robotic systems, cognitive and nano technologies, big data, virtual and augmented reality and smart grids have become an integral part of daily life today (Bahar et al., 2019; Yurt, 2021). Customers' experiences in both real and virtual settings have changed as a result of this technological shift, which has also given rise to reality virtuality technologies including virtual reality (VR), augmented reality (AR), and pure mixed reality (PMR) (Tom Dieck and Hang, 2022). Especially VR and AR technologies stand out as an important innovation and application for the tourism industry in the rapidly advancing digitalization process (Guttentag, 2010). These new generation technologies enable potential tourists to discover and experience tourism products and destinations in innovative ways. Innovative technologies such as VR and AR have the potential to increase the attractiveness of destinations by providing tourists with more immersive and interactive experiences.

As the tourism sector focuses on the direct tourist experience, it can quickly adapt to innovative technologies. VR and AR technologies with three-dimensional objects and virtual environment simulations are becoming increasingly important in the tourism sector (Demirezen, 2019). Today, when digitalization is accelerating, tourism businesses adopt these technologies to gain competitive advantage and increase customer satisfaction. In this context, this section discusses the role of VR and AR technologies in tourism, how they transform tourist experiences, and their contributions to the sector.

1 Asist., Prof. Karabuk University, Safranbolu ŞYD Vocational High School, Tourism and Hotel Management, Karabuk, Turkey isaugur@karabuk.edu.tr

Overview of VR and AR Technologies

Virtual Reality (VR) is defined as a human-computer interaction technology that allows individuals to interact and explore a three-dimensional computer-generated environment. This technology allows users to navigate and move in the virtual world independently from the real world, interact with various objects and even have real-time experiences in this environment (Siegrist et al., 2019). In other words, virtual reality provides a highly immersive interaction by providing an experience in which the user is completely surrounded by sensory stimuli in the virtual world and psychologically involved in the events of this world (Wang et al., 2021). This technology allows individuals to feel as if they are in a virtual environment by using advanced systems (Craig et al., 2009). In the tourism sector, it has been realized that virtual reality is an important factor affecting the purchase intention of tourists and this technology has started to be used effectively not only for entertainment purposes but also in sales and marketing processes (Yılmaz & Karamustafa, 2022). Augmented reality (AR) is defined as the integration of computer-generated graphics into the real world. Unlike virtual reality, augmented reality allows users to see the real world enriched with digital images (Billinghurst, 2002). In the tourism sector, augmented reality stands out as an important innovative application, especially in the processes of obtaining pre-booking information, exploring destinations and enriching the visitor experience (Cranmer et al., 2020). In this context, while enabling potential tourists to obtain more detailed information about destinations before travel, it also increases satisfaction by providing interactive and personalized experiences during travel. In addition, by integrating augmented reality into their sales and marketing strategies, tourism businesses both attract the attention of technology-sensitive guests and gain competitive advantage by adapting to the dynamics of the changing digital age (Yılmaz and Karamustafa, 2022).

Virtual reality (VR) and augmented reality (AR) technologies have the potential to create significant transformations in the field of tourism (Guttentag, 2010; Tussyadiah et al., 2018) as well as in many other fields such as artificial intelligence, health, telecommuting, e-commerce, digital marketing, education and training, marketing, online social networks, robotics, architecture, cultural heritage, sustainability (Özel & Uluyol, 2016; Wedyan et al., 2016). VR and AR technologies contribute to tourism in many areas such as hotel management, destination promotion, preservation of cultural heritage and enhancement of visitor experiences, and museums. These technologies transform the tourist experience by providing interactive content and virtual discovery opportunities and accelerate the digitalization process of the sector.

VR and AR Applications in Tourism

The development of VR and AR technologies in accommodation, food and beverage, travel and transportation businesses in the tourism sector has made the service processes of these businesses more efficient (Yurt & Sağır, 2023). It is observed that virtual and augmented reality applications have developed rapidly in recent years. The tourism sector is undergoing a transformation because to technologies like virtual reality and augmented reality. These technologies are changing how visitors engage with tourist attractions and providing previously unheard-of options for immersive experiences (Buhalıs et al., 2019). Initially guiding tourists through GPS applications for wayfinding purposes, these technologies have been enriched with functions such as obtaining information in travel and accommodation processes and exploring hotel and destination options in a virtual environment. Today, tourists can obtain interactive information about destinations through smart devices with augmented reality support, examine accommodation options with two or three-dimensional models, and make their cultural heritage experiences more impressive through virtual tours during museum visits. These developments both improve tourists' experiences in the tourism sector and provide businesses with the opportunity to offer more innovative and competitive services (Boone et al., 2016).

VR and AR Applications in Hotel Businesses

Virtual Reality (VR) and Augmented Reality (AR) offer unique experiences in hotel businesses, significantly influencing potential tourists' travel decisions and experiences. Nowadays, some hotel businesses have reached a level where they offer their guests the opportunity to feel the atmosphere of the real environment and introduce their accommodation areas in detail through virtual experiences. In this context, VR and AR technologies represent the beginning of a new era in the promotion and marketing processes of hotel businesses. VR and AR applications allow tourists to examine various hotel alternatives in detail in a virtual environment and shape their preferences accordingly without physically visiting them. In addition, potential tourists can make a more informed choice by viewing the available rooms, prices and views of the hotel business in three dimensions (Keckes & Tomicic, 2017; Kabadayı, 2020). Kabadayı (2020) stated in his study that Virtual (VR) and Augmented Reality (AR) applications used in hotel businesses are basically shaped around two main processes. The first is the applications aimed at increasing the motivation of potential tourists to travel. In this context, VR and AR technologies can positively influence individuals' travel decisions by promoting destinations and accommodation facilities visually and experientially.The second is the product experience stage, which allows for the prior experience of accommodation services. In this context, guests can explore the hotel rooms, facilities and services offered by the facility in three dimensions in a virtual environment, and even make an informed choice by comparing different room types and price options and have a better travel experience (Kabadayı, 2020). Thus, VR and AR applications are becoming important tools that strengthen the marketing strategies of hotel businesses and increase customer satisfaction. These technologies help potential customers make more informed decisions by experiencing accommodation options that meet their expectations in advance, while helping hoteliers gain a competitive advantage. In this context, global hotel chains such as Marriott, Hilton, Accor, Radisson and Best Western are developing innovative VR and AR-based applications for their guests.

For example, Marriott stands out with its "VRoom Service" application that allows guests to explore destinations in a virtual environment. As can be

Figure 1. Virtual Reality Room Tour

Source: mize. tech, 2025

seen in the picture on the side, the application allows guests to explore different destinations around the world using VR headsets in their hotel rooms. Similarly, Accor Hotels is investing in VR and AR technologies to enhance the customer experience. The hotel offers 360-degree virtual tours, allowing guests to evaluate their accommodation options in a more informed way. For example, the hotel offers 360° virtual tour experiences on its websites and sales platforms, allowing potential guests to explore rooms and facilities in a realistic way. This VR-supported content allows customers to examine the property in detail before booking, enabling them to make a more informed decision. In fact, it has been reported that hotels using 360° virtual tours get up to 50 % more views on their websites than traditional photos. With such VR hotel tours, Accor aims to both increase direct bookings and provide a powerful visual tool for remote sales teams (Globenewswire.com, 2025). In one of the applications Best Western hotels use VR technology as an effective marketing tool for the promotion of hotel rooms. Potential tourists can experience the services offered by the hotel with a virtual tour before making a reservation and choose the most suitable option for themselves. As can be seen, VR and AR technologies are being used in hotel businesses for both marketing strategies and enhancing the customer experience. In addition to digitizing promotion and booking processes, these technologies make the accommodation experience more interactive and personalized. Thus, hoteliers can offer a more realistic and customized experience to their guests while gaining significant advantages in both marketing and operational processes.

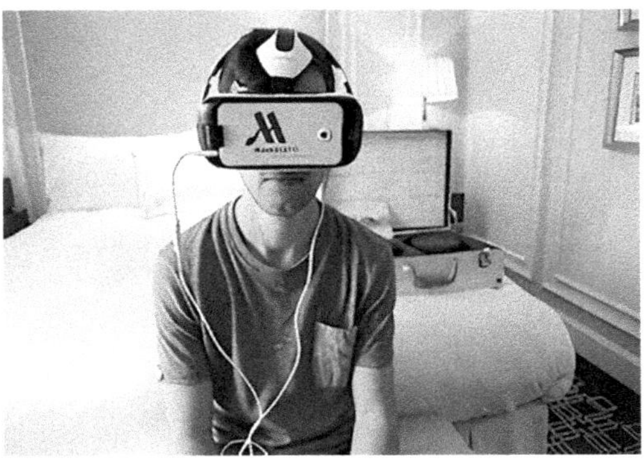

Figure 2. "VRoom Service"

Source: engadget.com, 2025

VR and AR Applications for Destinations

The attractiveness of a destination consists of its tangible and intangible cultural heritage elements and natural beauties. In the intensely competitive tourism sector, promotion and marketing activities of destinations play an important role in highlighting these attractions. Today, innovative technologies such as VR and AR make these promotional activities more effective and make the destination experience more interactive and accessible for tourists (Sarı & Kozak, 2005; Özdipçiner, 2010). In parallel with the continuous change and development of the tourism sector, destinations are using VR and AR applications to make it easier for tourists to explore the destination, allowing them to experience cultural and natural heritage elements in a virtual environment and thus gain competitive advantage.

Augmented reality (AR) applications have been developed to provide more comprehensive and interactive information about tourist destinations. The effective use of this technology by tourists contributes to enriching their experiences and increasing their overall satisfaction levels (Akkuş & Akkuş, 2018). The image in the figure shows how augmented reality technology works in tourism; a process in which digital information is added to the real world environment and this enriches the user experience. Virtual reality technology is becoming increasingly widespread in the tourism sector in Turkey; the 3D ancient city project, which offers the opportunity to visit the Red Courtyard, Zeus, Athena and Asclepius Temples in Bergama in a virtual environment, is an example of

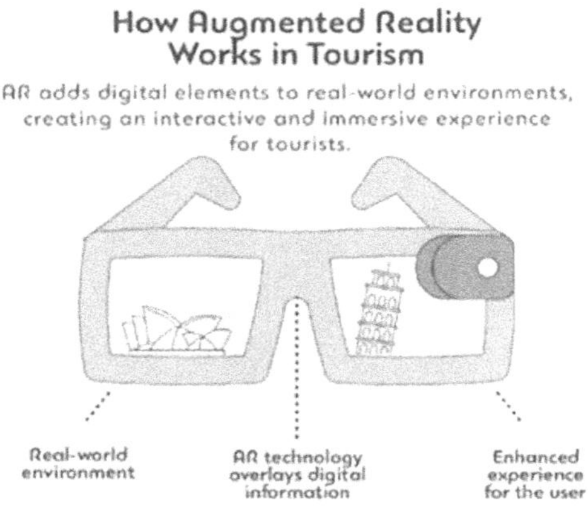

Figure 3. Destination Experience with AR

Source. mize. tech, 2025

the effective use of technology in the promotion of cultural heritage (Çilkin & Hamrişçi, 2017). Similarly, at the London World Travel Market in 2024, the island of Menorca promoted its touristic areas with VR technology and organized virtual tours and interactive events. VR and AR technologies contribute to destination promotion, cultural heritage preservation and sustainable tourism, while making tourism more inclusive for people with limited access (Ekici & Güven, 2017; Guerra, 2015).

VR and AR Applications in Cultural Heritage and Museums

Many cultural tourism organizations have started to make use of VR and AR technologies to enhance the experience of tourists. Especially for cultural heritage sites, augmented reality allows digital signage and content to be provided without compromising the original architecture or landscape (Jung et al., 2016). In areas such as the protection of cultural heritage and the enrichment of visitor experiences, virtual reality applications seem to play an important role both in the promotion of touristic places and in the decision-making processes of potential tourists. VR and AR technologies increase the number of visitors by making museum experiences more interactive and deepen learning by enriching the presentation of historical and cultural elements through digital animations (Hjalager, 2015; Woody, 2018). For example, the "StreetMuseum" application developed for the Museum of London offers visitors the opportunity to visualize historical images on real

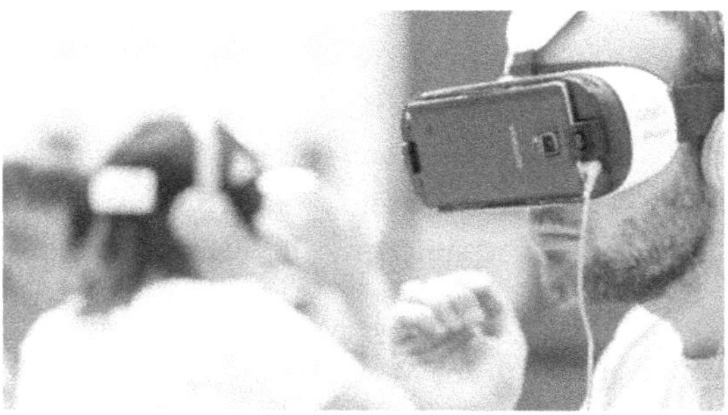

Figure 4. British Museum's Bronze Age Virtual Reality Project

Source: museumsandheritage.com, 2025

spaces, while an AR-based application for the National Palace Museum of Korea contributes to the transfer of cultural heritage by providing detailed information about the space. In this context, augmented reality is becoming an increasingly adopted technology in tourism by providing informative and interactive experiences in museums (Yovcheva, 2015).

Museum visitors' interaction with the exhibited artifacts is usually limited, but AR and VR technologies increase this interaction through 3D digital models, enriching the visitor experience and contributing to the delivery of cultural heritage to a wider audience (Wojciechowski et al., 2004) For example, with the Bronze Age VR Project of the British Museum, the artifacts of the period are presented in three dimensions, while the Turtle Terbiyecisi artifact at the Pera Museum has been transformed into an interactive experience with VR (Tanrıkulu & Karagöl, 2021). Many leading museums around the world, such as the National Museum of Singapore, Kennedy Space Center, Smithsonian National Museum of Natural History, British Museum and Pérez Art Museum, enrich the visitor experience by offering VR and AR-based exhibitions. In this context, augmented and virtual reality technologies are considered as an innovative tool that strengthens cultural interaction by eliminating spatial and physical constraints between museums and visitors (Billock, 2017; Woody, 2018; Quiroz, 2018).

As an innovative technology, VR and AR applications are increasingly being adopted in the tourism sector and enhance various touristic experiences. Especially in different tourism environments such as museums, exhibitions, historical and cultural heritage sites, it stands out as a tool that enriches the spatial perception of visitors, increases interaction and digitizes traditional information presentation methods (He et al., 2018). In the study conducted by Eryılmaz and Aydın (2020), it was determined that 15 of the 28 applications examined included virtual reality (VR), 11 included augmented reality (AR) and 2 included mixed reality solutions where both technologies were used together. It was stated that these technologies are applied in various tourism areas such as travel businesses, museums, hotels and transportation.

Results

As discussed in this chapter, applications in VR and AR technologies have a wide range of applications in the tourism sector, from accommodation businesses to destination marketing, from museum experiences to food and beverage sector, from food and beverage businesses to travel and transportation businesses.Technological transformation is of great importance in the tourism sector in order to increase efficiency, ensure customer satisfaction and strengthen the interaction between customers and businesses.

In a study conducted by Ergen (2020), it was stated that VR and AR applications stand out among the most preferred technological innovations in the tourism sector. VR and AR technologies play an important role in developing marketing strategies in the tourism sector and making customer experience interactive. These technologies, which are used in many fields from hospitality to museology, facilitate booking processes through virtual tours and 3D visuals and contribute to more effective promotion of destinations. While digitized artifacts in museums provide visitors with information and experience about the past through AR applications, tabletop interactions and multimedia menus in the food and beverage industry offer personalized and informed consumption. In the tourism sector, service quality stands out as a determining factor on behavioral outcomes such as customer satisfaction, loyalty and intention to revisit. As a matter of fact, Ünlü et al. (2024) revealed that customers' perception of service quality and the reasons for choosing an accommodation facility have significant and positive effects on their intention to revisit. Similarly, according to Buhalis et al. (2019), with the integration of emerging technologies into the tourism experience, innovative applications such as VR and AR strengthen the perception of service quality and have positive effects on customer loyalty. One of the most concrete indicators of this transformation is that 78% of hotels will include technological investments in their plans to increase operational efficiency in 2024 (Protel, 2024). Especially thanks to AR technology, tourists can virtually visit hotels, examine rooms and make more informed travel decisions by receiving interactive information about destinations (Tourism Agency, 2024). Recent studies show that VR technologies create an effective and innovative learning method by offering high school and university students, especially those studying tourism, the opportunity to reinforce theoretical knowledge with practical experiences in applied courses (Kumlu 2021; ClassVR, 2025). VR and AR technologies have become widespread in the global tourism market, playing an important role in marketing destinations and enriching the tourist experience (Cognitive Market Research, 2025). While North America and Europe are at the forefront of these technologies,. the Asia-Pacific region is also emerging as a rapidly developing market. In Turkey, AR/VR integration is increasing, but is still limited compared to developed countries (Demirezen, 2019).

As a result, the integration of VR and AR technologies in the tourism sector both increases the competitiveness of businesses and offers tourists more impressive, informative and memorable experiences. It is predicted that these technologies will develop further in the future and become much more widespread in the tourism sector. With the developing digital transformation process, the adoption of these innovations by tourism destinations and businesses will contribute significantly to the sustainable growth of the sector. Destination management organizations, travel agencies and tour operators should consider integrating augmented reality technology to provide tourists with more personalized and interactive experiences. Adoption of this technology offers the potential to respond more effectively to the ever-changing expectations and needs of tourists, while improving the customer experience and increasing the competitiveness of destinations.

References

Akkuş, G. and Akkuş Ç. (2018). Evaluation of mobile augmented reality applications used in historical touristic areas, Journal of Tourism and Gastronomy Studies, 6(1): 83–104.

Bahar, M., Yüzbaşıoğlu, N. Topsakal and Y. (2019). An Overview of Future Tourism Guidance in the Light of Smart Tourism and Super Smart Tourist Concepts. Journal of Travel and Tourism Research, 14: 72–93.

Billinghurst, M. (2002). Augmented reality in education, *New Horizons for Learning*, Seattle, USA, 1–5.

Billock, J. (2017). Five Augmented Reality Experiences That Bring Museum Exhibits to Life,

Boone, A. E., Foreman, M. H. ve Engsberg, J. R. (2016). Development of a novel virtual reality gait intervention, *Gait&Posture*. 52, 202–204.

Buhalis, D., Harwood, T., Bogicevic, V., Viglia, G., Beldona, S. and Hofacker, C. (2019), "Technological disruptions in services: lessons from tourism and hospitality," Journal of Service Management, Vol. 30 No. 4, pp. 484–506.

Craig, A. B., Sherman, W. R. ve Will, J. D. (2009). Developing virtual reality applications foundations of effective design, *Multiple: Morgan Kaufmann Publishers.*

Cranmer, E. E., tom Dieck, M. C. ve Fountoulaki, P. (2020). Exploring the Value of Augmented Reality for Tourism. *Tourism Management Perspectives*, *35*, 100672.

Demirezen, B. (2019). A Literature Review on the Usability of Augmented Reality and Virtual Reality Technology in Tourism Sector. International Journal of Global Tourism Research, 3(1), 1-26.

Ekici, R., & Güven, A. (2017). The Role of Virtual Reality Technology in Tourism Industry. Eurasian Academy of Sciences Journal of Social Sciences, 406–408.

Ergen, F. D. A Literature Review on the Cultural Applicability of Augmented and Virtual Reality Technologies in Isparta Province. Journal of Tourism Theory and Research,6(1), 62–74.

Eryılmaz & Aydin, R. (2020). The place of virtual reality and augmented reality technologies in tourism applications and marketing. International Journal of Rural Tourism and Development (IRTAD) E-ISSN: 2602-4462, 4(2), 9–25.Guerra, J. P., Pinto, M. M., & Beato, C. (2015). Virtual Reality-Shows a New Vision for Tourism and Heritage. European Scientific Journal, ESJ, 11(9), 49–54.

Guttentag, D. (2010). Sanal Gerçeklik: Turizm Uygulamaları ve Çıkarımları. Turizm İşletmeciliği, 31 (5), 637–651.

He, Z., Wu, L. and Li, X. (2018). When art meets tech: the role of augmented reality in enhancing museum experiences and purchase intentions, Tourism Management, 68: 127–139

Hjalager, A. M. (2015). 100 Innovations that transformed tourism, Journal of Travel Research, 54 (1): 3–21.

Jung, T., Dieck, T. M. C., Lee, H. ve Chung, N. (2016). Effects of Virtual Reality and Augmented Reality on Visitor Experiences in Museum, In Inversini, A. ve Schegg, R. (eds), Information and Communication Technologies in Tourism, New York: Springer International Publishing.

Kabadayı, M. (2020). Virtual reality and augmented reality applications in hotel businesses. Journal of Travel and Hotel Management, 17(3), 464–479.

Kečkeš, A. L., ve Tomičić, I. (2017). AugmentedReality in Tourism–Researchan Applications Overview. *InterdisciplinaryDescription of ComplexSystems: INDECS*, *15*(2), ss. 157–167.

Kumlu, S. T. (2021). The effect of virtual reality usage on learning motivation in tourism education (Master's thesis, Kocaeli University, Institute of Social Sciences).

Özdipçiner, N. S. (2010). Electronic Marketing in Tourism. Journal of Internet Applications and Management, 1(1), 5–22.

Özel, C., & Uluyol, Ç. (2016). Development of an augmented reality application and student opinions, Journal of Turkish Social Research, 793–823.Protel. (2024). *Otelcilik sektörü 2024 raporu: Teknolojik yatırımların etkisi*.

Quiroz, J. (2018). The 5 best museum AR/VR experiences this summer,

Sarı, Y. & Kozak, M. (2005). The Effect of the Internet on Tourism Marketing: A Model Proposal for Destination Web Sites, Journal of Akdeniz İ.İ.B.F., 248–271.

Siegrist, M., Ung, C., Zank, M., Marinello, M., … A. K.-F. research, & 2019, undefined. (n.d.). Consumers' food selection behaviors in three-dimensional (3D) virtual reality.

Tanrıkulu, B., & Karagöl, A. (2021). Virtual Reality Applications in the Museum: Turtle Trainer as a Case Study. New media electronic journal, 5(2), 95–111.

Tom Dieck, M.C. and Hang, D.I.D. (2022), "The role of immersive technology in customer experience management," Journal of Marketing Theory and Practice, Vol. 30 No. 1, pp. 108–119.

Tussyadiah, I. P., Wang, D., Jung, T. H., & tom Dieck, M. C. (2018). *Virtual reality, presence, and attitude change: Empirical evidence from tourism*. Tourism Management, 66, 140–154.

Ünlü, H., Kıvanç, M. İ., Nazikgül, M., & Apak, Ö. C. (2024). The Effect of Service Quality Perception and Reasons for Preference of Customers Staying in Public Guesthouses on Revisit Intention. Journal of Travel and Hotel Management, 21(3), 630–648.

Wang, Q. J., Escobar, F. B., Da Mota, P. A., & Velasco, C. (2021). Getting started with virtual reality for sensory and consumer science: Current practices and future perspectives. *Food Research International*, *145*, 110410.

Wedyan, F., Aloqily, I., Freihat, R. ve Wedyan, S. (2016). JoGuide: a mobile augmented reality application for locating and describing surrounding sites, *ACHI 2016: The Ninth International Conference on Advances in Computer-Human Interactions*, 88–94.

Wojciechowski, R., Walczak, K., White, M., ve Cellary, W. (2004). Building virtual and Augmented Reality museum exhibitions, Proceedings of the Ninth International Conference on 3D Web Technology-Web3D, 135–144

Woody, R. C. (2018). The Future of Augmented and Virtual Reality (AR & VR) in Museums.

Yılmaz, M., & Karamustafa, K. (2022). Augmented Reality Applications and Tourism. Journal of Travel and Hotel Management, 19(3), 399–422.

Yovcheva, Z. (2015). User-Centred Design of Smartphone Augmented Reality in Urban Tourism Context, (Published Dissertation Thesis), Bournemouth University.

Yurt, İ and Sağır, Y. E. (2023). The Effect Of Staff Behaviours in Food and Beverage Establisments On Customer's Eating and Drinking Experience. Gastroia: Journal of Gastronomy and Travel Research, 7(2), 448–460.

Yurt, İ. (2021). Society 5.0. In Ö. Bozkurt (Ed.), Entrepreneurship with examples - Transformation in the digital age (pp. 137–158). Nobel Academic Publishing.

Internet Resources

https://www.engadget.com/2015-09-28-vroom-service-marriott-hotel.html

https://www.globenewswire.com/newsrelease/2019/10/22/1933492/0/en/Accor-Recommends-Visiting-Media-s-Virtual-Sales-Marketing-Platform-to-Hotels Across-Europe-and-the-Americas.

https://museumsandheritage.com/advisor/insights/british-museum-uses-virtual-reality-to-explore-bronze-age-objects/?utm

https://www.classvr.com/vr-ar-resources/travel-and-tourism-vr-teaching-resources/

https://www.cognitivemarketresearch.com/ar-vr-in-travel-and-tourism-market-report?campaign_source=google_ads&campaign_name=cmr_Pmax_APAC_02

https://mize.tech/blog/4-well-thought-examples-of-virtual-reality-in-tourism/

https://mize.tech/blog/5-ways-augmented-reality-is-enhancing-the-tourism-experience/#conclusion

Tourist Guiding and Technology

Ayşe ARSLAN ÖZKAN[1]

Introduction

Tourism is a service industry, with humans at its core and a structure based on reciprocal communication. Although it has been stated that the use of technology would be limited due to these characteristics, recent changes in technology are rapidly and profoundly transforming this view (Çakmak ve Demirkol, 2019). Nowadays, developing information and communication technologies are widely used in the tourism industry, as in many other industries. Tourism industry, which has a more labor-intensive structure and where mechanization and automation are more limited, utilizes information and communication technologies to gain an advantage in the market due to increasing competitive conditions (Ünüvar, 2008: 597). Technology is not only a necessary tool in carrying out tourism activities and providing basic services to tourists, but it is also a critical factor in enhancing the marketing and service performance of tourism businesses and all other units (Noti & Llazo, 2016). Today, the use of digital technologies, which are either the tourist experience itself or an important complement, is increasingly seen as a necessary tool in tour management and enhancing communication with tourists in the field of tour guiding (Neuhofer & Buhalis, 2014).

Tour guides are skilled professionals in the tourism industry who assist tourists by offering informative insights about the locations they visit, guiding them throughout the experience, and maintaining close communication with them during tour programs (Chowdhary & Prakash, 2022). The use of technology by tourists during tours is increasing every day and tourist guides, who provide guidance services to them, must also adapt to these changes. In the literature, there are studies that evaluate the outcomes of technology use in tourist guiding in different ways. For example, El Kafy (2020) suggests that the use of technology will reduce the demand for tourist guiding, while (Abdelhamid, 2020) states that technology will help tourist

[1] Aydın Adnan Menderes University, Faculty of Tourism, Department of Tourism Guidance, ayse.arslan@adu.edu.tr, Orcid No: 0000-0001-5459-0147

guides become more equipped and qualified. So, understanding the effects of new technologies on tourist guiding is a critical issue for both academics and industry professionals.

Tourism and Technology

To investigate the role of technology in the modern tourism industry or system, a comprehensive model is required that offers a broader perspective on technology's role rather than focusing solely on industry-specific issues, particular technologies, or narrow concerns. Although it's useful to examine technology in relation to a specific element of the tourism industry (such as a tour operator, travel agent, or hotelier), a single issue (like environmental sustainability), or a particular technology (such as Computer Reservation Systems or CRS development), a more holistic view becomes increasingly valuable when evaluating technology's overall impact on tourism. The relationship between tourism and technology is one of mutual influence, rapidly evolving, and shaping both sectors. Technology impacts almost every stage of tourism and offers new possibilities for the industry. Here are some key ways in which tourism and technology are interconnected:

Recommendations shared on social media is very important in tourists' travel decision process (Arslan & Şimşek, 2022; Barbu et al., 2024). Tourists use technology at every stage of their travel such as making reservations (flights, hotels, restaurants, car rental), buying tours, sharing content on social media platforms, using review sites. They can easily search and purchase the services and products through platforms such as Booking.com, Expedia, Airbn, tripdavisor, etc. (Garcıa et al., 2022). Also, tourism businesses and destinations influence tourists' decision-making processes through technologies such as virtual reality (VR) and artificial reality (AR). These technologies offer virtual trips to tourists and enrich their experiences (Calisto & Sarkar, 2024; Özkul & Kumlu, 2019). Chatbots and virtual assistants provide 24/7, artificial intelligence can analyze customer requests and offer personalized suggestions or solutions to problems in real-time (Doğan & Bayar, 2024).

Technology has transformed the hospitality industry. Smart rooms, integrated with IoT devices, can be controlled by guests using their phones. Guests can adjust lighting, room temperature, and interact with in-room devices (Chang et al., 2022). Also, technology is used in airports to streamline passenger flow. Innovations like facial recognition, automated check-in kiosks, and luggage tracking systems have significantly increased the efficiency and speed of travel processes (Mrňa et al, 2021; Khooshechin Gilak, 2023). Additionaly, technology can play a role in sustainable tourism practices by enabling eco-friendly transport options, reducing energy consumption in hotels, or aiding in waste management. Moreover, technological advances could help address societal challenges like overtourism by monitoring visitor flows and helping with capacity management (Vargas-Sánchez, 2020; Erdoğan et al., 2022; Gangurde & Dsouza, 2023). Technology is used to monitor tourist flows and natural resource usage. This data helps optimize capacity management in tourist destinations and supports sustainable development policies (Wang, 2024).

Technology has transformed all aspects of the tourism industry, making it more efficient, accessible, and customer-oriented. Additionally, these technological advancements are transforming tourism into a more environmentally friendly, experience-centered, and digital industry. Innovations in technology continue to simplify travel for both travelers and tourism businesses, and they will play an increasingly significant role in the future.

Tourist Guiding and Technology

In recent years, technology has played a major role in tourism industry. Technology is effectively manifesting itself in many areas, from accommodation and transportation to reservation systems and resource management. As businesses adapt technological innovations into their operations, it is also essential for tourism employees to adapt to this change. In this section, tourist guides who practice their professions in the forefront of tourism and their relationship with technology are discussed.

Tourist guiding, although referred to by different names throughout history, is one of the oldest professions known to humanity (Khalil, 2021; Corfu, 2022), although its designations have changed over time and historical records frequently reference its existence. A tourist guide is a person who conveys the historical, natural and cultural values of the destination in the preferred language to the tourists she/he accompanies within the scope of the tour. They are one of the tourism employees who spend the longest time with tourists (ancient site, museum, restaurant, hotel, airport, vehicle). Tourist guides, who are in direct contact with tourists, encounter different tourist profiles during theirs tours. Therefore, in order to meet the expectations of tourists and to keep up with the age, tourist guides should embrace new technological innovations and adapt to changes. For example, tourists can access information about destinations before their travels. Therefore, this situation poses a challenge for tourist guides when dealing with tourists who think they know everything. An perception may emerge that these new technologies could take over the roles of tourist guides, such as language translation, direction information, general service information, and interpretation. This situation is also an important issue for tour operators, who emphasize the advantage of guided tours over non-guided tours. Indeed, the internet allows tourists to easily review tourist guides, sites and operators at no cost to themselves (Weiler & Black, 2014:167).

Many innovations such as the head-set application used in museums and archaeological sites, GPS-supported maps, smartphones, applications that provide written and audio narrations in museums and archaeological sites, developments in transportation systems, etc. are known to be effective in the way guides perform their profession (Çakmak & Demirkol, 2017:224). Besides, tourist guides use smartphone applications, microphones, wireless headphones, bluetooth systems, voice recorders during their tours. These technological devices help tour and time management (Kara et al., 2022). The use of these technologies can enhance the educational aspect of the experience. Guides can provide real-time information and context about the visited sites through applications like AR and VR, enriching their narratives, which makes the experience more informative (Zhang, 2024).

Geographic Information Systems (GIS) is one of the location-based information system. GIS, has been used in environmental conservation and wildlife protection, while its use in tourism planning, management, and marketing of destination products has been recent (Magige et al., 2020). GIS enables access to dynamic and evolving information based on location, such as nearby tourist attractions and services (accommodation, restaurants, entertainment and shopping centers, transportation services), weather, emergency services, currency exchange, mobility restrictions, and traffic. The adoption of this technology enhances the quality of the tourist experience while also impacting the sustainability of environmental resources (Kim et al., 2005). It can be considered as a tool that will make the work of tour guides easier in tour management. By improving time management and service quality, this system can enable tour guides to earn the appreciation of tourists, it can even lead to positive feedback, such as compliments and thanks.

Technology, in certain mandatory situations, has directed both tourists and tour guides towards alternatives. With COVID-19, travel restrictions and health concerns have led tourists to seek alternatives for their travels. During this period, virtual tours have become a popular option, allowing individuals to explore destinations safely from their homes (Patria et al., 2023). This has introduced the concept of the digital tourist (those engaging in this experience), while also bringing digital storytelling and virtual engagement to the forefront for tourist guides (Nazare et al., 2024).

As a result of technological innovation, today we can speak of four different types of tour guiding. This classification can be made according to the real or virtual form of the tour guide and the real or virtual environment in which the guide will deliver their services:

1. In Real Life (IRL) Tour Guiding: It is the type of guiding that combines a human tour guide with a real workplace environment. For example, tour guiding in Ephesus.
2. Augmented Reality (AR) Tour Guiding: In this type of guiding, the workplace is physical, but the tour guide is virtual. Listening to the narration of a virtual tour guide with the help of an app at the Taj Mahal. A hologram explaining the Alexander Sarcophagus at the Istanbul Archaeology Museum.
3. Mixed Reality (MR) Tour Guiding: In this type of tour guiding, a real environment is combined with a virtual environment to create a new setting. It allows users to naturally interact with both electronically produced objects and real objects. For example, experiencing the environment designed with wearable technologies, such as walking on Mars.
4. Virtual Reality (VR) Tour Guiding/Avatar Guide: In this type of guiding, both guide and environment is virtual. Avatars are the first form of tourist guides created with virtual reality technology (Aboelmagd, 2023).

According to De Le Harpe and Sevenhuysen (2020), combining technology with the tourist guiding profession can offer new experiences for tourists in the digital age. By using technology, tours and activities can be designed to encourage personalized tourist experiences. Seyitoğlu et al., (2025), reveal that technology can impact tourist guiding in dual ways. While technology can enhance service quality and help improve experiences, technologies such as robot tourist guides, VR, and AR are seen as threats and are believed to have the potential to negatively affect service quality and tourist experiences.

While technology and apps provide tourists with a lot of information and convenience, they have some limitations. Many scholars agree that tourist guides offer a richer experience for tourists. The ability of a tourist guide to add their own interpretation, personality, and enthusiasm to their narrations is highly valuable for tourists. Tourist guides also help tourists engage with the local environment, serving as a bridge for interaction, which is something many tourists seek. The experience, charisma, and personalisation of the tour provided by a tourist guide go far beyond what an app can offer. The profession of tour guiding is in competition with technological innovations. Therefore, in many studies, technology is considered both a supportive tool and a challenging competitor. Reed (2015) summarizes this situation as follows:

Apps can answer tourists' questions on many topics, but not every question from tourists is likely to be satisfactorily addressed by apps. For example, emergencies related to health. Tourist guides, by profession, are trained to handle emergencies, and therefore can manage such situations based on their experience and knowledge. Tourist guides are superior to apps when it comes to responding to tourists' special requests. In solving problems and meeting requests, tourist guides can observe and understand tourists, making changes to the tour without them even realizing it. For example, preparing a surprise birthday cake or offering the opportunity to have a romantic time by catching the sunset. On the other hand, apps provide tourists with accurate information and are cost-effective. It provides detailed information about tourist attractions, helps travelers who organize their own trips, and offers opportunities to add/create different places and activities beyond the guided tour programs. While tourist guides can provide guidance to a limited number of tourists, apps can cope with large number of tourists at the same time with many different languages. Additionally, it provides communication and interaction opportunities among group members. Tourists can share their experiences with their social network, comments and feedback on their posts provide tourists with a broader perspective on the destination and its facilities

(Reed, 2015). Additionally, data privacy and security is one that should not be taken lightly. Many applications can be used by entering personal data. As tourists share personal data through apps and online platforms, concerns about data misuse and breaches have increased. Indeed, tourism businesses and tour guides must leverage technology while considering these concerns from the perspective of tourists (Wuenderlich et al., 2015). Despite its great advantages, technology is unlikely to replace the elements of emotion and personal communication in tours (Sotohy, 2020). Tour guides are significantly superior to robot guides and apps in terms of emotional intelligence, problem-solving ability, consideration of special requests, mastery of local values, and conveying them (Seyitoğlu et al.,2025).

Conclusion

Tourism industry is quite dynamic, and as a result, it undergoes technological changes and innovations. In recent years, tourists have been benefiting from technological tools such as AR, VR, and robot guides, mobile applications, social media, review sites, at all stages of the travel. They want to experience technology-enhanced tourism experiences and personalized ones. Businesses and employees are expected to adapt to these changes and improve themselves. Tour guides are undoubtedly an important profession group that needs to evolve in this regard.

The impact of rapidly developing information-communication and artificial intelligence technologies on the tourist guiding profession is one of the topics examined by both academia and the industry. These technologies offer both opportunities and challenges, influencing various aspects of the profession, such as tour planning, real-time communication, and enhancing the overall tourist experience. While they can improve service quality and efficiency, they also pose potential threats to traditional guiding roles. Therefore, it is essential for tourist guides to adapt to these technological advancements to stay relevant and enhance their services in the digital age.

Tourist guides benefit from various technologies pre, during, and post tours. Before the tour, they use technology to gather information about destinations and tourist groups, as well as to communicate with stakeholders and tourists. During the tour, tools like communication devices, microphones, navigation apps, and social media platforms help improve the tour experience, assist lost tourists, and validate the information provided. After the tour, technology helps establish connections for future tours, allows guides to identify areas for improvement, and enables continued communication with people in different destinations. Additionally, feedback from tourists on social media plays a key role in helping guides, travel agencies, and tour operators improve their services (Karamustafa & *Kılıçhan, 2024:640–643).* It is important for tourist guides to have their own web pages and you tube pages in terms of their recognition. Thus, as their visibility increases, they can communicate directly with potential tourists and increase the number of tours. Türkmen & Özcan (2024) state that WhatsApp groups and social media are highly effective in helping tourist guides find jobs.

Some studies show that technology is a supportive tool for tour guides (Sotohy, 2020). However, as a result of changes in tourism trends, especially after the Covid-19 period, the shift towards individual travel and a decrease in the demand for services from tourist guides has made it possible for tourists to organize their own tours by making greater use of technology. This changing situation has raised concerns that, in the future, it may threaten the profession of tour guides (Nazlı, 2020; Kavak et al., 2022; Karamustafa & Kılıçhan, 2024; Seyitoğlu et al., 2025). Tourist guides can further improve themselves through technology, plan and manage their tours much more effectively, and market themselves on a wider range of platforms. Additionally, they can use negative feedback from tourists to improve their services. They can make efforts to correct their shortcomings and mistakes. By maintaining the new friendships they form during the tours through technology, they can work towards reuniting in future tours and encourage tourists to recommend them to others. After poorly conducted tours, tourist guides can use technology to respond quickly and constructively to negative comments on social media, turning a negative situation into a positive outcome.

To sum up, use of technology provides numerous benefits for tourists guides. If tourist guides adopt these technological services and use them while practicing their profession, they can make a positive contribution to their careers. It facilitates the professional development of tour guides and helps them stay up-to-date. It offers tour guides an opportunity to expand their business scope or increase their customer base. Technology helps tour guides reshape and empower their roles, while also making cultural experiences more interactive, educational, and exciting. Tourist guides can use technology as a tool to help interpret natural and cultural heritage sites through new experiential and intellectual approaches. By using technologies such as AR and VR, they can make sightseeing more appealing and offer more engaging experiences. They can promote themselves through social media and online platforms, and follow feedbacks. Thus, they take the opportunity to improve themselves and build stronger relationships with tourists. Tourist guides should adopt these technologies and use them as a tool to sustain their profession and enrich tourists' experiences.

References

Areni, C. S., & Kim, D. (1993). The influence of background music on shopping behavior: classical versus top-forty music in a wine store. *Advances in Consumer Research*, 20 336–340.

Aboelmagd, A. (2023). Emerging technology trends in tour guiding: Virtual and distance tour guiding, *Research Journal of the Faculty of Tourism and Hotels Mansoura University* Vol No. 13, 341–370.

Abdelhamid, M. (2020). Using 3d hologram technology (3dht) in the distance learning program to enhance the professional skills of tour guidance undergraduate students. Journal of Association of Arab Universities for *Tourism and Hospitality*, 18(3), 17–34. https://doi.org/10.21608/jaauth.2020.109878.

Arslan, A., & Şimşek, G. (2022). Gezginlerin Seyahat Deneyimlerini Sosyal Medyada Paylaşma Motivasyonları. Turizm Akademik Dergisi, 9(2), 379–392.

Barbu, F. S., Sanda, G., Rusu, C-M., Balint, P-G. (2024). *The Impact of Social Networks on the Choice of Tourist Destinations*, Proceedings of the International Conference on Business Excellence 18(1):2327–2343.

Calisto, M., L., & Sarkar, S. (2024). A systematic review of virtual reality in tourism and hospitality: The known and the paths to follow, *International Journal of Hospitality Management*, Volume 116, 103623.

Chang, Y. S., Cheah, J. H., Lim, X. J., Morrison, A. M., & Kennell, J. S. (2022). Are unmanned smart hotels du jour or are they here forever? Experiential pathway analysis of antecedents of satisfaction and loyalty. *International Journal of Hospitality Management*, 104, 103249.

Chowdhary, N., & Prakash, M. (2022). *Interpreration and tour leadership: principles and practices of tour guiding.* Britain, Oxford: Goodfellow Publishers.

Corfu, G. T. (2022). Tourist Guide: Everything you need to know about this Ancient Profession, Retrieved from https://corfuguidedtours.com/tourist-guideeverything- you-need-to-know-about-this-ancientprofession/. Last accessed 20 March 2025.

Çakmak, T. F., & Demirkol, Ş. (2017). Teknolojik gelişmelerin turist rehberliği mesleğine etkileri üzerine bir swot analizi. *Journal of Social Sciences Institute/Sosyal Bilimler Enstitüsü Dergisi, Yıl: 7, Cilt: 7, Sayı: İktisat Özel Sayısı, 221–235.*

De Le Harpe, M., & Sevenhuysen, K. (2020). New technologies in the field of tourist guiding: threat or tool?, *Journal of Tourismology*, Vol. 6 No. 1, pp. 13–33, doi: 10.26650/jot.2020.6.1.0009.

Doğan, S., & Bayar, S. B. (2024). Hospitality and tourism marketing in an artificially ıntelligent world. *Ondokuz Mayıs Üniversitesi Karadeniz Turizm Araştırmaları Dergisi* 2 (1), 51–72.

El Kafy, J. H. A. (2020). Challenges facing tour guide profession and their ımpacts on the egyptian guides performance. *Journal of Association of Arab Universities for Tourism and Hospitality* (JAAUTH), 19(3), 113–130.

Erdoğan, S., Gedikli, A., Çevik, E., İ., Erdoğan, F. (2022). Eco-friendly technologies, international tourism and carbon emissions: Evidence from the most visited countries, *Technological Forecasting and Social Change*, 180(7):121705 DOI:10.1016/j.techfore.2022.121705.

Gangurde, D. & Dsouza, C. C. (2023). A study on the influence of ecofriendly technologies in the hospitality industry, *International Journal of Advanced Research in Science, Communication and Technology*, 465–469.

Garcıa, G, M., Dos Anjos, S., J., G., Doğan, S., (2022). Online travel agencies and their role in the tourism industry, *Advances in Hospitality and Tourism Research (AHTR)*, Vol. 10 (3) 361–386, DOI:10.30519/ahtr.865546.

Kara, D., Kurt Yılmaz, B., Güler, M. E. (2022). Turist rehberlerinin teknoloji kullanımına yönelik algıların belirlenmesi üzerine nitel bir araştırma, *Journal of Yasar University*, 17/66, 517–537.

Karamustafa, K. & *Kılıçhan. B.* (2024). Bilgi-iletişim ve yapay zekâ teknolojileri turist rehberliği mesleğinin geleceği için fırsat mı yoksa tehdit mi?, *Journal of Gastronomy, Hospitality and Travel*, 7(2), 634–645.

Kavak, M., Emir, O., & Arslantürk, Y., (2024). A new alternative in the field of tour guiding emerging after covıd-19: Online guided virtual tour. *Journal of Tourism, Leisure and Hospitality*, vol.6, no.1. 59–71.

Khalil, H. (2021). History of the tourist guide in egypt, *The Scientific Journal of the Faculty of Tourism and Hotels, Alexandria University, Vol. 18,* (Issue 1), 121–131. doi:https://doi.org/10.21608/thalexu.2021.59183.10.44.

Khooshechin Gilak, M. (2023). *Systematic mapping review of facial recognition applications in airports,* The 14th International Conference on Innovation and Research in Engineering Sciences.

Kim, J.W., Kim, C. S., Gautam, A., Lee, Y. (2005). *Location-based tour guide system using mobile gıs and web crawling.* In: Kwon, YJ., Bouju, A., Claramunt, C. (eds) Web and Wireless Geographical Information Systems. W2GIS 2004. Lecture Notes in Computer Science, vol 3428. Springer, Berlin, Heidelberg. https://doi.org/10.1007/11427865_5

Magige, J. M., Jepkosgei, C., Onywere, S. M. (2020). *Use of GIS and remote sensing in tourism*. In: Xiang, Z., Fuchs, M., Gretzel, U., Höpken, W. (eds) Handbook of e-Tourism. Springer, Cham. https://doi.org/10.1007/978-3-030-05324-6_118-1.

Mrňa, D., Badánik, B., Novák, A. (2021). Internet of things as an optimization tool for smart airport concept, *European Transport \ Trasporti Europei*, Issue 82, Paper n° 6, ISSN 1825-3997.

Nazare, A.-K.; Moldoveanu, A; Moldoveanu, F. (2024). Virtual journeys, real engagement: Analyzing user experience on a virtual travel social platform. *Information*, 15, 396.

Nazlı M. (2020). The Future of tourist guidance concerning the digital technology: A comparative study, *International Journal of Contemporary Tourism Research*, Vol 4: No: 1, p. 66–78, doi: 10.30625/ijctr.692463.

Neuhofer, B., & Buhalis, D. (2014). Technology enhanced tourism experience, Retrieved from https://www.researchgate.net/publication/272566581_ Technology_enhanced_tourism_experience.

Noti, E., & Llazo, E. (2016). The use of computer technology in albanian tourism enterprises a way to ıncrease the performance. *International Journal Ecosystems and Ecology Sciences (IJEES),* 6(1), 89–94.

Özkul E., Kumlu S. T. (2019). Augmented reality applications in tourism, *International Journal of Contemporary Tourism Research*, Vol 3: No: 2, p.107–122, doi: 10.30625/ijctr.625192.

Patria, T., Hidayah, N., & Suherlan, H. (2023). Effect of immersive experience on repurchase intention of virtual heritage tours among gen-z in Indonesia. *E3s Web of Conferences*, 388, 04013. https://doi.org/10.1051/e3sconf/202338804013.

Reed, T. (2015). Can apps replace tourist guides? Daily Southern African Tourism Update. https://www.tourismupdate.co.za/article/can-apps-replace-tourist-guides. Last accessed 20.03.2025.

Seyitoğlu, F., Atsız, O. & Akyuz, M. (2025). The role of technology in tourist guiding: future career, service quality, marketing and tourist experiences, *Consumer Behavior in Tourism and Hospitality*, Vol. 20 No. 1, pp. 57–74. https://doi.org/10.1108/CBTH-05-2024-0169.

Sotohy, H.T. (2020). New trends in tour guiding, the guide faces technology 'applied study to selected sites in Egypt, *Journal of Association of Arab Universities for Tourism and Hospitality*, Vol. 19 No. 3, pp. 35–47.

Türkmen, S. Özcan, B. (2024). Turist rehberlerinin iş (tur) alma biçimleri, *Balıkesir Üniversitesi Sosyal Bilimler Enstitüsü Dergisi*, *27(52), 691–704.*

Ünüvar, Ş. (2008). Turizm sektöründe bilgi iletişim teknolojileri kullanımı. *Selçuk Üniversitesi Sosyal Bilimler Meslek Yüksekokulu Dergisi*, *10*(1–2), 597–618.

Vargas-Sánchez, A. (2020). *Overtourism and smart cities: Present and future*. In J. A. Oskam (Ed.), The overtourism debate (pp. 267–278). Emerald Publishing Limited.

Wang, L. (2024). Enhancing tourism management through big data: Design and implementation of an integrated information system, *Heliyon*, Volume 10, Issue 20, 38256.

Weiler, B., & Black, R. (2014). Tour guiding research: ınsights, ıssues and ımplications (62 ed.). Channel View Publications.

Wuenderlich, N., Heinonen, K., Ostrom, A., Patricio, L., Sousa, R., Voss, C., & Lemmink, J. (2015). Futurizing" smart service: implications for service researchers and managers, *Journal of Services Marketing*, Vol. 29 Iss 6/7 pp. 442–447.

Zhang, J. (2024). Design and implementation of a tour guide system based on augmented reality technology for tourist attractions. *Advances in Transdisciplinary Engineering.* 241–249, https://doi.org/10.3233/atde240393.

Postmodernism, Tourism and Transformation in Marketing Approach

Fuat ATASOY[1], İlker ÖZTÜRK[2], Emre TANKUŞ[3]

Introduction

In the contemporary era, there have been substantial shifts in economic, cultural, and social dynamics. These changes have surpassed the structure of modernism, which was shaped within the framework of certain patterns and rules, and have exposed the flexible, fragmented, and pluralistic structure of the postmodern period. Postmodernism has had a profound impact on the consumption habits, identity construction, and lifestyles of individuals, and consequently, it has engendered substantial transformations in tourism and marketing approaches. Significant changes in the consumption habits and lifestyles of individuals have been particularly marked during this period of accelerating globalization and technological development. The traditional understanding of tourism has been transformed by postmodernist concepts such as individualization, experience-oriented consumption, and storytelling (Urry, 1990).

These changes have had a profound effect on the tourism sector and the understanding of marketing (Baudrillard, 1998). The postmodern transformation in tourism underscores the notion that tourists no longer merely visit a destination but also consume the meanings and experiences offered there. In lieu of conventional package tours, tourists now gravitate toward personalized, experience-focused, and meaning-laden travel (Pine & Gilmore, 1999). This study focuses on the transformation of postmodernism in tourism

1 Dr., Ankara University Beypazari Vocational School, Department of Hotel-Catering and Catering Services, atasoyf@ankara.edu.tr
2 Assist. Prof. Dr. Sivas Cumhuriyet University, Cumhuriyet Vocational School of Social Sciences, Department of Hotel-Catering and Catering Services, ilker5885@gmail.com
3 PhD(c), Kastamonu University, Faculty of Tourism, Tourism Management, etankus@sinop.edu.tr

and marketing understanding, changes in consumer behavior, and examples of postmodern tourism in the tourism sector.

Postmodernism: Definition and Evolution

Different fields, including linguistics and cultural studies, interpret postmodernism in a variety of ways. This intellectual movement, which emerged in the aftermath of modernism, garnered particular attention in the mid-20th century, particularly with regard to its approach to art, culture, and society. In defining postmodernism, one can see the traces of a period in which the basic principles of modernity are rejected and the concepts of reality, universality, and absolute truth are questioned (Featherstone, 2007). While modernity was shaped by the Enlightenment and the Industrial Revolution in the West, postmodernity questions the contradictions and absolute truths of modernism as a contrast to this process. The notion of postmodernism emerged in the 1930s under the term "deconstruction" as a reaction to modernism, but it was popularized by artists, writers, and critics in the 1960s and subsequently (Kohler, 1977; Hassan, 1985). This paradigm shift, which can be seen in architecture, music, literature, and the visual arts, is part of an ideological shift toward a way of thinking that values independence and local values. This challenges the usual "high art" understanding of modernism (Türköne, 2005:497; cited in Çağlar, 2008:371). The fundamental tenets of postmodernism encompass pluralism, locality, variability, and relativism. By challenging the tenets of modernism, including universality, reason, and scientific truths, postmodernism has profoundly altered our understanding of reality and truth (Möngü, 2013). While a diversity of views exists among theorists regarding postmodernism, there is a general consensus that postmodernism repudiates the metanarratives of modernism, supplanting them with novel concepts such as multiplicity, diversity, and contingency (Bauman, 2005; Lyotard, 1984). Postmodernism is defined as an intellectual movement that transcends the limits of modernist thought, where individual expressions, simulations, and pluralism are at the forefront. In the postmodern era, there has been a transformation in which social structures and norms have been disintegrated and universal rationality has been replaced by individual understandings of morality. Postmodern culture is characterized by the prioritization of difference and diversity, with the definitive norms of the past being superseded by a more flexible and pluralistic structure.

Postmodern Marketing

The transition from agriculture to industrial production has precipitated profound transformations in the conceptualization of marketing, particularly with the escalating significance of information and technology. The socio-cultural and economic transformations witnessed throughout the last century have precipitated significant shifts in societal structures, with the advent of information technologies serving to enhance individual agency. These changes have precipitated a shift from mass production and consumption patterns to market conditions based on individual consumer behavior. Consequently, the traditional marketing approach has evolved into a modern marketing approach that prioritizes customer satisfaction and quality. The advent of the internet has rendered e-commerce an indispensable facet of marketing (Babacan & Onat, 2012). Postmodern marketing, in contrast to modern marketing, examines consumer behavior from a standpoint that is more individualized and multifaceted. In contrast to modern marketing, which often makes generalizations about consumers, postmodern marketing acknowledges each individual consumer as a unique entity. The impact of postmodernism on marketing is characterized by an emphasis on consumers' emotional responses rather than the functional characteristics of products. Postmodern marketing advocates a 'both/and' approach, as opposed to the 'either/or' approach that is characteristic of modern marketing. This shift in perspective deconstructs the rigid boundaries that characterized traditional marketing, replacing them with a more flexible and engaging approach (Tek, 1997:58–59; cited in Varinli, 2012:18). In contemporary consumer culture, the focus has shifted from mere products to the experiences they engender. In this context, marketing strategies are becoming increasingly personalized, experience-based, and entertaining. Concrete manifestations of this phenomenon include theme parks, shopping centers, and simulations (Güneren & Karakuş, 2015). In addition, consumers adopt multiple identities and images in combination rather than a single identity. This paradigm shift in consumer behavior has prompted a shift in marketing strategies, which are now directed toward niche and specialized markets, as well as bricolage markets (Brown, 1993). The fragmentation of language, information, and social structures has resulted in the diversification, flexibility, and personalization of marketing strategies. This has resulted in an increase in personalized products (Thomas, 1997). Consequently, this paradigm shift has given rise to the emergence of personalized products and markets that are now exclusively managed by individual marketing strategies (Babacan & Onat, 2012). Research demonstrates that this approach fosters a more creative, artistic, and dynamic marketing paradigm. Strategies shape advertisements by encouraging consumers to think and establish emotional bonds. In postmodern marketing, the consumer is regarded as a constituent element in the unceasing production and dissemination of images and symbols, as opposed to being the focal point of the marketing strategy (Fırat, Dholakia & Venkatesh, 1995).

Postmodern Tourism and Post Tourist

The elite spearheaded the development of pre-modern tourism, characterized by a hedonistic framework. The advent of the Industrial Revolution signaled the ascendency of a modern tourism paradigm, concomitant with a proliferation of travel and communication opportunities, thereby enabling widespread participation in tourism. In the contemporary era, there has been an observable rise in the tendency toward individualization, accompanied by the emergence of virtual experiences and meta-reality as key components of the tourism landscape. The mounting environmental concerns and the emergence of mass tourism have catalyzed the evolution of postmodern tourism that is oriented toward sustainability (Kozak, Evren, & Çakır, 2013). Postmodern tourism is shaped in the context of tourist motivation, experiences, and environmental factors (Uriely, 1997). As an alternative to mass tourism, increased sensitivity to the protection of natural life and interest in past lives have led to the importance of authenticity and simulation. Themed centers, in this context, have emerged as significant cultural hubs, redefining areas lacking tourist attraction (Featherstone, 2007). Postmodern tourists seek to avoid risks by opting for the copy instead of the real, and they want to have tourist experiences without physical effort (Varinli, 2012). The advent of communication technologies has facilitated enhanced access to information, elevated service expectations, and fostered a shift toward eco-tourism (Urry, 1990). Consequently, mass tourism is being superseded by alternatives such as culture, adventure, and rural tourism (Rooney, 2011:20–21, cited in Kozak, Evren, and Çakır, 2013:17). The ease of access to information, influenced by the media, has accelerated consumer movements, and the transience of fashion and the shortening of the experience period have led tourism to gain a more aesthetic dimension (Urry, 1995).

Postmodern Tourism Examples

Postmodern tourism represents a paradigm that surpasses conventional tourism, providing distinct experiences. Kozak and colleagues (2013:11) say that the rise of tourism categories based on nature, culture, education, and personal interests that allow for real experiences is a sign of postmodern tourism. Space tourism, thermal parks, and hotels that have become popular because of technological progress are some examples. Postmodern tourism is characterized by a consumer-centric and experiential approach. Mustonen (2006:164) asserts that voluntary tourism exemplifies one of the most commendable forms of travel. Voluntourism encompasses activities designed to aid environmental research or alleviate poverty, in which tourists engage. This category exists at the intersection of contemporary backpacker tourism and conventional pilgrimage tourism, exhibiting elements of postmodernism. Blom (2000:32) asserts that death-themed tourism occupies a significant position within postmodern tourism. Historically, death, fear, and unusual experiences were unappealing to tourists; however, in the postmodern era, these themes have gained popularity. Instances like "Dracula-Land," envisioned in the Transylvanian Alps of Romania, and "Jack the Ripper" in England exemplify postmodern tourism's fascination with unconventional and remarkable experiences. Heritage tourism serves as another illustration of postmodern tourism. Nuryanti (1996:250) asserts that postmodern tourists seek self-discovery through the cultural values and symbols present in historical sites. Heritage tourism emphasizes universal phenomena, including traditions, cultures, and religions, showcasing the historical and cultural legacies of various societies to visitors. The Taj Mahal in India and the pyramids of Egypt are notable heritage sites that draw tourists. The connection between postmodern tourism and cultural spaces is acquiring a new dimension. Cultural institutions in Amsterdam, including the Rembrandt House and the Rijksmuseum, have evolved from simple exhibition venues for art into spaces for social interaction and entertainment. Richards (2002:1057) examines the influence of cultural venues on tourism amid this era of heightened cultural consumption. Museums now serve as venues for public enjoyment and commerce. The Sabancı Museum, through events such as "Taxi Driver Day" and "Neighbor Day," aimed at engaging the local community, exemplifies the postmodern transformation of museums. Thematic restaurants play a significant role in the entertainment industry and exemplify experiential marketing within postmodern tourism. For instance, Rainforest Cafe establishments replicate a tropical rainforest ambiance, providing patrons with a unique experience (Williams, 2002:211). The Ritz Carlton Hotel employs design and scenography principles to provide its guests with profound emotional and physical experiences (Michelli, 2009; cited in Güzel, 2014:9). These hotels offer tourists more tailored and memorable experiences, distinct from conventional hotel stays. The architectural tourism in Northern Cyprus distinctly embodies the influences of postmodernism. Themed hotels in the Bafra region arose as a reaction to mass tourism and are predominantly designed with classical Western elements (Besim, Kiessel, and Kiessel, 2010:11-120). The Luxor Casino & Hotel in Las Vegas is a prominent example of the postmodern era. This hotel, designed in the form of an Egyptian pyramid, is embellished with ancient Egyptian symbols and embodies contemporary casino culture. In the context of postmodern tourism, tourist experiences extend beyond physical spaces. Bicycle tourism exemplifies the postmodern quest for individual and multifaceted experiences. Aşan and Kozak (2015:271) assert that cyclists are driven by experiences including adventure, skill acquisition, and personal challenge. Such experiences suggest that the postmodern tourist seeks both a physical and a social expedition. CitizenM Hotels also enhance the experiential dimension of postmodern tourism. By creating communal spaces for guest interaction, they enable individuals to customize their hotel experiences (Güneren and Karakuş, 2015:483). These hotels cater to consumers' experiential demands and provide a service distinct from the conventional hotel model.

Results

Postmodernism has transformed the comprehension of tourism and marketing, resulting in a diversification of consumer behaviors, the reconfiguration of cultural identities, and the rise of an experience-oriented consumption culture. This comprehension motivates tourists to not only explore a destination but also to engage with its meanings, emotional experiences, and cultural wealth. This transformation enables the tourism sector to become more adaptable, customized, and focused on experiences. Tourism destinations and enterprises are catering to the varied requirements of travelers by providing specialized, tailored travel experiences rather than conventional mass tours. In the postmodern era, authentic, sustainable, and culturally significant tourism concepts have become increasingly important (Güneren and Karakuş, 2015). The incorporation of technologies like virtual and augmented reality has enhanced tourists' experiences, rendering them more enriching and multifaceted. This transformation has significantly impacted marketing strategies. A shift has occurred from traditional product-centric marketing to strategies that prioritize emotional connections, highlighting storytelling and cultural identities. Postmodern marketing is influenced by strategies that honor individual identities and provide each consumer with a tailored experience. Consumers now prioritize not only the acquisition of a product but also the experience it provides (Yıldız, 2021). This circumstance has resulted in marketing endeavors becoming increasingly innovative, artistic, and dynamic.

In conclusion, postmodernism has arisen in tourism and marketing as both a consumption process and a complex phenomenon encompassing cultural discoveries and individual experiences. The tourism sector is devising innovative and original strategies to adapt to this new paradigm, generating opportunities for individuals to engage in deeper, more meaningful, and personalized experiences. This transformation denotes an ongoing and dynamic process necessitating the reconfiguration of both destination management and marketing strategies.

References

Aşan, K. ve Akoğlan K., M. (2015). Postmodern Turist Deneyimi ve Bisiklet Turizmi, International Journal of Eurasia Social Sciences, 6(21):265–288.

Babacan, M. ve Onat, F. (2002). Postmodern Pazarlama Perspektifi, Ege Akademik Bakış Dergisi, 2(1):12–20.
Baudrillard, J. (1995). Simulacra and Simulation Translated: Sheila Faria Glaser. University of Michigan Press.
Bauman, Z. (2005). Modernidad y ambivalencia (Vol. 44). Anthropos Editorial.
Blom, T. (2000). Morbid Tourism - A Postmodern Market Niche with an Example from Althorp, Norwegian Journal of Geography, 54:29–36.
Brown, S. (1993). Postmodern Marketing? European Journal of Marketing, 27(4):19–34. doi.org/10.1108/03090569310038094.
Çağlar, N. (2008). Postmodern Anlayışta Siyaset ve Kimlik, Süleyman Demirel Üniversitesi İktisadi ve İdari Bilimler Fakültesi Dergisi,13(3):369–386
Featherstone, M. (2007). Consumer Culture and Postmodernism.2nd Edition.Sage Publications.
Fırat,A.F.,Dholakia,N. and Venkatesh,A.(1995).Marketing In Postmodern World, European Journal of Marketing, 29(1):40–56, www.emeraldinsight.com/doi/abs/10.1108/03090569510075334, 01.09.2017
Gilmore, J., & Pine, B. (1999). The experience economy. HighBridge.
Güneren, E. ve Karakuş, Y. (2015). Postmodern Pazarlama.İçinde Turizm Pazarlamasında Güncel Yaklaşımlar. (Editörler: Burhan KILIÇ ve Zafer ÖTER). İstanbul: Beta Basım Yayım Dağıtım A.Ş., ss.454–487.
Güzel, F.Ö.(2014). Postmodern Çağ Tatil Deneyimlerinde Hazcı Tepki Basamağı Çıktısı Olarak Temalı Oteller, Uluslararası İktisadi ve İdari İncelemeler Dergisi, Yıl:7, (13):1–16
Hassan, I. (1985). The culture of postmodernism. Theory, Culture & Society, 2(3), 119–131.
Kohler, K. J. (1977). Einführung in die Phonetik des Deutschen. Berlin: Erich Schmidt.
Kozak, A.M., Evren, Ş. ve Çakır, O. (2013). Tarihsel Süreç İçinde Turizm Paradigması, Anatolia: Turizm Araştırmaları Dergisi, 24(7):7–22.
Kozak, A.M., Evren, Ş. ve Çakır, O. (2013). Tarihsel Süreç İçinde Turizm Paradigması, Anatolia: Turizm Araştırmaları Dergisi, 24(7):7–22.
Lyotard, J. F. (1984). The Post Modern Condition: A Report on Knowledge, Translation from the French by Geoff Bennington and Brian Massumi, Theory and History of Literature, Vol:10. Manchester University Press.
Möngü, B. (2013). Postmodernizm ve Postmodern Kimlik Anlayışı, Atatürk Üniversitesi Sosyal Bilimler Enstitüsü Dergisi, 17 (2): 27–36.

Mustonen, P. (2006). Volunteer Tourism: Postmodern Pilgrimage? Journal of Tourism and Cultural Change, 3(3):160–177.
Nuryanti, W. (1996). Heritage and Postmodern Tourism. Annals of Tourirm Research, 23(2):249–260.
Richards, G. (2002). Tourism Attraction Systems Exploring Cultural Behavior, Annals of Tourism Research, 29(4):1048–1064.
Thomas, J.M. (1997). Consumer market research: does it have validity? Some postmodern thoughts, Marketing Intelligence and Planning, 15(2):54–59.
Uriely, N. (1997). Theories of Modern and Postmodern Tourism, Annals of Tourism Researh, 24(4):982–985.
Urry, J. 1990. The Tourist Gaze. London: Sage.
Urry, J. (1995). Consuming Places. London and New York: Routledge.
Varinli, İ. (2012). Pazarlamada Yeni Yaklaşımlar. Ankara: Detay Yayıncılık.
Williams, A. (2002). Understanding the Hospitality Consumer. Butterworth Heinemann Publications.
Yıldız, E. (2021). Postmodern Pazarlama Yaklaşımıyla Modern Pazarlama Yöntemleri: Güncel Şirket Uygulamaları. Süleyman Demirel Üniversitesi Sosyal Bilimler Enstitüsü Dergisi, 34, 123–140.
Yücel B., D., Kiessel, M., & Tozan Kiessel, A. (2010). Postmodernist hotel-casino complexes in Northern Cyprus. METU JFA, 1:103–123.

Motivations of Luxury Tourism

Merve DOĞAN[1]

Introduction

Consumption is a two-ended phenomenon. While one end is the fulfillment of basic demands and needs of an individual, the other end is the fulfillment of hedonic and status acquisition needs. Although tourism essentially includes activities that are beyond compulsory needs of an individual, there are two similar ends of the spectrum in touristic consumption as well. On the one hand, there is desire to get away from the routines of daily life and relaxation, on the other hand, there is desire to buy a touristic product that represents status of touristic consumer against others in a society and personal satisfaction stems from that desire. In the past a few decades, consumer choices have been affected by the transition to an experience-based economy rather than just buying products or services (Pine & Gilmore, 2011). Namely, consumers are searching for experiences to fulfil their needs and desires. The essence of luxury is to demand something that is not needed (Som & Blanckaert, 2015). Furthermore, the concept of luxury has undergone a transformation from owning luxury goods and brands to acquiring experiences recently. Since tourism is a set of experiences that are not a necessity for survival, there is a close relationship between luxury and tourism.

While the global economic uncertainty and increase in inflation, which emerged especially after Covid-19 pandemic, many sectors was negatively affected, the luxury market has been upswing. It is estimated that the world luxury market will reach 530–570 billion Euros by 2030, more than double its size in 2020 (https://www.luxurydaily.com/clarifying-luxury-brand-values-and-positioning-for-an-evolving-marketplace/). Likewise, luxury tourism is one of the fastest-growing market segments in the global tourism industry. Additionaly, it has the advantage of being little affected by the global economic downturn (Trihas, Panagiotaki & Kyriakaki, 2020). It was estimated that the global luxury travel market generated a revenue of

[1] PhD., Bilecik Seyh Edebali University, Faculty of Applied Science, Department of Tourism Management, merve.dogan@bilecik.edu.tr

USD 1,339,737.1 million in 2023 and is expected to reach USD 2,268,239.8 million by 2030. The market is expected to grow at a CAGR (Compound Annual Growth Rate) of 7.8% by 2030. (https://www.grandviewresearch.com/horizon/outlook/luxury-travel-market-size/global).

The current picture of the global luxury tourism market and its expected upward trend in the near future encouraged us to prepare this chapter. In this chapter, concept of luxury, luxury consumption and luxury tourism from different perspectives are presented. In order to provide better understanding to why people prefer luxury tourism, motivations behind luxury consumption are discussed.

The Concept of Luxury and Luxury Consumption

Luxury is defined as "non-essential items or services that contribute to luxurious living; an indulgence or convenience beyond the indispensable minimum" (Webster, 2002). Heine (2012) also defines the concept of luxury as high-priced, rare, status-providing, desirable and non-ordinary. Similarly, Kapferer, Klippert & Leproux (2014) proposed that luxury means rare experiences and hedonic emotions beyond the necessities of life. On the other hand, Berry (2016) characterizes luxuries as objects of desire that provide pleasure. Furthermore, there are several definitions regarding luxury in the literature. Some consider it in terms of higher price, some in terms of symbolic meaning, some in terms of uniqueness and some approach from hedonic feelings or desire.

Although luxury is widely seen as a tangible reflection of material wealth, it has evolved into a phenomenon that perception of luxury may differ for individuals and cultures or perceptions of luxury may also change over time (Ergüven, 2020). Actually, the meaning of luxury is highly relative that each individual may attach different meaning since it is a dynamic concept. Features that someone perceives those as luxury may be perceived as ordinary or exaggerated by another person. Similarly, things that are considered luxurious for a period of time, may become ordinary in the future period of time. From cultural perspective, it was discussed previously how luxury consumption motivations and practices differ across countries, cultures as well (Wong & Ahuvia, 1998; Hennigs, Wiedmann, Klarmann, Strehlau, Godey & Pederzoli, 2012). Based on self-concepts of different cultures as such independent self or interdependents self, meaning attributed to luxury products might differ. For example, since Asians' interdependent self, they pay a great deal of attention to possessions that are both public and visible, such as luxury designer clothing and accesorise, expensive cars, jewelry so on. However, Westerns' self is independent and it is inner, personal oriented mostly (Markus&Kitayama, 1991). Western people consume luxury goods to satisfy themselves rather than satisfy others.

Kapferer (1998) suggested that luxury consumption is multi-dimensional construct. In a similar vein, Wiedmann, Hennigs & Siebels, (2009) proposed that luxury value consists of different types of values: Financial value, functional value, individual value and social value. Price is positioned within the scope of financial value. In the context of luxury consumption, a more expensive product means better quality and a better indicator of status. Usability, quality and uniqueness values are positioned within the scope of functional value. Usability refers that luxury consumption subjectively respond to the needs (symbolic) of the individual. Quality value refers luxury products have a superior quality. Uniqueness value relates individuals may experience feeling of exclusivity through luxury products. On the other hand, individual value covers self-identity, hedonic and materialistic values. Individuals prefer luxury brands which are congruence with their ideal self. In terms of hedonic value, it is assumed that luxury products provides more hedonic utility rather than functional utility. Materialistic value refers that materialistic-oriented consumers prefers luxury products which have symbolic signal. Finally, social value consists of conspicuousness value and prestige value. In terms of these values, consumers can use luxury products in a conspicuous way and send messages to others in order to associate themselves with reference groups. According to Vigneron & Johnson (2004), luxury brands are perceived in two different aspects: Non-personal-oriented perceptions and personal-oriented-perceptions. While non-personal-oriented perceptions cover conspicuousness, uniqueness, quality; personal-oriented-perceptions cover hedonism and extended self.

From traditional view, conspicuous consumption of luxury is forefound. That is, luxury is seen as elitist, exclusive, symbolic, prestigious, expensive, and attached firmly to the products and their attributes. (Hennigs et al., 2012). However, there is a shift within recent decades regarding the meaning of luxury from wealth and status symbol to experience and indulgence

(Yeoman, 2010). That is to say, new luxury is defined more from the experience perspective; much more related to experience and authenticity in contrast to traditional understanding (Tynan, McKechnie & Chhuon, 2010; Yeoman, 2010; Correia, Kozak& Del Chiappa, 2020).

From experiantal view, luxury tourism was defined as an experience in which touristic consumers obtain different benefits and memorable experiences with touristic products&services (Ariffin, Maghzi, Soon & Alam, 2018). The investments in accommodation, spa, food&beverage services by leading luxury brands of fashion and cosmetic sectors such as Bvlgari, Louis Vuitton (LV), Christian Dior (CD), Gucci, Coach, Ralph & Lauren (RL) are suitable evidences that the meaning of luxury has evolved from owning luxury products to experiencing luxury. For example, there are Bvlgari Hotels & Bvlgari Restaurants in many cities worldwide (https://www.bulgarihotels.com/en_US/). Recently, LV has invested in the hospitality sector that The Louis Vuitton Hotel will open in 2026 at Paris (https://www.thehoteltrotter.com/the-louis-vuitton-hotel-in-paris-all-we-know-about-this-future-beacon-of-luxury/). Moreover, CD has luxury spa centers in Dubai, Doha, Paris, The French Riviera and Scotland (https://www.dior.com/en_int/beauty/store-page-folder/home-diorspa.html). Gucci, RL and Coach offer opportunity to experience luxury through dining and coffee shops in numerous locations in the USA, Europe and Asia.

Based on Iloranta's (2022) comprehensive literature review, there are 4 different perspectives within the luxury tourism literature: *Consumer consumption-oriented perspective* which is fed from Veblen's conspicuous consumption (Bagwell & Bernheim, 1996). Accordingly, individuals view luxury tourism as an indicator of wealth and differentiation from other people. This perspective focuses on buying/purchasing side of luxury tourism and considers tourists' luxury consumption habits, revisits, repurchases, reactions for prices as well. *Brand management oriented perspective* defines luxury tourism based on traditional meaning of luxury. In other words, this approach considers luxury tourism more as a product than an experience, Research more focuses on brand such as brand image, brand equity, brand attachment and tourism products such as luxury destinations or hotels with specific tangible and intangible attributes. However, researches based on *service marketing perspective* focuses on more service, employee interaction with tourists. As well known tourism is a service based industry itself and service is always a part of luxury

(Kepferer, 2014). Namely, service marketing perspective is the most common approach within luxury tourism research. Service quality, customer loyalty, satisfaction, trust and customer retention, employe and customer interactions are the most common topics based on the service marketing perspective. Finally, *consumer behavioural-oriented perspective* defines luxury tourism not only means of status but also ideal values and experiences as different needs of consumers (Kiessling, Balekjian & Oehmichen, 2009; Hemetsberger, von Wallpach & Bauer, 2012). This perspective approaches luxury tourism standing behavioral practices, motivation, experiental consumption.

Luxury in tourism may take different forms. Luxury tourism products/services are often associated with *unique destinations* such as, Maldives, Bora Bora, Amalfi Coast, Monaco, Dubai; *customized services* such as spa & wellness, butler services in hotels, personalised services; *exclusive& unique experiences* such as luxury shopping, golf, high-end cruise trips, gourment culinary; *transport services* such as first class flight, luxury train, private yacht and *accommodation services* such as private villas, luxury hotels&resorts, deluxe or ultra all inclusive concepts (Petroman, 2021; https://www.grandviewresearch.com/industry-analysis/luxury-travel-market).

Motivations of Luxury Tourism

Tourism demand depends on two factors. Enabling factors, such as income, growing prosperity and leisure time and causing factors, such as changes in taste, inclinations, interests and felt needs (Theuns, 2014). In terms of luxury tourism demand, as income and prosperity grows, causing factors are evolving to exclusive needs, need of uniqueness, hedonic desire so on. From a consumer behavior perspective, consumers seek for luxury products, services or experiences basically driven by intrinsic and extrinsic motivations. Whereas intrinsic motivations refer to an emotional value, self-fulfilment and pleasure of individuals; extrinsic motivations refer to a search for signalling something to others as such social status, prestige. Thus, extrinsic level relates to social and interpersonal aspects of luxury consumption. (Tsai, 2005; Correia et al., 2020). Extrinsic motives fall under the theories of status consumption (Veblen, 2005), which state that individuals purchase luxury goods to display their wealth and status. Based on the tourism literature, tourist motivations and choice of a destination are shaped by push and pull factors. While pull factors are related with the attractiveness of a visiting destination and its tangible features, push factors are psychological and socio-psychological needs that motivate a tourist to travel. In other words, push factors refer to the intrinsic desires of the tourist (Crompton, 1979). In this regard, desire of luxury consumption or luxury consumption tendency (Dogan, Ozkara & Dogan, 2022) can be assumed as one of the push factors on choosing a tourism product/service or destination. However, what are the triggers behind consuming luxurious in the tourism context?

In general, prices or affordability can be considered as one of the pull factors for a destination or a facility; however luxury tourists seek the best-quality tourism products and the most complete spectrum of services rather than the best price (Yang & Mattila, 2017). Particularly, expensiveness is important in terms of symbolic meaning as such indicators of status, prestige and quality. According to Kapferer (1998), luxury brands plays a role as a carrier of meaning and enable consumers to differentiate themselves, to have a higher status than others and to attribute a different meaning to themselves than others (Heine, 2012). To metaphorize, luxury consumption is a kind of non-verbal communication, through which people communicate to their social environment about their status, financial situation, taste of living etc. Since the price paid for a good or a service is a kind of signal for a prestige and a social class (Amaldoss&Jain, 2005), luxury tourists are looking for new experiences as well as a dream vacation, and they are willing to pay high prices for services that will set them apart from other tourists in terms of status (Hall, Harrison, Weaver & Wall, 2013). Regarding to identity-signaling perspective, consumers sometimes make choices which diverge them from others (Berger & Heath, 2007). Luxury tourism can be considered as one of the best instruments for that.

Rarity is considered luxury's fundamental code (Phau & Prendergast, 2000). Namely, uniqueness needs can provide explanation for luxury consumption motivation (Tian, Bearden & Hunter, 2001). In the context of tourism, luxury is identified with experiencing luxury rather than having luxury items (Özkoç & Çalışkan, 2018). Online reviews of luxury travellers show that majority care about a memorable stay which supports experience-oriented operations in luxury tourism activities (Arıcı & Bayram, 2022). In a nutshell, perception of luxury is not only about facilities, making customers feel unique and exclusive should become a priority for all luxury tourism sector. Luxury experiences provide to consumers something out of the ordinary, which will envelop and engage with the senses to support a unique experience (Harkison, Hemmington & Hyde, 2018). Luxury corresponds to unique personlized experiences in the tourism industry and tourists expect unique products or services in return for higher prices (Heine, 2012). Customization plays an important role for luxury travelers in order to orchestrate their own unique travel experiences.

Hedonic consumers also prefer luxuries for their positive emotions rather than their rational attributes. Luxury brands provide hedonic pleasures to consumers (Tsai, 2005). The hedonistic nature of luxury consumption reveals consumers' hedonic pleasures. Consumers' hedonic pleasures might signal their social class such that consumers could diverge themselves from others through their specific hedonic pleasures (Dogan et al.2020). Features such as limited capacities or quantities, limited editions, higher prices, difficulty of access cause people who own luxury goods to feel a strong sense of pleasure. The more limitations or difficulties, the higher level of pleasure and satisfaction consumers feel by using that luxury brand/product (Heine, 2012). Therefore, luxury tourism facilities such as hotels, restaurants with high prices and limited capacity, luxury destinations that are difficult to reach and expensive (e.g. Malvides, Caribbean) can be considered as pull factors in terms of tourists' hedonic pleasure motivation.

Furthermore, bandwagon effect may play a trigger role for participating luxury tourism. *Bandwagon effect* referred to the extent to which the demand for a commodity is increased due to the fact that others are also consuming the same commodity (Leibenstein, 1950, p.189). Accordingly, bandwagon oriented luxury consumption is the tendency to purchase luxury products or services with motivation to purchase products purchased by people who are desired to be associate (Kastanakis & Balabanis, 2012). Bandwagon effects in tourism travel decisions have been already investigated by Boto-García & Baños-Pino, (2022) and it was illustrated that bandwagon effect as a social influencer is an important determinant of individual travel decisions. Through media innovations, interventions and expansions, media-driven is one aspect of luxury today (Cristini, Kauppinen-Räisänen, Barthod-Prothade & Woodside, 2017). In this regard, social media sharings (such as post, video or short video) of rich celebrities or influencers may play a bandwagon role in terms of their luxury travel or luxury experiences. Their sharings related luxury vacations can play a driven role for their followers/fans. In particular, it has been revealed that interdependent self-construal compared to independent self feeds bandwagon antecedent more (Kastanakis & Balabanis, 2014). On this basis, professionals from interdependent-self oriented cultures or countries should consider the current difference for the marketing strategy of luxury tourism products/services.

Result

It is possible to consider luxury tourism in two different aspects: Luxury tourism as a segment in tourism market and luxury tourism as a type of tourism. The market segment aspect is based on the fact that higher expectations and expenditures of luxury tourists, also higher investments in order to produce high quality services and experiences. On the other hand, luxury tourism is a type of alternative tourism which include unique and customized experiences such as exclusive accommodation, spa&wellnes, culinary, shopping, golf, cruise, yachting, safari, gambling etc. Since tourism is a service industry, no matter which aspect you approach, experience is the key factor for luxury tourism. Luxury tourism is a set of exclusive, unique experiences in exchange with higher prices. According to Correie et al. (2020), luxury experiences are mostly driven by social traits such as social value, conspicuousness and uniqueness. On the other hand, Dogan et al.(2020), evaluate luxury consumption tendency as a trait and define it as consumer's intrinsic tendency toward luxury consumption. It is a multi-dimensional construct composed of uniqueness, expensiveness, symbolic meaning, arbitrary desire and belonging to an exclusive minority. Since luxury is a complex and multi-dimensional construct, destination marketers and hospitality managers must take into account triggers of luxury consumption which help them to define market segments and marketing strategies properly. Furthermore, tourism practitioners should plan or run marketing operations for luxury consumer segments regarding their motivators. Particularly, social media contents or advertising messages for luxurious tourists can be created congruence with luxury consumption drivers.

References

Amaldoss, W., & Jain, S. (2005). Pricing of conspicuous goods: A competitive analysis of social effects. *Journal of Marketing Research*, 42(1), 30–42.

Arıcı, S. & Bayram, M. (2022). Lüks Turizm in *Örnek Olaylarla Niş Turizm*, ed. Canan Tanrısever, Detay Yayıncılık:Ankara, Pages: 287–309
Ariffin, A. A. M., Maghzi, A., Soon, J. L. M., & Alam, S. S. (2018). Exploring the influence of hospitality on guest satisfaction in luxury hotel services. *E-review of Tourism Research*, 15(1), 1–20.
Bagwell, L. S., & Bernheim, B. D. (1996). Veblen effects in a theory of conspicuous consumption.*The American Economic Review*, 349–373.
Berger, J., & Heath, C. (2007). Where consumers diverge from others: Identity signaling and product domains *Journal of Consumer Research*, 34(2), 121–134.
Berry, C. J. (2016). Luxury: A dialectic of desire?. *Critical Luxury Studies: Art, Design, Media*, 47–66.
Boto-García, D., & Baños-Pino, J. F. (2022). Social influence and bandwagon effects in tourism travel. *Annals of Tourism Research*, 93, 1–15.
Correia, A., Kozak, M., & Del Chiappa, G. (2020). Examining the meaning of luxury in tourism: A mixed-method approach. *Current Issues in Tourism*, 23(8), 952–970.
Cristini, H., Kauppinen-Räisänen, H., Barthod-Prothade, M., & Woodside, A. (2017). Toward a general theory of luxury: Advancing from workbench definitions and theoretical transformations. *Journal of Business Research*, 70, 101–107.
Crompton, J.L. (1979). Motivations for pleasure vacation. *Annals of Tourism Research*, 6(4), 408–424.
Dogan, V., Ozkara, B. Y., & Dogan, M. (2020). Luxury consumption tendency: conceptualization, scale development and validation. *Current Psychology*, 39, 934–952.
Ergüven, M. H. (2020). Lükse kavramsal bir yaklaşım. *Jass Studies-The Journal of Academic Social Science Studies,* 80 (2), 435–446.
Hall, C. M., Harrison, D., Weaver, D., & Wall, G. (2013). Vanishing peripheries: Does tourism consume places? *Tourism Recreation Research*, 38(1), 71–92.
Harkison, T., Hemmington, N., & Hyde, K. F. (2018). Luxury accommodation–significantly different or just more expensive? *Journal of Revenue and Pricing Management*, 17(4), 231–243.
Heine, K. (2012). The concept of luxury brands. *Luxury Brand Management*, 1(2), 193–208.
Hemetsberger, A., von Wallpach, S., & Bauer, M. (2012). "Because I'm Worth It" - Luxury and the Construction of Consumers' Selves. *Advances in Consumer Research*, *40*, 483–489.

Hennigs, N., Wiedmann, K. P., Klarmann, C., Strehlau, S., Godey, B., Pederzoli, D. (2012). What is the value of luxury? A crosscultural consumer perspective. *Psychology and Marketing*, 29(12), 1018–1034.
Iloranta, R. (2022). Luxury tourism – a review of the literature. *European Journal of Tourism Research* 30, 3007
Kapferer, J. N. (1998). Why are we seduced by luxury brands?. *Journal of Brand Management*, 6(1), 44–49.
Kapferer, J. N., Klippert, C., & Leproux, L. (2014). Does luxury have a minimum price? An exploratory study into consumers' psychology of luxury prices. *Journal of Revenue and Pricing Management*, 13, 2–11.
Kastanakis, M. N., & Balabanis, G. (2012). Between the mass and the class: Antecedents of the "bandwagon" luxury consumption behavior. *Journal of Business Research*, 65(10), 1399–1407.
Kastanakis, M. N., & Balabanis, G. (2014). Explaining variation in conspicuous luxury consumption: An individual differences' perspective. *Journal of Business Research*, 67(10), 2147–2154.
Kiessling, G., Balekjian, C., & Oehmichen, A. (2009). What credit crunch? More luxury for new money: European rising stars & established markets. *Journal of Retail and Leisure Property*, 8(1), 3–23.
Leibenstein, H. (1950). Bandwagon, snob, and Veblen effects in the theory of consumers' demand. *The Quarterly Journal of Economics*, 64(2), 183–207.
Markus, H. R., & Kitayama, S. (1991). Cultural variation in the self-concept, in *The self: Interdisciplinary approaches,* Eds.Jaine Strauss, George R. Goethals, Springer: New York, pages: 18–48.
Özkoç, A.G. & Çalışkan, N. (2018). Lüks Turizm in *Turizmde Güncel Konu ve Eğilimler, Eds.* Aydın, Ş. & Boz, M., Detay Yayıncılık: Ankara.
Petroman, C. (2021). Types of luxury tourism consumers. *Quaestus*,18, 312–320.
Phau, I., & Prendergast, G. (2000). Consuming luxury brands: the relevance of the 'rarity principle'. *Journal of Brand Management*, 8(2), 122–138.
Pine, B. J., & Gilmore, J. H. (2011).*The experience economy*. Harvard Business Press.
Som, A., & Blanckaert, C. (2015). *The road to luxury: The evolution, markets, and strategies of luxury brand management*. John Wiley & Sons.
Theuns, H. L. (2014). Is tourism a luxury: Can we do without? *Tourism Recreation Research*, 39(2), 221–233.

Tian, K. T., Bearden, W. O., & Hunter, G. L. (2001). Consumers' need for uniqueness: Scale development and validation. *Journal of Consumer Research, 28*(1), 50–66.

Trihas, N., Panagiotaki, G., and Kyriakaki, A. (2020). Insights into luxury tourism: the case of Elounda, Crete. *Journal of Tourism, Leisure and Hospitality*, 2(2), 62–70.

Tsai, S. P. (2005). Impact of personal orientation on luxury-brand purchase value: An international investigation. *International Journal of Market Research*, 47(4), 427–452

Tynan, C., McKechnie, S., & Chhuon, C. (2010). Co-creating value for luxury brands. *Journal of Business Research*, 63(11), 1156–1163.

Veblen, T. (2005). *Conspicuous Consumption*. Penguin:UK.

Webster's Third New International Dictionary (2002). Merriam-Webster, New York.

Wiedmann, K.-P., Hennigs, N., & Siebels, A. (2009). Value-based segmentation of luxury consumption behavior. *Psychology and Marketing*, 26(7), 625–651.

Wong, N. Y., & Ahuvia, A. C. (1998). Personal taste and family face: Luxury consumption in Confucian and Western societies. *Psychology & Marketing*, 15(5), 423–441.

Yang, W.,& Mattila, A.S. (2017). The Impact of Status Seeking on Consumers' Word of Mouth and Product Preference: A Comparison Between Luxury Hospitality Services and Luxury Goods. *Journal of Hospitality & Tourism Research* 41 (1), 3–22.

Yeoman, I. (2010). The changing behavior of luxury consumption. *Journal of Revenue and Price Management*, 10(1), 47–50.

https://www.grandviewresearch.com/horizon/outlook/luxury-travel-market-size/global (Access: 18.03.2025)

(https://www.luxurydaily.com/clarifying-luxury-brand-values-and-positioning-for-an-evolving-marketplace/ (Access:18.03.2025)

https://www.bulgarihotels.com/en_US/ (Access:15.03.2025)

https://www.thehoteltrotter.com/the-louis-vuitton-hotel-in-paris-all-we-know-about-this-future-beacon-of-luxury/ (Access: 15.03.2025)

https://www.dior.com/en_int/beauty/store-page-folder/home-diorspa.html (Access:15.03.2025)

Ethical Dilemma of Climate Challenge in the Tourism Industry

Onur Şevket YILDIZ[1], *Gamze ÇOBAN YILDIZ*[2]

Introduction

The tourism industry is arguably an industry with great potential for economic development. In that regard, creation of new jobs and investment opportunities sound quite pleasing to both developed and developing countries. However, the impact of tourism activities on nature could be quite complicated and difficult to manage. More challengingly, some tourism activities solely depend on the use of natural assets. The recent common worries about climate change – though mostly about its consequences rather than the sensitive and responsible approach to nature – triggered many thoughts on how to minimize it and manage its impacts. Although the experts' warnings about climate change have a longer history, actions have only recently started to be applied. From a business perspective, some reluctance about these actions is understandable since a commercial activity is basically a way of making profit by managing costs and income. When climate-friendly applications cost, then any business would reconsider whether to apply them unless there is a legal obligation to do so. Moreover, businesses challenging in a tough market where profitability is a harder task may not be able to adopt a longer-term vision since they are too busy with survival in the short term. Thus, sustainability, both for the business and environment, could be ignored. Governments which depend on the tax income financially and are responsible for the welfare of the people are hesitating to intervene this situation since they have to create a

1 PhD., Assistant Professor of Tourism Management, Nevsehir Haci Bektas Veli University, Faculty of Tourism, Department of Tourism Management, osyildiz@gmail.com
2 PhD., Assistant Professor of Tourism Management, Nevsehir Haci Bektas Veli University, Faculty of Tourism, Department of Tourism Management, gamzecoban_85@hotmail.com

desirable investment atmosphere and a job market big enough to employ the relevant population. In such a complicated or dialectical situation, it is, indeed, very difficult to make a choice between climate fight and stability based on the conventional economic truth. Transformation needed for resolution is not less challenging than that. The truth of business, or the economic truth, is based on a commercial approach but it is not limited with just the rules of economy. Moral values of societies or the common ethical understanding of an occupation do play their roles. In that regard, businesses are reacting to moral understandings of societies and some adopt these values sincerely whereas some do this just in appearance (Besio & Pronzini, 2014). However, it is safe to argue that the ethical values are likely to be ignored when they clash with the economic interests in a commercial setting. It does not mean that businesses always decide ethically when there is no such clash. In other words, ethical dilemma is not the sole dilemma; but, it is a major one. The very base of the ethical dilemma lies in the contradiction between ethical understanding and economic system. Any of these two has to be modified when there is a contradiction between these two major factors. However, changing the economic system(s) might be too challenging if one is to avoid damaging the welfare of any party. On the other hand, reformations are clearly needed and economic systems should continue to evolve (supposing that they are evolving and doing this in a good way). Focusing on the ethical side might even be a prerequisite for the economic transformation.

The existence of different views on the issue increases the chance of business worlds' adopting a functional approach, as similarly argued by Stern (2014a). Although some approaches translate into a non-reconcilable situation between climate change focus and business focus due to the role they attain to moral philosophy (Jamieson, 1996; Stern, 2014a), there could be some way for solution. For example, if one is to use moral philosophy to justify something, then he can justify climate action rather than profit maximization. As Garvey (2009) states, ethical aspect of the climate change issue is still obscure despite the clarity of horrible impacts of climate change. In order to discuss the ethical dilemmas the tourism industry is facing, it is useful to clarify climate change issues first.

Climate Change and the Tourism Industry

Climate change is quite a contemporary topic which is recently being discussed in many areas. The broad spectrum of climate change impacts and factors makes it relevant to anything. Specifically, the industries are re-questioned in terms of their contribution to climate change factors and the required transformations to make them climate-friendlier. Climate change which naturally occurs throughout the ages or due to natural factors are out of the ethical discussion this chapter makes. The anthropogenic climate change, which is the climate change caused by human activities – especially industrial activities –, is the main discussion point here. As their technological capabilities enhance, human beings are gaining more power which may be more destructive to the nature. However, philosophical approaches throughout history questioned the relevant issues, which also hints that human beings can also change the way they do things for the sake of minimizing their impact on nature. Moreover, the fact that their existence depend on the nature is a motivator for solving (or managing) climate change issue even if they adopt a 'selfish' (or speciesist) approach. Climate change – in the context of contemporary general debate – could be briefly defined as the change in the climate which is triggered by anthropogenic factors. The natural change in the climate might also constitute many threats on lives but the industrial capabilities of human beings started to make such a big impact that interdisciplinary researchers, governments, and international organizations mostly focus on the climate change which is caused by humans. Technically, climate change is defined as the long-term shifts in climate (UN, 2025). In order to name these shifts as climate change, they should be statistically significant (Türkeş, 2008). Human-induced climate change, as especially observed in greenhouse gas emissions and their impacts, (Türkeş, 2008) might even be the sole subject of ethical discussions since it is the direct responsibility of human beings.

The tourism industry is sometimes regarded as a clean(er) industry in terms of its impact on nature. However, this is not always true. Same complexity is valid for the tourism industry's climate-friendliness. Moreover, some activities included in tourism, such as transportation, accommodation, and etc., are those major carbon emitting activities (Michailidou et al., 2016, Yildiz & Coban Yildiz, 2023). Land use issues are also quite relevant to the relation between climate change and tourism (Gössling & Peeters, 2015). Another interesting point about the tourism industry is that those tourism activities based on nature could both protect or destroy natural assets. For example, tourism can be a motivation for preserving a forest due to the fact that it is a tourism asset. On the other hand, building a tourism facility, a huge hotel or a couple of cottages, in the middle of a forest is quite likely to harm it. Despite the fact that the tourism industry contributes to climate change factors, no major transformation was achieved to make it considerably climate-friendlier. Economically, the tourism industry is considered as a good venue for investment and employment opportunities. Managing a (fast) growing industry with a focus on climate-friendliness might be a harder challenge than managing a stable one. Ethical discussion, then, gains more complexity.

Ethical Dilemma in the Tourism Industry

Arnold & Bustos (2005) discuss the relative weaknesses of consumers in their critical analysis of Bowie's (1990) views in 'Money, Morality, and Motor Cars'. This weakness is illustrated both in terms of consumers' power to influence decision makers (or consumers' participation opportunities in decision making) and their lack of technical knowledge on climate change factors. When consumers are willing to decide in a climate-friendlier manner, the perplexity of the issue and consumers' level of knowledge create some dilemmas (Juvan & Dolnicar, 2014; Feucht & Zander, 2018). The responsibility of businesses and consumers may be discussed with a view of 'who is responsible' but that might be too simplistic. Economic systems, cultures, and many other major factors dictate (or force) both businesses and individuals (consumers, employees, investors, and etc.) to behave in a specific way or within some boundaries. In the tourism industry, there are many other complexities. For example, the natural assets used for tourism purposes are sometimes the product themselves. Sea, sun, and sand tourism is based on the relevant resources and replacement of the product with another alternative may be practically impossible. It is not the case in some other industries, such as energy production. Energy can be produced from renewable (possibly more climate-friendly) or conventional resources (which are likely to generate more emissions). Of course, there are limitations, such as technical capabilities, costs, and etc., in those industries, too.

When the very nature of the relevant tourism product is a determinant of the used resources (as in the sea, sun, sand tourism mentioned above), it might be more challenging to transform it into a climate-friendlier way – if it does not turn into a Gordian knot. Thus, ethical dilemma is less challenging in planning phase; but this applies only the very first planning phase (such as in the creation of a new destination). In other words, planning with a climate focus could prevent the ethical dilemma in practice. However, the tourism industry has already matured at some level and the discussion on the ethics in practice is likely to get more intense. These do not mean that planning phase do not include any dilemma. First of all, planning is the start of 'preferences' and 'decisions'. In that sense, planning could also be the base of those dilemmas. Focusing on practice is just more sensible due to tourism system's maturity. If the case is not a very new destination, then the problems 'at hand' are the point of discussion. The above mentioned consumers-businesses issue is basically a practical discussion. The system that force them to behave in the way they do includes planning discussions. In short, ethical dilemmas are present in both planning and execution phases. The perplexity of those dilemmas mostly depends on the case.

Since the science already clarified what the human beings have to do technically to minimize climate change, the big question is now about how human beings should transform almost everything. As also noted by Broome (2008), the ethical question begins when there are conflicting interests. The current state of the tourism industry offers many examples of conflicting interests. For example, luxury tourism is quite a big market, thus a job creator and an investment opportunity but many of luxury tourism activities have a negative impact on nature and climate. From a point of view, we can easily decide that we should limit (or even ban) luxury tourism (e.g. limit the number of golf hotels, yachts, etc.). However, we cannot ignore that these decisions will harm many people's interests. Those who find jobs, those who make profits, those who gain tax income (and its benefits), and many others will have to find some alternative. In practice, we cannot magically cover all those 'costs'. This ethical dilemma could lead the decision makers (at least those who believe that transformation is possible) to search for alternatives of each and every activity. In other words, the efforts could be on the transformation of today's tourism activities without disturbing those who have an interest. On the other hand, this 'soft transformation' approach may be disregarding future generations' interests. Technically, the later we change into a climate-friendlier way of life, the harsher impact the future generations will have to deal with. From a broader perspective, the climate injustice is just another injustice and in ethical terms, it is useful to focus on eliminating injustice and harm by considering the end-results (in this case climate impacts) (see e.g. Grasso, 2013 and Barry et al., 2013). Therefore, the 'higher maturity of the tourism industry' or 'the currently established system of interests' cannot be an excuse for softer transition.

Another dilemma is that if the tourism industry is to undergo a sharp and sudden transformation, then human beings should find ways of not creating new ethical dilemmas; if it is possible at all. Sudden changes and current worries may force people to decide quicker. The bravery aspect of new decisions could be overly emphasized. These all constitute new ethical risks in the tourism industry. For example, luxury tourism could be transformed into a climate-friendlier form by even changing the perception of luxury. However, it is clear that it will probably take some considerable time to alter the 'culture of luxury' – created through many years or even ages – which has a major impact on perceptions. As Ballantyne et al. (2025) discuss, the socio-cultural aspect of tourist behavior and preferences should not be ignored.

From the point of view that 'the rich' has a higher impact on climate (Stern, 2014a), it is arguable that a lifestyle change in the wealthier classes is needed for a climate-friendlier global system. However, a more sensible approach states that a fairer distribution of resources would provide a higher success (see e.g. Mittler, 2014). This, at least as a starting point, could be a crucial development in climate change mitigation efforts. However, in both cases – when the wealthier class change their lifestyle or when the resources are distributed in a fairer way all around the world and among people –, there will be consequences for the tourism industry. First, the tourist profile as consumers will change. Therefore, former might be better in short term since the present wealthier classes are expected to change into more climate-sensitive people. The latter might be more challenging to manage since there will be many new potential tourists who may focus on new experiences with a high motivation that may overshadow climate-sensitivity. However, in the latter case, if the tourism industry is able to offer climate-friendlier products to these new tourists, this can even shift the perception of tourism in a better (climate-friendlier) way. Moreover, the new climate-friendlier tourism market will have more depth, which means an economically healthier tourism market. In this scenario, business ethics should also evolve if the tourism industry is to achieve a permanent positive change. Fairer distribution of wealth could also alter the transportation patterns of the world. As transportation activities are responsible for a big share of emissions, more tourism activity between neighboring countries could create a healthier tourism flow in terms of climate change. Rather than carrying the tourists from the developed part of the world to everywhere, greener transportation systems could be built between neighboring countries; but, miles per tourists and total miles for global tourism are difficult to estimate.

A better (fairer and more egalitarian) distribution of world resources and the benefits derived from them could also eliminate the individual vs. society dilemma. Competing interests are more difficult to ethically manage, thus less competing interests means an easier work for climate mitigation and adaptation. However, the competing interests are not only a matter of discussion for the living population but they are also present between the present and future generations (Garvey, 2009). Thus, thoughts on intergenerational ethics might help in that regard (see e.g. Davidson, 2015). When the conflict is between the present and future generations, then it could even be more difficult to ethically solve the issue since the future generations cannot even express themselves whereas the present ones can even decide and apply something else. The 'good' behavior could be less likely in such a case, as Garvey (2009) illustrates in the example of temporal version of prisoner's dilemma.

Not limited to the tourism industry, there is a tendency (or an actual situation) that industrial discussions on the impact of climate change focus too much on the economical side of the issue and neglect the fact (or idea) that lives actually matter (Garvey, 2009). As Jamieson (1992) points, it can also be interpreted as the hegemony of the economic view. From a point of view, the industry is just an industry and focusing on the economic aspect is understandable. However, this results in more difficulty in the creation of a humane (and non-speciesist) transformation of the global economy. There is even some financial cost-based justification of inaction about the climate issue, which is somehow, as Garvey (2009) states, vicious. Thus, there is the need of a humane focus for the issue. The tourism industry is still labor-intensive and there is still plenty of interpersonal communication in many tourism activities. This may be a chance for changing the intense economic focus of climate change debate and moving the discussion into a wider-perspective ethical domain. The tourism industry could even take a leading role in terms of this positive change. Moreover, the level of moral maturity human beings reached could be a determinant of moral perspective they are to adopt regarding climate change-tourism industry relation. Garvey's (2009) explanation starting with biocentrism and reaching through considering the planet as a whole gives the hints of this moral maturity. Indeed, there are philosophical thoughts that could well lead the human beings to perceive the entire universe as a whole (see e.g. DesJardins' (2023) explanations on biocentrism and environmentalism). On the other hand, the moral maturity that we refer is based on the idea that as people question, their thoughts will evolve in 'the right' direction which will eventually lead them to a level in which they show great respect to everything that exists. This assumption might be too optimistic but this can give a brief road-map to those who are trying to change the industries, including the tourism industry, into more climate-friendlier ones. In the case of tourism, even some climate-friendlier hospitableness approaches (which could be adapted to the hospitality industry) can be developed. For example, locals may plant a tree for the honor of each visitor and do this with the visitor. During this sort of activities, tourists and locals could be informed and reminded about the serious climate threats and what to do about them.

Right to travel (or right to participate in a tourism activity) is certainly another matter of ethical discussions in tourism. Every tourism activity creates some impact on climate. The source of some ethical dilemma is due to the scarcity of resources (Patil & Nahata, 2024). The discussion on 'who has the right to use which resources?' is important but it is far from solving the dilemma. Use of resources and the purpose they are used for are justified by various thoughts. Human beings may prefer to use the resources for a group, for financial value creation, for every person, for every creature, for the whole universe, for joy and entertainment or they can avoid using these resources for the sake of future generations, for the sake of a healthier nature, or just because they think they do not have the right to use them. The tourism industry may somehow justify or rationalize the use of resources for tourism purposes but that does not ensure a 'fair' tourism system.

The tourism industry, unlike some other industries such as food, is more open to discussion since tourism is not a prerequisite for our existence. Unless we find a natural way to feed billions of people on earth, we have to keep the food industry. However, we do not necessarily need the tourism industry for our existence (supposing that we have to ensure human existence but even this assumption is speciesist). This is another ethical dilemma that the tourism industry faces. Beside the potential and actual benefits of tourism activities to human beings (such as recreation, better understanding of other cultures etc.), the tourism industry should either stress the significance of these benefits more or find some more ways to enhance the well-being of people, nature, and etc. through tourism. Otherwise, in case of a cost-benefit analysis for the sake of climate, the tourism industry may find itself in a group of 'secondary importance'. Of course, it is difficult to estimate whether people will reach (or have reached) a stage where they have to eliminate some of their industrial activities for the sake of climate change mitigation and whether they will decide to eliminate the tourism industry in such a case.

The Dilemmas, Solutions, and Perplexities

The challenge in the decision making about climate change action is that it is quite a complicated issue and one cannot see the actual results that are to happen in the future, thus, have to rely on the forecasts. In other words, the perception of climate change and its seriousness is something to be dealt for a successful transition to a climate-friendlier system (Stern, 2014b). Same is true for tourism policy making and tourism industry practices. Moreover, even when the seriousness of the impacts are clear, the tourism industry cannot give up, for example, transportation. Transformation of the whole passenger transportation is, on the other hand, an issue which cannot be solved by the tourism industry alone. Transforming tourism accommodation is also challenging since it requires quite a lot of investment (or giving up a bulk of current investment). On the other hand, greener accommodation and transportation opportunities, especially when they also reduce the costs of tourism businesses, are quite attractive. Climate-focused preference of consumers might be a major force to transform the industry. Approaches such as responsible tourism also constitute some opportunities for a smooth transition if the ethics of responsibility is well established (Cheer et al., 2021).

The volume of tourism activities is something difficult to manage. The ever-growing number of travels, tourists, stays, and etc. could seriously undermine, or delay, climate efforts. Possible efforts on limiting these numbers is quite controversial in terms of human rights, let alone their 'impossibility' considering the economic truth. However, policies aimed at optimal distribution of tourist flows in terms of their climate impact is not unrealistic. The tourism industry could direct the marketing efforts with this end in view. For example, tourism businesses in a destination could enhance their joint marketing efforts to attract potential tourists from nearby regions (to limit the transportation emissions) although this would require the presence of potential demand there.

The need for a major transformation of the industry seems quite clear (Stern, 2014b). However, this is the main ethical dilemma. Every economic system establishes some values. If one is to make a big change in that system or change the system altogether, then the ethical conflicts begin. Majorly changing the perceptions is not an easy task to achieve. However, the tourists as the active players in the tourism industry could make a facilitating impact on this sort of transformation. Either by choosing the climate-friendlier tourism products or protesting the other ones, they can force the businesses to operate in a climate-friendlier manner. Consumer demands are hard to reject in today's world (supposing that the market is competitive); thus, any positive change in consumer demands could result in a smoother transition to a climate-friendlier economy (see e.g. Goodwin & Francis, 2003). The cost of climate-friendlier production is another determinant in that regard since businesses will be willing to meet consumers' demands but the costs associated to achieving this goal should be within reasonable limits so that they can offer these products still at affordable prices.

Those natural resources used for tourism purposes is another issue. Natural resources are scarce and they are quite vulnerable in such a climate crisis. Moreover, climate-focused decisions about these scarce resources might trigger more ethical dilemmas considering vulnerable communities (Byskov et al, 2021). Thus, sustainable use of these resources is a task harder than ever. Responsible tourism practices may help for the sustainability to some extent.

Destinations might be competing for attracting visitors but they will also ideally try to attract those with higher sense of responsibility. Those tourists with higher sense of responsibility are likely to fit in climate-friendly tourism activities better. Thus, ethical dilemmas can be minimized.

The negative reactions to tourism activities, such as anti-tourism, due to the problems faced because of over-tourism, might help to develop a different ethical understanding of tourism. This could well change the perception of tourism from an industrial/consuming activity to a totally new state such as a way of communication, a social experience, and etc. The commercial side of tourism will probably still exist in these activities but the developed products could reflect the 'new idea of tourism'. In that case, climate-friendliness should also be embedded in this new understanding of tourism. On the other hand, over-tourism is not always perceived as something unpleasant by all stakeholders (Dodds & Butler, 2019). From the definition and scope of over-tourism to its metrics and management, there are many issues (Goodwin, 2017) which are essentially ethical. Thus, the debates on over-tourism issues are quite likely to continue for a long time.

Since climate change is a global issue, international collaborative projects among the tourism industry are vital (Coban Yildiz & Yildiz, 2024). This is also likely to trigger ethical questions such as 'who will do what and for whom'. The tourism industry's advantage, in that regard, is that tourism is already an international domain. Many tourism businesses have been experiencing international communication, which provides them with a collaborative ethical/mental background.

Economies, industries, and businesses should ideally focus on sustainable economic benefit. However, if they are fighting for survival then they focus on immediate economic gains. This is probably a bigger problem in developing countries or regions and in times of crisis (see e.g. Coban Yildiz & Yildiz, 2024). Thus, it is quite likely that the ethics of sustainability, also for the sake of climate-friendliness, will continue to be controversial.

Conclusion

As the threats of climate change become clearer, businesses are expected to consider them and act responsibly. However, the production and consumption activities and the profit maximization-focus usually result in more impact on nature. Indeed, the industrial activity is the major anthropogenic factor of climate change which, dialectically, became a serious threat to many industries such as tourism. Although there is such a consensus on what to do to minimize climate change, many of the necessary acts are difficult to apply due to ethical or financial reasons. The clash between the truth of business and environmental protection, including climate considerations, creates an ethical dilemma especially in the tourism industry. The tourism industry creates many jobs and offers many opportunities especially for the developing countries. On the other hand, transportation and accommodation facilities cause major CO_2 emissions. Moreover, the tourism industry uses many natural assets. The ethical dilemma usually arises when the financial interests – or the truth of business world – contradicts with minimizing environmental footprint and transitioning to a climate-friendlier system. The power of individuals, businesses, and governments are all limited; however, adoption of a common ethical understanding which prioritizes climate considerations is not impossible. The tourism industry with that great number of people, as consumers, employees, local people, and employers, involved in the actions has its advantages as well as disadvantages. Dilemmas will probably exist in any decision but it is not to reject that climate-friendlier advancement is possible in tourism.

References

Arnold, D. G. & Bustos, K. (2005). Business, ethics, and global climate change. *Business & Professional Ethics Journal, 24*(1/2), 103–130. https://doi.org/10.5840/bpej2005241/26

Ballantyne, A.G., Petersen, L.K., Fals, K.B., & Lysgaard, J.A. (2025). Climate-Friendly Vacations and Tourism. *Wiley Interdisciplinary Reviews: Climate Change, 16*(e929), 1–9. https://doi.org/10.1002/wcc.929

Barry, J., Mol, A.P.J., & Zito, A.R. (2013). Climate change ethics, rights, and policies: an introduction. *Environmental Politics, 22*(3), 361–376. https://doi.org/10.1080/09644016.2013.788861

Besio, C. & Pronzini, A. (2014). Morality, ethics, and values outside and inside organizations: An example of the discourse on climate change. *Journal of Business Ethics, 119*, 287–300. https://doi.org/10.1007/s10551-013-1641-2

Broome, J. (2008). The ethics of climate change. *Scientific American, 298*(6), 96–102. https://doi.org/10.1038/scientificamerican0608-96

Byskov, M.F., Hyams, K., Satyal, P., Anguelovski, I., Benjamin, L., Blackburn, S., Borie, M., Caney, S., Chu, E., Edwards, G., Fourie, K., Fraser, A., Heyward, C., Jeans, H., McQuistan, C., Paavola, J., Page, E., Pelling, M., Priest, S., Swiderska, K., Tarazona, M., Thornton, T., Twigg, J., & Venn, A. (2021) An agenda for ethics and justice in adaptation to climate change, *Climate and Development, 13*(1), 1–9. https://doi.org/10.1080/17565529.2019.1700774

Cheer, J. M., Ting, H., & Leong, C-M. (2021). Responsible tourism: A new era of responsibility. *Journal of Responsible Tourism Management, 1*(1), 1–17. https://doi.org/10.47263/JRTM.01-01-01

Coban Yildiz, G. & Yildiz, O.S. (2024). A Debate on Turkey Tourism's Adaptation to Climate Change. In: Remoaldo, P., Lopes, H., Ribeiro, V., Alves, J. (Eds.) *Tourism and Climate Change in the 21st Century*. Advances in Spatial Science. Springer, Cham. https://doi.org/10.1007/978-3-031-59431-1_9

Davidson, M. D. (2015). Climate change and the ethics of discounting. *Wiley Interdisciplinary Reviews: Climate Change, 6*(4), 401–412. https://doi.org/10.1002/wcc.347

DesJardins, J.R. (2023). Biocentrism. *Encyclopedia Britannica*. https://www.britannica.com/topic/biocentrism

Dodds, R. & Butler, R. (2019). The phenomena of overtourism: A review. *International Journal of Tourism Cities, 5*(4), 519–528. https://doi.org/10.1108/IJTC-06-2019-0090

Feucht, Y. & Zander, K. (2018). Consumers' preferences for carbon labels and the underlying reasoning. A mixed methods approach in 6 European countries. *Journal of Cleaner Production, 178*, 740–748. https://doi.org/10.1016/j.jclepro.2017.12.236

Garvey (2009). *The ethics of climate change: Right and wrong in a warming world.* Bloomsbury Publishing: London and New York.

Goodwin, H. (2017). The challenge of overtourism. *Responsible Tourism Partnership, Working Paper 4*, 1–19.

Goodwin, H. & Francis, J. (2003). Ethical and responsible tourism: Consumer trends in the UK. *Journal of Vacation Marketing, 9*(3), 271–284. https://doi.org/10.1177/135676670300900306

Gössling, S. & Peeters, P. (2015). Assessing tourism's global environmental impact 1900–2050. *Journal of Sustainable Tourism, 23*(5), 639–659. https://doi.org/10.1080/09669582.2015.1008500

Grasso, M. (2013). Climate ethics: with a little help from moral cognitive neuroscience. *Environmental Politics, 22*(3), 377–393. http://dx.doi.org/10.1080/09644016.2012.730263
Jamieson, D. (1992). Ethics, public policy, and global warming. *Science, Technology, & Human Values, 17*(2), 139–153.
Jamieson, D. (1996). Ethics and intentional climate change. *Climatic change, 33*(3), 323–336. https://doi.org/10.1007/BF00142580
Juvan, E. & Dolnicar, S. (2014) Can tourists easily choose a low carbon footprint vacation?, *Journal of Sustainable Tourism, 22*(2), 175–194. https://doi.org/10.1080/09669582.2013.826230
Michailidou, A. V., Vlachokostas, C., & Moussiopoulos, N. (2016). Interactions between climate change and the tourism sector: Multiple-criteria decision analysis to assess mitigation and adaptation options in tourism areas. *Tourism Management, 55*, 1–12. https://doi.org/10.1016/j.tourman.2016.01.010
Mittler, D. (2014). The changing ethics of climate change. *Ethics & International Affairs, 28*(3), 351–358. https://doi.org/10.1017/S0892679414000422
Patil, R. R. & Nahata, R. (2024). Ethical issues Climate Change. *Journal of Epidemiology and Public Health, 2*(1), 1–12.
Stern, N. (2014a). Ethics, equity and the economics of climate change paper 1: Science and philosophy. *Economics & Philosophy, 30*(3), 397–444. https://doi.org/10.1017/S0266267114000297
Stern, N. (2014b). Ethics, equity and the economics of climate change paper 2: Economics and politics. *Economics & Philosophy, 30*(3), 445–501. https://doi.org/10.1017/S0266267114000303
Türkeş, M. (2008). Küresel iklim değişikliği nedir? Temel kavramlar, nedenleri, gözlenen ve öngörülen değişiklikler. İklim Değişikliği ve Çevre, 1(1), 26–37.
United Nations (2025). *What is climate change?* https://www.un.org/en/climatechange/what-is-climate-change
Yildiz, O. S. & Coban Yildiz, G. (2023, 26–28 October). Climate adaptability for tourism in Rocky Cappadocia: A model proposal. 3rd International Congress of New Generations and New Trends in Tourism (pp. 151–166). Sakarya University of Applied Sciences, Sakarya, Türkiye.

Sustainable Sports Tourism and Event Management

Selda KOCAMAZ ADAŞ[1], *Fatih TURGUT*[2]

Introduction

Sustainable sports tourism and event management are two complementary fields. Integrating these areas with sustainability principles contributes to reducing environmental impacts, supporting local economies and fulfilling social responsibilities. Event management and sports tourism play an important role in this process by acting together and contributing to the development of sustainable sports tourism. In this context, considering sports tourism and event management in terms of sustainability is of great importance for both the future of the sector and the society. This book chapter explains the historical development of sports tourism, its types, economic impacts, environmental and social impacts, the concept of event management, and discusses how sustainability principles can be integrated into sports tourism and event management.

Sports Tourism

Sports tourism refers to the travels made by people to participate in, watch or engage in sports-related activities. This concept is often associated with visits to cities, countries or regions where major sporting events are held. However, sports tourism is not limited to professional sports events; It covers a much wider range such as amateur and educational sports activities, nature sports and health tourism. In recent years, sports tourism has become an important sector in terms of both economic contributions and cultural interaction. Definitions of sports tourism by different researchers are given:

---

1 PhD., Istanbul Aydın University, Faculty of Sport Sciences, Department of Sport Management, seldaadas@aydin.edu.tr

2 PhD., Istanbul Aydın University, Faculty of Sport Sciences, Department of Sport Management, fatihturgut@aydin.edu.tr

- ***Gibson (1998)*:** Leisure-based trips that take people away from their homes temporarily to participate in, watch, or visit physical activities.
- ***Hazar (2007)*:** Sports tourism is the tourism movement of people to participate in or watch sports events.
- ***Pigeasou (2004)*:** These are travels made by people to destinations and regions for non-commercial reasons in order to popularize sports culture.
- ***Gammon & Robinson (1997)*:** It is a combination of sports and tourism that involves actively or passively participating in sports activities during travels outside the region of permanent residence.
- ***Dogu (2006)*:** It is defined as an activity in which individuals benefit from tourism services in addition to participating in sports.
- ***Kurtzman (2005)*:** It is a type of tourism that emerges by using tourism as a tool to participate in sports activities.

Historical Development of Sports Tourism

The origins of sports tourism date back to ancient times. Major sporting events such as the Olympic Games held in Ancient Greece required both participants and spectators to travel from different regions, and thus the first examples of sports tourism emerged. As a matter of fact, it is known that travelling and accommodation activities were carried out during the ancient Olympics (Weed & Bull, 2004). The foundations of modern sports tourism were laid in the 19th century with the spread of major sports organisations, especially in Europe and North America. Since the late 1970s, the sector has grown rapidly and started to attract more attention as sports became a global industry. International events such as the Olympics, World Cup and Formula 1 have played an important role in the development of sports tourism. Sports tourism is categorised in different ways in the literature. These categories vary according to the purpose of the sport and the scope of the activities. The main types of sports tourism are as follows:

- ***Event-Based Sport Tourism***: It covers travelling to watch major sports organisations. Events such as the Olympics, World Cup and Formula 1 races are the most well-known examples of this type (Brouwers et al., 2012).

- ***Active Sports Tourism*:** This type means that people travel to experience a particular sport. Activities such as mountaineering, skiing, golf and cycling can be given as examples of this type of sports tourism (Weed, 2009).
- ***Nature Sports Tourism*:** It refers to tourism in which nature-related sports are performed and one is in touch with nature. Such sports include activities such as trekking, mountaineering, surfing and canoeing (Hudson, 2003).
- ***Health and Wellness Tourism*:** In recent years, health-oriented sports tourism has become an important area. Sports-related health activities, therapy and wellness-oriented holidays constitute health and wellness tourism (Foley et al, 2012).

Event Management

Event management is an interdisciplinary field that covers the processes of planning, organizing and executing events organized for individual or corporate purposes. Event management targets not only the logistics elements of an organization, but also the experience of the participants and the achievement of the event's purpose. According to Bowdin et al (2023), event management is "a professional process that includes the planning, implementation and evaluation of all types of events and has both operational and strategic dimensions."

Event management is the process of planning, organizing and managing large organizations, festivals, races and sporting events. Event managers coordinate the entire process from beginning to end, creating the necessary infrastructure, attracting participants and audiences, managing sponsorships and ensuring the events run smoothly.

Stages of Event Management

Managing sporting events involves many complex steps to organize a successful event. The event management process generally consists of the following stages:

- **Planning Process and Strategic Management**
  According to Allen et al (2005), the planning process in event management is the key to the success of the event, and in this process, goals must be

clearly defined, effective use of resources and detailed logistics planning must be made. First of all, the goals of the event are determined. Who is the target audience? What kind of audience will the event appeal to? Which sport or type of event will be organised? The answers to these questions lay the foundation for all planning. Getz (2007) views event management as not only an operational process but also a strategic tool used to achieve organizational goals. In this context, he emphasized that events should be integrated with areas such as branding, marketing and public relations.

- **Budgeting and Finance**
  A budget is determined for the event. This budget is tried to be balanced with sponsorships, ticket sales, media rights and other sources of income. In addition, the necessary financial sources for the event should be determined and sponsorship agreements should be made when necessary. A strong logistics planning is required for the events to be held successfully. Elements such as stadiums or event areas, stage setups, security measures, healthcare services and transportation should be arranged in detail.
- **Marketing and Communications**
  An effective marketing strategy must be created to publicize the event and attract participants. Social media, advertisements, sponsorships and public relations campaigns are critical at this stage.
- **Sponsorships and Collaborations**
  Sponsorships play a big role in financing the event and increasing brand value. Event management must make the right sponsorship agreements and manage these sponsorships effectively. The day of the event is a time when all plans will be put into action and everything must run smoothly. All stakeholders such as participants, spectators, media, security and volunteers must work in a coordinated manner. The event manager must ensure that everything happens smoothly and on time. Bowdin et al (2023) state that event management should focus not only on the organizational process but also on the experience that participants will gain from the event. In this context, it is important that the events are designed to be innovative and meet the expectations of the participants.
- **Evaluation and Reporting**
  According to Abson et al (2012), evaluating activities is critical to measuring success and creating learning opportunities for future

organizations. Participant feedback and measurable success criteria should be taken into account in this process. Post-event, organizers evaluate the event and draw lessons for future events. Participant feedback, financial reports, and media coverage are used to measure the success of the organization.

The Relationship Between Sports Tourism and Event Management

Sports tourism includes travel for the purpose of participating in, watching, or engaging in sports-related activities. Major events such as football matches, Olympics, World Cup, World Championships, Formula 1 races are the main examples of sports tourism. These activities can stimulate local economies by increasing tourist attraction, but can also lead to environmental degradation, conflicts with local people and depletion of resources. Event management covers the processes of planning, organizing and managing large organizations such as sports. These processes include all the logistics, marketing and organizational strategies necessary for the successful realization of the events. Aligning event management with sustainability principles is critical for long-term success.

The relationship between sports tourism and event management is a complementary structure. While sports tourism triggers economic growth by attracting major events, event management ensures that this process runs smoothly. However, without integrating both areas into sustainability principles, it will be difficult to achieve long-term success. Sports tourism and event management can develop strategies that support each other by integrating sustainability principles.

Concept of Sustainability

In the Our Common Future report published by the United Nations World Commission on Environment and Development (Brundtland Commission), the concept of sustainability was first discussed in detail in 1987. In this report, sustainability was defined as meeting the needs of the current generation and not jeopardising the capacity of future generations to meet their own needs (Aksu, 2011). Over time, this approach has evolved into a multifaceted perspective that encompasses economic, social, cultural and technological dimensions rather than being limited to the environmental dimension.

One of the main goals of sustainability is to protect natural resources and leave a healthy environment for future generations (Yücel et al., 2018). Therefore, it is emphasised that sustainability principles should be adopted and integrated into practices in dynamic areas such as sport (Lindsey, 2008). In addition, in order for sustainable development policies to be successful, environmental, economic and social elements must be evaluated in harmony (Nemli, 2001).

Sustainability is the combination of efforts to ensure a better future in the environmental, economic and social fields. Sports tourism and event management, in this context, is a growing sector and provides significant economic benefits. However, these sectors not only provide economic gain, but also bear great responsibility in terms of environmental impacts and social responsibilities. This chapter examines the relationship between sustainable sports tourism and event management, focusing on how these two fields can support each other to achieve a more sustainable sports management.

- ***Sustainable sports tourism*** is an approach that not only provides economic benefits but also takes environmental and social impacts into consideration. This includes factors such as protecting natural resources, preventing environmental pollution and supporting economic benefits for local communities.

Sustainable sports tourism is shaped by the following basic principles:

- ***Environmental Impact Reduction*:** Practices such as energy efficiency, waste management and water saving are important to minimize the negative effects of sports events on the environment.
- ***Supporting Local Economies*:** Major sporting events support regional development and increase revenues for local businesses(Jago et al, 2010).
- ***Social Responsibility and Participation*:** Participation of local people in events and promotion of local cultures constitute the social responsibility dimension of sports tourism.
- ***Sustainable event management*:** It takes into account environmental, economic and social factors at every stage of activities. This means setting

and implementing sustainability goals from planning to completion of events. Sustainability of events can be ensured by the following elements:

- ***Logistics and Infrastructure Planning*:** In areas where events are held, energy efficiency, optimization of water use and sustainable transportation must be provided.
- ***Education and Awareness Raising*:** By making attendees aware of environmental impacts, event organizers can deliver a more sustainable event experience. It is important to provide training to participants on topics such as waste recycling, energy saving and protection of natural resources.
- ***Social Responsibility*:** Social benefits can be achieved through strategies such as economic empowerment of local communities, promotion of social equality and protection of cultural heritage.

Economic Impacts of Sports Tourism

Sports tourism plays an important role in global economic development. Sports events organized around the world directly contribute to local economies by attracting tourists to these regions. Sports tourism, which is an indispensable part of the tourism industry, offers serious economic opportunities. The International Olympic Committee and the World Tourism Organization state that the interest in sports tourism is increasing day by day and this trend is accelerating. As emphasized by Hall (1992), supporting economic development is one of the most important functions of sports tourism. According to TURSAB's 2014 Sports Tourism Report, the economic size of the sports tourism sector worldwide has reached 1.2 trillion dollars. This sector has a share of 180 billion dollars in tourism revenues. While the general tourism sector is growing by 4–5% annually, the growth rate of sports tourism has reached 14% (TURSAB, 2014). Mega sports events such as the Olympic Games, World Championships, World Cup, European Championships, Mediterranean Games and Islamic Games provide significant economic and employment contributions to the country, region or city where they are held. In addition, events such as World Tour, National League, Formula 1 and Grand Slam are among the organizations that provide economic benefits (Hall & Piggin, 2001; Turgut, 2022; Preuss, 2013). Such large sports organizations attract many tourists to the regions where they are held. Incoming tourists directly contribute to the regional economy by staying in local hotels, eating at restaurants, shopping and participating in various events. These expenditures allow local trade to flourish and economic development to accelerate. Both large-scale mega sports events and smaller-scale sports tourism events offer important opportunities for the development of sports and tourism (Allen et al, 2000).

Environmental and Social Impacts of Sports Tourism

The environmental impacts of sports tourism appear to be more evident in areas where large events are held. Environmental impacts include factors such as depletion of natural resources, waste management problems, environmental impacts of transportation, and degradation of local ecosystems. Therefore, sustainable management of sports events is of great importance. In order to minimize the effects of sports events on the environment, it is necessary to provide energy saving, water management, recycling and environmentally friendly transportation options. In addition, organizing the areas where events are held without harming the environment is of great importance for the protection of nature. (Weed, 2009; Beaumont & O'Gorman, 2013) Sports tourism also affects social responsibilities and relations with local people. While major sporting events can promote local cultures, they can also lead to conflicts with local people. Organizing events in harmony with the local community is an important factor in preventing social problems (Foley et al, 2012).

Sustainable Sports Tourism and Event Management

In recent years, some studies have examined the relationship between environmental sustainability and sport, with a particular focus on sport tourism (Mallen, 2018). These studies have focussed more on the environmental impacts of destinations rather than the environmental impacts of sporting events (Moyle et al, 2018). In addition, literature reviews on the relationship between sport, tourism and environmental sustainability have examined the link between sport organisations and environmental sustainability (Mallen, 2018; Trendafilova & McCullough, 2018) and investigated the effects of leisure activities on sustainability (Ardoin et al, 2015; Vaugeois et al, 2017).

The need to prioritise environmental issues in future tourism research is particularly emphasised by Buckley et al. Sports tourism, as a rapidly growing sector (Alexandris and Kaplanidou, 2014), needs further research in terms of environmental sustainability (WTO, 2019). In this context, the current study aims to contribute to sports tourism management and develop environmental sustainability strategies. In particular, the ecological behaviour of sport participants and spectators (Lin and Lee, 2020; Trail and McCullough, 2020), strategies to reduce the environmental impacts of nature-based sport activities (Wolf et al., 2019) and sport events (Babi et al., 2019) are the focus of research in this field. Sports tourism should ensure that local people benefit from the events and strengthen local economies. Supporting local businesses, creating employment and supporting social projects contribute to the sustainability of sports tourism (Brouwers et al, 2012).

Results

Although sports tourism is a rapidly developing field, more research is needed on many issues. Research on the long-term effects of sports tourism on local people is limited. Studies on this subject can show how activities can be made more efficient in terms of social responsibilities and cultural interaction (Hall & Piggin, 2001). Research examining the environmental impacts of sporting events in more depth may contribute to the development of more sustainable event management strategies (Beaumont & O'Gorman, 2013). More studies should be conducted on the impact of technology on sports tourism, especially on the provision of environmentally friendly transportation options, the digitalization of events and the use of environmental monitoring tools (Weed, 2009).

Sports tourism is a sector that provides great economic benefits, but its environmental and social responsibilities must be taken into consideration. While organized large events strengthen local economies, they can also cause environmental and social problems. Therefore, managing sports tourism in a sustainable manner is of great importance not only for the future of the sector but also for the preservation of environmental and social balances. Sustainable sports tourism is based on the understanding of minimizing environmental impacts, developing strategies that benefit local people and providing long-term economic benefit. To reduce the environmental impact of sporting events, strategies such as recycling waste, energy efficiency practices, reducing carbon footprint and promoting environmentally friendly transportation options should be implemented. Sustainability is the combination of efforts to ensure a better future in the environmental, economic and social fields. In this context, sports tourism and event management, in addition to being an important economic sector, have a great responsibility in terms of environmental impacts and social responsibilities. As a result, sports tourism and event management, acting together, can contribute to a more sustainable future and play an important role in this process. In this context, strengthening the relationship between sustainable sports tourism and event management is of critical importance for both the development of the sector and the welfare of the society. New regulations and improved strategies in this field will be important steps to ensure both environmental and social sustainability.

References

Abson, E., Bladen, C., Kennel, J., & Wilde, N. (2012). Events Management: An Introduction. Oxon: Routledge.

Aksu, C. (2011), Sürdürülebilir Kalkınma ve Çevre, Güney Ege Kalkınma Ajansı.

Alexandris, K., & Kaplanidou, K. (2014). Marketing sport event tourism: sport tourist behaviors and destination provisions. Sport Marketing Quarterly, 23(3), 125+. https://link.gale.com/apps/doc/A383574164/AONE?u=anon~c149070&sid=googleScholar&xid=e5486ac3

Allen, G., Davis, K., Goodale, T., Hutanu, A., Kaiser, H., Kielmann, T., ... & Ullmer, B. (2005). The grid application toolkit: toward generic and easy application programming interfaces for the grid. Proceedings of the IEEE, 93(3), 534–550. Doi:10.1109/JPROC.2004.842755

Allen, J., O'Toole, W., McDonnell, I. & Harris, R. (2000). Festival & Special Event Management. Sydney: John Wiley & Sons Australia Ltd. https://doi.org/10.4324/9781003044963

Ardoin, N. M., Wheaton, M., Bowers, A. W., Hunt, C. A., & Durham, W. H. (2015). Nature-based tourism's impact on environmental knowledge, attitudes, and behavior: a review and analysis of the literature and potential future research. Journal of Sustainable Tourism, 23(6), 838–858. https://doi.org/10.1080/09669582.2015.1024258

Babí, J., Inglés, E., & Soler, S. (2019). Trail races in protected mountain areas and their effects on sustainable development. Journal on Protected Mountain Areas Research and Management, 11, 18–26. http://epub.oeaw.ac.at/eco.mont https://dx.doi.org/10.1553/eco.mont-11-2s18
Beaumont, N., & O'Gorman, K. (2013). "Sport Tourism and the Environment: Sustainable Tourism Development." Annals of Tourism Research, 42, 109–123. 10.1080/14775085.2016.1220863
Bowdin, GA, Allen, J., Harris, R., Jago, L., O'toole, W. & McDonnell, I. (2023). Events Management. Routledge.
Brouwers, J., De Bosscher, V., & Sotiriadou, P. (2012). Sport Management Review.
Buckley, R., Gretzel, U., Scott, D., Weaver, D., & Becken, S. (2015). Tourism megatrends. Tourism Recreation Research, 40 (1), 59–70. https://doi.org/10.1080/02508281.2015.1005942
Collins, A., Jones, C., & Munday, M. (2009). Assessing the environmental impacts of mega sporting events: Two options?. Tourism management, 30(6), 828–837. https://doi.org/10.1016/j.tourman.2008.12.006
Doğu, G. (2006). Sporun sosyal ve ekonomik faydaları. Sağlıklı Kentler Birliği Toplantısı, 27, 28.
Foley, M., McGillivray, D., & McPherson, G. (2012). Event policy: From theory to strategy. Routledge. https://doi.org/10.4324/9780203806425
Gammon, S., & Robinson, T. (1997). Sport and tourism: A conceptual framework. journal of sport Tourism, 4(3), 11–18. https://doi.org/10.1080/10295399708718632
Getz, D. 2007. Event studies: Theory, research and policy for planned events, Oxford: Elsevier.
Gibson, H. J. (1998). Sport tourism: a critical analysis of research. Sport management review, 1(1), 45–76. DOI: 10.1016/S1441-3523(98)70099-3
Hall, C. M. (1992a). Hallmark Tourist Events: Impacts, Management and Planning. London: Belhaven Press.
Hall, C. M., & Piggin, R. (2001). Tourism and World Heritage in OECD countries. Tourism Recreation Research, 26(1), 103–105. https://doi.org/10.1080/02508281.2001.11081182
Hazar, A. (2007). Spor ve turizm. Ankara: Detay Yayıncılık, 80–82.
Hudson, S. (2003). Sport and adventure tourism (pp. xxi+-324). New York: Haworth Hospitality Press.
Jago, L., Dwyer, L., Lipman, G., van Lill, D., & Vorster, S. (2010). Optimising the potential of mega-events: an overview.

International Journal of Event and Festival Management, 1(3), 220–237. DOI 10.1108/17852951011078023
Kurtzman, J. (2005). Sport tourism categories, Journal of Sport and Tourism, 10:1, 15–20. https://doi.org/10.1080/14775080500101502
Lin, Y. H., & Lee, T. H. (2020). How do recreation experiences affect visitors' environmentally responsible behavior? Evidence from recreationists visiting ancient trails in Taiwan. Journal of Sustainable Tourism, 28(5), 705–726. https://doi.org/10.1080/09669582.2019.1701679
Lindsey, I. (2008). Conceptualising sustainability in sports development. Leisure studies, 27(3), 279–294. https://doi.org/10.1080/02614360802048886
Mallen, C. (2018). The robustness of the sport and environmental sustainability literature and where to go from here. In B. McCullough & T. B. Kellison (Eds.), Routledge handbook of sport and the environment (pp. 11–35). New York: Routledge.
Moyle, B.D., Hinch, T.D., & Higham, J.E.S. (2018). Sports tourism and sustainable destinations. New York: Routledge.
Nemli, E. (2001). Çevreye duyarlı yönetim anlayışı. İstanbul Üniversitesi Siyasal Bilgiler Fakültesi Dergisi, (23–24), 211–224. https://dergipark.org.tr/en/download/article-file/5457
Pigeassou, C. (2004). Contribution to the definition of sport tourism, Journal of Sports and Tourism, 9 (3), 287–289. https://doi.org/10.1080/1477508042000320205
Preuss, H. (2013). The contribution of the FIFA world cup and Olympic games to the green economy. Sustainability (Switzerland), 5(8), 3581–3600. https://doi.org/10.3390/su5083581
Trail, G. T., & McCullough, B. P. (2020). Marketing sustainability through sport: Testing the sport sustainability campaign evaluation model. European Sport Management Quarterly, 20(2), 109–129. http://dx.doi.org/10.1080/16184742.2019.1580301
Trendafilova, S. & McCullough, B. P. (2018). Environmental sustainability scholarship and sport sector endeavours: A rapid review of the literature. Persuasive Social Science, 4(1), 1467256. https://doi.org/10.1080/23311886.2018.1467256
Turgut, F. (2022). Spor Turizmi. Spor Bilimleri IV, DİNÇ, Z. F. (Ed.). Akademisyen Kitabevi.
TURSAB Spor Turizmi Raporu, (2014). Erişim: 18 Şubat 2025, https://www.tursab.org.tr/dosya/12195/tursab-spor-turizmiraporu_12195_5670173.pdf

Vaugeois, N., Parker, P., & Yang, Y. (2017). Does leisure research contribute to sustainability? A systematic review of the literature. Leisure/Loisir, 41(3), 297–322. https://doi.org/10.1080/14927713.2017.1360151

Weed, M. (2009). "Sport Tourism: Participants, Policy and Providers." Elsevier. DOIhttps://doi.org/10.4324/9780080942117

Weed, M. and Bull, C. (2004) Sports Tourism: Participants, Policy and Providers. Oxford: Elsevier.

Weed, M. E., & Bull, C. J. (1997). Integrating sport and tourism: A review of regional policies in England. Progress in tourism and hospitality research, 3(2), 129–147.

Wolf, I. D., Croft, D. B., & Green, R. J. (2019). Nature conservation and nature-based tourism: A paradox?. Environments, 6(9), 104. https://doi.org/10.3390/environments6090104

World Commission on Environment and Development. (1987). Our common future. New York: Oxford University Press.

World Tourism Organisation. (2019). Sports tourism and sustainable development goals. https://webunwto.s3.eu-west-1.amazonaws.com/s3fs-public/2019-09/sporttourismandsdgs.pdf

Yücel, A. S., Eskiyecek, C. G., Çiftçi, İ., Tan, Ç., Savucu, Y., & Türkmen, A. (2018). Lisans Öğrencilerinin Sporda Sürdürülebilirliğe İlişkin Görüşlerinin Değerlendirilmesi (Firat Üniversitesi Örneği). Mukaddime, 9(2), 261–292. https://dergipark.org.tr/en/download/article-file/575686

Absence Tourism: Conceptual Framework, Impacts, and Future Perspectives

Arif Nİhat SAMUR[1], *Sinan GÖÇ*[2]

Introduction

Absence tourism, entails visiting locations that are "endangered" or "altered" owing to natural disasters, climate change, armed conflicts, or anthropogenic environmental degradation. Nonetheless, absence tourism ought not to be regarded solely as a travel phenomena, but rather as a notion that embodies human existential anxieties, collective memory, and ecological awareness. The initial instances of absence tourism were documented in antiquity by travellers intrigued by the "traces of destruction." Submerged by the eruption of Mount Vesuvius in AD 79, Pompeii was frequented by the aristocracy and explorers of the Roman Empire shortly after the catastrophe. Archaeologist Mary Beard observes that Pompeii, seen as a "time capsule frozen at the moment of death," elicited both trepidation and reverence among ancient visitors. The Egyptian pyramids and Mesopotamian ruins have been characterised in Hellenistic travellers' writings as embodying a "nostalgic longing for lost civilisations."

During the Middle Ages, the pilgrimages of Christians to sacred locations like Jerusalem, where Jesus was purportedly crucified, evolved into a perpetually fluctuating "tourism of destruction" as a result of wars and invasions. During the Crusades, European knights interpreted the ruins in Anatolia and the Middle East as "evidence of divine punishment," illustrating the convergence of loss tourism and religious reasons (Riley-Smith, 2005). In the Renaissance, fascination with old Roman and Greek ruins led to the emergence of a trend termed "ruin tourism." During the 18th century, the Grand Tour tradition facilitated European aristocrats'

---

1 PhD., Nevşehir Hacı Bektaş Veli University, Kozaklı Vocation School, Health Tourism Management, asamur@nevsehir.edu.tr

2 PhD., Nevşehir Hacı Bektaş Veli University, sinangoc@nevsehir.edu.tr

visits to sites such as Pompeii, the Acropolis of Athens, and the Roman Forum. The trips were driven by both a historical curiosity and a romantic viewpoint that highlighted the "melancholic beauty of the ruins." The Industrial Revolution, conversely, dramatically altered the dynamics of ruin tourism. The rapid urbanisation and industrialisation of the 19th century rendered the degradation of the natural environment evident. The Great Stink of 1854 in London exemplified the early recognition of the necessity to safeguard natural resources due to the pollution of the Thames River by industrial effluents. The classification of regions such as Yosemite Valley in the United States as "national parks" exemplified the intricate dynamics of humanity's connection with nature: amidst the rapid pace of industrialisation, areas at risk of obliteration were being conserved (Nash, 2001).

The 20th century saw the establishment of absence tourism in its contemporary form. The two World Wars, the Holocaust, and nuclear catastrophes rendered human-induced devastation evident on a global scale. Locations such as the Auschwitz-Birkenau Concentration Camp, the Hiroshima Peace Memorial, and Chernobyl have emerged as emblems of "dark tourism." Lennon and Foley (2000) characterise the visit to these locations as a "caution regarding the vulnerability of modernism." The 1986 Chernobyl accident marked a pivotal moment that transformed the relationship between absence tourism and popular culture. Three decades post-disaster, visits to the area regained prominence following HBO's Chernobyl series. In 2023, 40% of the 124,000 visitors to the region were influenced by this series (Smith & Doe, 2023). This scenario underscores the media's pivotal role in absence tourism. The COVID-19 pandemic has induced significant alterations in the tourism sector, with absence tourism also experiencing its part of this transition. From 2020 to 2022, as physical visits declined owing to travel limitations, virtual tours and augmented reality (AR) applications gained popularity. In 2021, virtual tours of Chernobyl attracted over 2 million online participants (Digital Tourism Report, 2022). Post-pandemic, it has been noted that travellers have gravitated towards journeys centred on "the quest for meaning" and "resilience." For instance, 67% of visitors to Glacier National Park expressed the "desire to preserve these locations for future generations" (National Park Service, 2023).

The Conceptual Framework of Absence Tourism

To understand the concept of absence tourism, it is crucial to analyse the fundamental ideas related to this type of tourism. Initially, it is crucial to examine the impact of the concept of "absence" on tourism and the categorisation of these experiences. In tourism, deprivation refers to individuals deliberately forgoing contemporary amenities to seek experiences defined by scarcity and hardship (Anderson, 2020). A important concept is "authenticity." The concept of authenticity, as defined by MacCannell (1976), clarifies tourists' intrigue with genuine and untainted encounters. In the context of absence tourism, the quest for authenticity relates to travellers' desire to evade the artificiality of modern life and partake in experiences considered "genuine" and "unrefined" (Wang, 1999). The concept of the "experience economy" is essential for understanding deprivation tourism. The experience economy theory, articulated by Pine and Gilmore (1999), asserts that customers are increasingly pursuing meaningful and personalised experiences rather than merely acquiring items or services. Deficient tourism, as a component of the experience economy, enables tourists to attain personal growth through experiences characterised by obstacles and deprivation (Smith & Puczkó, 2010).

Absence tourism conceptually interacts with dark tourism, historical tourism, and eco-tourism. Nonetheless, the delineations between these categories are ambiguous. These notions are assessed within a comprehensive framework that includes the reinterpretation of history, ecological awareness, and calamities as components of a tourism experience. The indistinct boundaries between these categories require a reevaluation of absence tourism from an alternative perspective. Lennon and Foley (2000) characterise dark tourism as "visiting places of death and tragedy," while they assess absence tourism from a more expansive viewpoint, encompassing not only tragedy but also themes of environmental degradation and cultural erosion. The burgeoning eco-tourism initiatives in the Amazon Rainforest seek to enhance awareness of biodiversity depletion while bolstering local economies (Brondizio et al., 2019).

Stone (2006) characterises gloomy tourism as "the engagement with locations linked to death, suffering, and violence." Locations such as Auschwitz and the September 11 Memorial fall within this category. This form of tourism involves individuals visiting locations associated with historical traumas, including battlefields, genocide sites, and areas affected by natural disasters. Dark tourism is primarily influenced by the endeavour to commemorate and comprehend horrific historical occurrences. This initiative connects the past to the present with the concept of time and prompts visitors to recall historical events. This procedure is typically predicated on a linear conception of time, wherein an event that transpired in the past is regarded as immutable within a designated temporal framework. Absence tourism encompasses not just human-induced disasters but also locations that have been lost due to natural phenomena.Heritage tourism pertains to the conservation and display of cultural, historical, and archaeological heritage. Structures, old towns, museums, and traditional lifestyles designated as global heritage by UNESCO are the essential components of heritage tourism. This form of tourism is primarily intended to conserve the past and transmit it to future generations (Timothy & Boyd, 2003). Heritage tourism, like to dark tourism in its concentration on the past, underscores the perpetuation of cultural values. Nonetheless, the notion of time is also examined sequentially; it is regarded as immutable in a historical context and is conveyed to the present solely as a story or tangible vestige. Heritage tourism generally emphasises conserved or repaired sites, whereas absence tourism pertains to locations under risk of vanishing or undergoing transition.Eco-tourism is a form of tourism that is founded on sustainability and environmental consciousness. Tourists in natural settings have a sense of maintaining ecological equilibrium. This form of tourism has typically emerged as a response to modern society' estrangement from nature (Weaver, 2001). In eco-tourism, the concept of time is anchored in the continuity of the natural environment from the past through the present and its prospects for the future. Thus, this form of tourism is structured within a specific timescale, and the transformation of nature is analysed from a linear temporal viewpoint. Eco-tourism emphasises the preservation of the natural environment, while absent tourism underscores the degradation of both natural and cultural assets. The non-linear experience of time is the primary characteristic that differentiates absence tourism from other analogous notions. It confronts the past. It evokes memories of the devastation and losses endured in history. A traveler in Chernobyl encounters a historical confrontation by observing the remnants of the 1986 nuclear disaster (Smith & Doe, 2023). It pertains to comprehending the current situation. Visitors comprehend modern environmental and cultural challenges by exploring locations at risk of extinction. A tourist witnessing the retreating glaciers in Glacier National Park perceives the concrete impacts of climate change (Lemieux et al., 2018). It is a prediction of the future. Absence tourism acts as a cautionary signal to travellers over the future. A tourist exploring the Amazon Rainforest anticipates the future consequences of deforestation (Brondizio et al., 2019). It constitutes an existential contemplation. This impression prompts visitors to contemplate their mortality and humanity's connection to the earth. A tourist visiting Pompeii contemplates the future of their own society upon witnessing the devastation of an old civilisation (Beard, 2008). It constitutes a shared obligation. The non-linear experience of time amplifies visitors' consciousness of shared responsibility. A tourist at the Great Barrier Reef discovers the causes of coral bleaching and engages in environmental activism (Spalding et al., 2021).

Absence tourism thematically interacts with both memory tourism and loss tourism. These interconnected frameworks aim to enhance historical awareness and safeguard collective memory (Jones, 2020). Memory tourism emphasises the role of physical sites in identity development, whereas absence tourism presents the contradiction that the disappearance or inaccessibility of locations may serve as its main allure (Garcia, 2022). Absence tourism is a complex phenomena that entails travelling to locations characterised by experiences of "non-being" or "deprivation," therefore questioning conventional tourism models. This conceptual framework extends beyond mere physical absence to incorporate cultural, historical, ecological, and even metaphysical dimensions. Rooted in the philosophical concept of nothingness, absence tourism facilitates an existential confrontation with void spaces, compelling participants to engage in profound ontological questioning. The observatories in the Atacama Desert of Chile provide travellers the chance to encounter the universe's boundless expanse (Guzmán, 2021). Aligned with Bauman's (2000) concept of "liquid modernity," absence tourism represents the quest for meaning in a contemporary society characterised by ambiguity. Post-industrial cities such as Detroit epitomise the decline of capitalism, with tourists perceiving this decline as a "documentary reality" (Edensor, 2005). In Europe and the United States, deprivation tourism is frequently linked to "nostalgia" and "aesthetic melancholy." The deserted mining town of Charleroi in Belgium attracts photographers (Van der Auwera, 2017). Poverty tourism in Africa is linked to the quest for post-colonial identity.

The Kolmanskop Ghost Town in Namibia epitomises the forsaking of the German colonial heritage (Ndlovu, 2021). In Japan, the notion of "ma" (space) is a cultural principle that encompasses architecture and art. The dry stone garden at Ryoan-ji Temple in Kyoto enables visitors to appreciate the "beauty of emptiness" (Kawabata, 2022). In the contemporary world, the swift globalisation and homogenisation of lifestyles have prompted tourists to seek authentic experiences that have become obsolete (Cohen, 1988). In this sense, absence tourism signifies travellers' pursuit of a "lost paradise" (MacCannell, 1999). Tourists seek to engage with both the present and the remnants of the past.

Absence tourism exemplifies contemporary society' endeavours to reconfigure their connection with history. In this environment, absence tourism functions as a mechanism for individuals to enhance their quest for identity and belonging (Brown, 2019). Absence tourism warrants a rigorous examination. This form of tourism may occasionally result in the romanticisation of history and the distortion of factual events. Locations such as battlefields or sites of natural disasters may be marketed as tourist attractions, potentially neglecting the emotional consequences they entail (White, 2021). Absence tourism influences group identity. The remnants of the Berlin Wall draw tourists as an emblem of the Cold War (Garcia, 2022). The commercialisation of such sites may undermine their credibility. The "Rose of Sarajevo" in Bosnia, although a poignant reminder of wartime suffering, may unsettle local residents due to the insensitivity exhibited by tourists (Causevic & Lynch, 2011).

The theoretical underpinnings of absence tourism are interconnected with postmodern tourism theories; it has arisen as a reaction to the rampant consumption, technological dependency, and environmental deterioration engendered by contemporary existence. Postmodern tourism highlights travellers' pursuit of authentic experiences, their inclination to engage with many cultures, and their aspiration to transcend conventional tourist activities (Urry, 1990). In this perspective, absence tourism is regarded as a facet of postmodern tourism, addressing tourists' pursuit of authentic and distinctive experiences (Richards & Wilson, 2006). In the postmodern era, individuals' want to evade the pressures of urban existence, re-establish a connection with nature, and pursue a simplistic lifestyle constitutes the fundamental impetus for this sort of tourism (Cohen & Cohen, 2012). The increasing global focus on sustainability and environmental consciousness has been a significant driver for the emergence of absence tourism (Gössling et al., 2009). This type of tourism is fundamentally rooted in minimalist ideas and voluntary simplicity. The minimalist paradigm opposes hyper-consumerist modernity by promoting the rejection of non-essential luxury and emphasising the fulfilment of basic human needs (Schor, 1998). In this context, absence tourism represents the quest for experience alternatives to traditional tourism, notably emphasising low-consumption behaviours, harmonious coexistence with natural ecosystems, and the promotion of simple living. (Gössling, 2006).

To mitigate the adverse effects of absent tourism on destination communities, the implementation of sustainable tourist practices is becoming increasingly essential. This method prioritises two objectives: promoting socioeconomic advancement for local communities while concurrently safeguarding cultural heritage and natural ecosystems (Hall, 2011). The efficient sustainable management of absence tourism is essential for mitigating its potentially disruptive impacts on community structures (Weaver, 2001). For absence tourism to expand responsibly, future advancements must emphasise two essential components: the significant involvement of local stakeholders in decision-making processes and the equitable allocation of tourism-generated earnings. Strategic actions, including the integration of indigenous viewpoints in tourism planning and the execution of comprehensive cultural preservation projects, can substantially improve the long-term sustainability of absence tourism. (Gössling, 2006). Furthermore, it is essential to routinely perform socioeconomic and cultural evaluations to assess the effects of tourism on local community frameworks (Richards & Wilson, 2006).

Results

Absence tourism reflects the intricate emotional, intellectual, and social requirements of individuals. This phenomenon presents an opportunity to address the past and safeguard the future, while also posing ethical and ecological risks. The tourist sector must establish transparent policies to uphold this equilibrium and inform travellers about the "responsibility of travel."Absence tourism provides a vital insight on humanity's connection with the Earth. This phenomena highlights the impermanence of natural and cultural legacy, compelling us to assume responsibility for creating a more sustainable environment for future generations. Absence tourism serves as a substantial economic catalyst for destination areas, delivering essential cash that bolsters local economies and conservation initiatives. Nonetheless, these financial advantages are counterbalanced by significant ecological costs and sustainability issues. This tourism model offers quantifiable economic benefits while simultaneously presenting significant dangers and concealed costs. These difficulties appear in four principal dimensions: overtourism demands, inequitable wealth distribution, ecological deterioration, and infrastructure overextension. To tackle these intricate difficulties, location managers and tourist operators must establish thorough, transparent governance frameworks and execute evidence-based sustainable practices. Absence tourism has become a vital competitive differentiation for locations in the modern tourism environment. To attain market success, locations must strategically use Porter's three generic strategies: product differentiation, cost leadership, and market focus. Moreover, digital transformation and new experiential design have become essential instruments for improving destination competitiveness in this specialised area. In the future, sustainable development principles and stringent ethical frameworks will serve as the core pillars of competitive strategy in absence tourism, necessitating that destinations reconcile commercial objectives with environmental stewardship and socio-cultural preservation. Absence tourism yields both beneficial and adverse repercussions regarding its economic, social, and cultural implications. This type of tourism can generate economic benefits for destinations; nevertheless, it also presents ethical dilemmas, such as the commodification of poverty, violations of local inhabitants' privacy, and the intensification of social inequalities. The view of poverty as a source of amusement or experience by tourists raises ethical concerns with contemporary tourism.

The "ruin tours" in Detroit face criticism for romanticising poverty. Likewise, tours conducted in the radioactive zones of Fukushima obscure the trauma experienced by the local population (Yamaguchi, 2021). Tourists visiting Greenland to observe the receding glaciers contribute to carbon emissions via their flights. This scenario presents the danger of climate tourism "engineering its own extinction." Cultural tours featuring Indigenous culture in Australia may constitute "cultural appropriation" if executed without the involvement of Indigenous communities. Virtual Chernobyl tours in the Metaverse may diminish the gravity of real tragedy, transforming it into digital entertainment (Lee, 2023). Nonetheless, international collaboration is essential to create an ethical framework for absence tourism on a global scale. The 2021 UNESCO climatic Change and World Heritage Guide advises the use of climatic scenarios into tourism planning. Absence tourism exemplifies humanity's dual capacity for destruction and protection. This phenomena, extending from antiquity to the digital era, compels individuals to confront history and reconsider the future. To ensure the sustainability of this invitation, the tourism sector must reevaluate its ethical parameters and prioritise the need of local communities and ecosystems.Absence tourism not only delivers personal experiences but also confers multiple advantages to communities. This form of tourism promotes engagement with local populations, facilitating better understanding of their lifestyles and cultures (Mowforth & Munt, 2015). This therefore enhances tourism's impact on local economies and promotes cultural sustainability (Richards & Wilson, 2006).

Absence tourism is a significant form of tourism that fosters economic development on both local and global scales. Data from the World Tourism Organisation and the World Bank indicate that this form of tourism enhances local revenues, generates jobs, and fosters infrastructure development. Nonetheless, meticulous planning and administration are essential for the longevity of this tourism model. When properly managed, absence tourism can yield substantial benefits for local economies and the global tourism sector. The escalating climate change and environmental concerns are prompting tourists to seek vacation options that minimise carbon footprints and are ecologically sustainable. In this setting, absence tourism emerges as a form of tourism that corresponds with environmental sustainability objectives (Gössling et al., 2009). The extensive use of absence tourism may significantly mitigate the environmental implications of the tourism industry. Absence tourism can be carefully employed to mitigate the strain of overtourism on tourist locations and to bolster local economy. This form of tourism facilitates more sustainable management, particularly in places with high visitor traffic (Honey, 2008). Consequently, absence tourism can be regarded not merely as a form of tourism but also as a tactic that bolsters sustainable development objectives. Absence tourism's influence on the social frameworks of destinations can enhance the economic autonomy of the local populace while simultaneously augmenting external reliance. Reliance on tourism revenue might render local economies unstable and susceptible to economic swings (Weaver, 2001). The incorporation of the local populace into the tourism industry may result in the disregard of traditional occupations and crafts (Cohen, 2012)

References

Anderson, J. (2020). Minimalist tourism: An exploration of self-imposed scarcity in travel experiences. Tourism Review, 75(3), 456–472
Bauman, Z. (2000). Liquid modernity. Polity Press.
Beard, M. (2008). Pompeii: The life of a Roman town. Profile Books
Brown, T. (2019). Globalization and the rise of absence tourism. Journal of Travel Research, 58(4), 678–692
Brown, L. (2021). Submerged heritage: Tourism and the Titanic. Maritime Tourism Review, 7(3), 112–125
Brondizio, E. S., Settele, J., Díaz, S., & Ngo, H. T. (2019). Amazon deforestation and eco-tourism. Science, 365(6452), 345–348
Causevic, S., & Lynch, P. (2011). Phoenix tourism: Post-conflict tourism role. Annals of Tourism Research, 38(3), 780–800.
Cohen, E. (1988). Authenticity and commoditization in tourism. Annals of Tourism Research, 15(3), 371–386.
Cohen, E., & Cohen, S. A. (2012). Current sociological theories and issues in tourism. Annals of Tourism Research, 39(4), 2177–2202.
Digital Tourism Report. (2022). Virtual tours and post-pandemic tourism.
Edensor, T. (2005). Industrial ruins: Space, aesthetics and materiality. Berg Publishers.
Garcia, M. (2022). Virtual absence: Technology and memory in tourism. Tourism Geographies, 24(2), 210–225.
Gössling, S. (2006). Tourism and Global Environmental Change: Ecological, Social, Economic and Political Interrelationships. Routledge.
Gössling, S., Peeters, P., & Scott, D. (2009). Consequences of climate policy for international tourist arrivals in developing countries. Third World Quarterly, 30(7), 1209–1227.
Guzmán, A. (2021). Stargazing in the Atacama: Tourism and existential void. Astro-Tourism Review, 12(4), 112–125.
Hall, C. M. (2011). Policy learning and policy failure in sustainable tourism governance: From first-and second-order to third-order change? Journal of Sustainable Tourism, 19(4–5), 649–671.
Jones, L. (2020). Memory and tourism: The role of absence. Annals of Tourism Research, 82, 102–115.

Kawabata, Y. (2022). The philosophy of "ma" in Japanese tourism. Asian Cultural Studies, 19(3), 45–60.
Kumar, R., & Aladin, N. (2020). Ecological disasters and tourism: The case of the Aral Sea. Environmental Tourism Journal, 12(4), 234–250.
Lee, S. (2023). Virtual voids: Tourism in the Metaverse. Future Tourism Trends, 7(1), 88–104.
Lennon, J., & Foley, M. (2000). Dark tourism: The attraction of death and disaster. Cengage Learning.
Lemieux, C. J., Groulx, M. W., Bocking, S., & Beechey, T. J. (2018). Last chance tourism: Adaptive responses to vanishing landscapes. Tourism Geographies, 20(5), 780–794.
MacCannell, D. (1976). The tourist: A new theory of the leisure class. University of California Press.
MacCannell, D. (1999). The Tourist: A New Theory of the Leisure Class (2nd ed.). University of California Press.
Mowforth, M., & Munt, I. (2015). Tourism and Sustainability: Development, Globalisation and New Tourism in the Third World. Routledge.
Nash, R. (2001). Wilderness and the American mind. Yale University Press.
National Park Service. (2023). Visitor demographics and motivations. https://www.nps.gov
Ndlovu, D. (2021). Post-colonial absence in Namibian tourism. African Heritage Journal, 14(2), 88–102.
Urry, J. (1990). The Tourist Gaze: Leisure and Travel in Contemporary Societies. Sage.
Brown, T. (2019). Globalization and the rise of absence tourism. Journal of Travel Research, 58(4), 678–692. doi:10.1080/14616688.2019.1630675
Pine, J. B., & Gilmore, J. H. (1999). The experience economy: Work is theater & every business a stage. Harvard Business Press.
Richards, G., & Wilson, J. (2006). Developing creativity in tourist experiences: A solution to the serial reproduction of culture? Tourism Management, 27(6), 1209–1223.
Riley-Smith, J. (2005). The Crusades: A history. Bloomsbury Academic.
Schor, J. B. (1998). The Overspent American: Why We Want What We Don't Need. Harper Perennial.
Seaton, A. V. (1996). Guided by the dark: From thanatopsis to thanatourism. International Journal of Heritage Studies, 2(4), 234–244.
Smith, M. K., & Puczkó, L. (2010). Health and wellness tourism. Routledge.

Smith, J., & Doe, A. (2023). Post-pandemic trends in dark tourism: A case study of Chernobyl. Journal of Tourism Studies, 15(2), 45–60.
Stone, P. R. (2006). A dark tourism spectrum: Towards a typology of death and macabre-related tourist sites. Tourism: An International Interdisciplinary Journal, 54(2), 145–160.
Thompson, R. (2023). Climate change and park visitation. Journal of Environmental Management, 299,
Thompson, J. L., & Haas, G. (2023). From spectators to advocates: Climate change engagement through nature-based tourism. Environmental Psychology and Social Ecology Review, 45(3), 112–129.
Timothy, D. J., & Boyd, S. W. (2003). Heritage tourism. Pearson Education.
UNESCO. (2023). Climate change and world heritage: A practical guide. https://whc.unesco.org
UNESCO. (2023). World Heritage and sustainable tourism programme. https://whc.unesco.org
Van der Auwera, S. (2017). The aesthetics of decay in Charleroi. European Urban Photography, 9(1), 22–35.
Wang, N. (1999). Rethinking authenticity in tourism experience. Annals of Tourism Research, 26(2), 349–370
Weaver, D. B. (2001). Ecotourism as mass tourism: Contradiction or reality? Cornell Hotel and Restaurant Administration Quarterly, 42(2), 104–112.
Weaver, D. B. (2001). The encyclopedia of ecotourism. CABI Publishing.
White, R. (2021). Ethical considerations in absence tourism. Tourism Management, 83, 104–118.
Yamaguchi, T. (2021). Fukushima and the ethics of absence tourism. Japanese Journal of Ethical Tourism, 14(3), 77–89.

Experience Innovation in Tourism: Virtual Reality (VR) and Augmented Reality (AR) Applications

Yasemin TEKİN[1]

Introduction

Technological innovations play a crucial role in shaping the tourism sector in the contemporary business environment. Digitalization, which influences various processes from production to service delivery, enables the creation of distinctive touristic products and services, offering guests unique experiences. By integrating traditional tourism activities with technology, businesses leverage digitalization to enhance their offerings, delivering guests an innovation-oriented experience enriched with memorable moments.

Grounded in technological advancements that reshape the services offered by tourism businesses in the contemporary era, this study aims to compile and assess Virtual Reality (VR) and Augmented Reality (AR) applications, key drivers of experience innovation in tourism. In addition, the study seeks to contribute to tourism science and industry stakeholders by offering innovative recommendations.

The research data were obtained through document analysis, literature review, and an examination of websites and social media platforms. The findings suggest that VR and AR applications, which enhance experience innovation for guests, offer substantial benefits to tourism businesses across various aspects.

Literature Review

Innovation was first defined by economist Joseph Alois Schumpeter as "the introduction of a novel product or a new variation of an existing product, the implementation of new production and sales methods, the expansion into new markets, the discovery of new raw material sources, the restructuring

[1] PhD., Balıkesir University, Tourism Faculty, Department of Torism Management, yasemin.tekin@balikesir.edu.tr

of enterprises, and the driving force of the economy and development" (Śledzik, 2013).

The continuous advancement of technology has made innovation and digitalization integral to tourism activities. The differentiation of touristic products, services, and activities through technological support and experience innovation serves as a strategic tool for businesses seeking success and sustainability under highly competitive market conditions.

Experience innovation refers to the practice of enhancing consumers' interactions with existing products and services to make them enjoyable, pleasant, and reliable. The goal of experience innovation is to cultivate admiration for the product or service offered, giving consumers the impression that "this is unavailable elsewhere." In the context of tourism, experience innovation involves the design of traditional tourism activities or leveraging tourism potential to create unforgettable experiences for guests (Soylu, Ceylan & Özdipçiner, 2019).

Providing unforgettable experiences is central to the tourism industry. Extraordinary tourist experiences can be delivered through exceptional services at luxury venues, gourmet meals prepared by top chefs, or even by offering unique experiences such as camping in the Sahara Desert (Pizam, 2010).

Virtual Reality (VR) provides users an immersive, virtual experience through a head-mounted device, while Augmented Reality (AR) overlays virtual images onto real-world environments. Both VR and AR serve as promotional and marketing tools that positively influence consumer purchase intentions (Chen, Wu & Lai, 2023) and are employed to create and deliver interactive, visually engaging, and memorable experiences (Chen, Wu & Lai, 2023; Nayyar, Mahapatra, Le & Suseendran, 2018). VR technology provides tourists with a bird's-eye view of a destination, allowing them to notice and experience its details in a virtual environment, whereas AR enables users to plan their trip, explore the destination in advance, and access location-specific information (Nayyar, Mahapatra, Le & Suseendran, 2018). While VR is typically utilized off-site, either before or after a visit, AR is employed on-site, during the actual visit, to enhance the experience (Yılmaz & Karamustafa, 2022).

Virtual Reality (VR) technologies provide tourism professionals with a wide range of applications in areas such as tourism policy planning, marketing, attractions, leisure and heritage tourism, and tourism education (Huang, Backman, Backman & Chang, 2016). As an effective marketing tool, VR enables tourism businesses and destinations to present their products and services in vivid and memorable ways, reaching potential global audiences beyond traditional markets. The immersive nature of VR experiences substantially enhances the perceived value of a tourism destination, leading to increased visitor numbers and tourism revenue (Jinathissa, Manjaree & Langseth, 2025).

With interactive VR devices, users can navigate and control their movements in a multidimensional virtual world, experiencing it as if they were physically present an imagined environment, rather than merely viewing it on a flat screen (Xi & Hamari, 2021). Through virtual tourism enabled by VR, visitors can explore a variety of destinations, including space and fictional, dangerous, or restricted areas. Additionally, VR allows users to access advance information about a trip, helping to alleviate uncertainty before embarking on the actual journey (Chen, Wu & Lai, 2023).

Virtual Reality (VR) brings immersive environments to life, allowing users to virtually explore destinations and experiences from their current location, thereby influencing their travel decisions (Jinathissa, Manjaree & Langseth, 2025). While VR cannot fully replace real tourism activities, it offers significant benefits, particularly in addressing the economic challenges faced by developing countries. Moreover, VR serves as an alternative method for individuals with disabilities or health conditions to explore global tourism resources (Chen, Wu & Lai, 2023). According to a 2017 report, the VR industry is projected to reach USD 692 billion by 2025, with widespread applications across various fields, including education, medicine, healthcare, architecture, sports, television, film, and music (Nayyar, Mahapatra, Le & Suseendran, 2018).

The origins of Virtual Reality (VR) devices date back to 1962 with the creation of the "Sensorama Simulator," which allowed users to experience riding a motorcycle in a virtual environment, complete with realistic movements, sounds, smells, and airflow. Since then, VR technology has undergone significant transformations. Today, driving and flight simulators are featured in Disneyland and various amusement parks, while multidimensional cinemas provide similar immersive experiences (Nayyar, Mahapatra, Le & Suseendran, 2018).

Augmented Reality (AR), enhancing real-world scenes with computer-generated images and additional data for interpretation and entertainment, was first recognized in 1999, initially emerging through cinema technology. AR, which can be experienced via smartphones or camera-mounted glasses, is an effective tool for enhancing tourism experiences, events, and museum visits. It provides visitors with the ability to explore beyond temporal, spatial, and linguistic limitations (Chen, Wu & Lai, 2023; Hjalager, 2015; Yılmaz & Karamustafa, 2022). Moreover, AR offers several business advantages, including promotion, encouraging purchases, providing a competitive edge, and improving efficiency and performance (Demirezen, 2019).

While traditional methods, such as web pages or printed brochures, require one to imagine a hotel, virtual tours now allow potential guests to explore hotels, restaurants, spas, and fitness centers in detail. For travelers planning their trips, the food and beverage offerings and entertainment options at a destination are crucial. Before making a travel decision, tourists can take a virtual tour of a restaurant, choose their preferred table setting, explore virtual menus in multiple language options, and engage in an interactive experience through Augmented Reality (AR) technology. Similarly, AR technology allows users to view hotel rooms, including their size, layout, view, and floor plans. In today's context, where sustainable tourism is a priority, environmentally conscious hotels use virtual platforms to showcase their green building materials, herb gardens, and sustainable practices, thereby fostering guest loyalty (Nayyar, Mahapatra, Le & Suseendran, 2018).

Tuscany+, the first smartphone application designed for tourism and utilizing Augmented Reality (AR) technology, offers an interactive and real-time guide. The app enhances the visitor experience by providing information on physical sites visible through the smartphone camera. Another globally recognized example of AR in tourism is the game Pokémon GO. In this AR-based game, players navigate the real world while holding their smartphones, discovering and capturing virtual characters, known as "pocket monsters," at their current locations, and collecting points (Nayyar, Mahapatra, Le & Suseendran, 2018). In addition, "PokéStops" within the game have been strategically placed at tourist destinations. Tour operators have developed new tours incorporating these PokéStop points, allowing participants to engage in an educational yet entertaining experience as they collect Pokémon, blending learning with enjoyment (Nayyar, Mahapatra, Le & Suseendran, 2018).

The Metaverse, a concept that enables users to connect in both real and virtual environments through technologies such as Virtual Reality (VR), Augmented Reality (AR), artificial intelligence, avatars, and holograms, derives its name from the combination of “meta” and “verse,” meaning “beyond the universe.” It represents a parallel and virtual universe designed to extend physical spaces, products, and services. The Metaverse consists of immersive, three-dimensional worlds that offer tourists a unique and captivating opportunity to explore destinations from their location within a virtual environment (Himanshu, Aggarwal, Gupta, Dhingra & Kumar, 2025). Virtual technologies such as VR, AR, and the Metaverse are powerful tools for destination marketing, encouraging potential tourists to visit these destinations by offering immersive experiences (Huang, Backman, Backman & Chang, 2016).

Applications in Tourism Businesses

The COVID-19 pandemic has accelerated the integration of innovative tourism practices with technologies such as Virtual Reality (VR) and Augmented Reality (AR), offering alternative tourism opportunities during crises like the pandemic (Roman, Kosiński, Bhatta, Niedziółka & Krasnodębski, 2022). The digital transformation and technological adaptation within the hospitality industry, driven by COVID-19, have led to the introduction of numerous innovations, including virtual tours, augmented reality content, mobile check-in processes, self-check-in kiosks, digital concierge services, and smart rooms (Jinathissa, Manjaree & Langseth, 2025). While virtual tourism cannot fully replace traditional tourism, 360-degree virtual tours have proven to be a valuable tool in helping individuals reduce stress during the pandemic by offering a form of escape while staying at home, as well as alleviating some of the challenges faced by the tourism industry (Chen, Wu & Lai, 2023; Yılmaz & Karamustafa, 2022).

The COVID-19 pandemic has played a significant role in the emergence and development of virtual tourism. Driven by the influence of virtual technologies, virtual tourism represents a new form of tourism that offers economic benefits to the tourism sector (Chen, Wu & Lai, 2023). Virtual tourism refers to activities undertaken by individuals in a digitally created three-dimensional environment, designed for leisure and entertainment, outside of their typical living space (Roman, Kosiński, Bhatta, Niedziółka & Krasnodębski, 2022).

International hotel chains such as Starwood, Hyatt, and Crowne Plaza utilize virtual worlds to promote and market tourism destinations. Similarly, Walt Disney's Virtual Magic Kingdom is designed to encourage visits to the company's theme parks (Huang, Backman, Backman & Chang, 2016).

In collaboration with Samsung, Marriott hotels introduced the "VRoom Service" project, offering guests virtual trips to destinations such as China, Rwanda, and Chile. Through this application, guests can also place room service orders. Additionally, Marriott, Samsung, and Framework collaborated on the creation of virtual reality postcards, which provided fun and interactive experiences through 360-degree and three-dimensional techniques (Durmaz, Bulut & Tankuş, 2018).

At Le Petit Chef restaurants, which operate in 89 locations worldwide, including the Sheraton Istanbul City Center Hotel and Antalya (https://lep etitchef.com), Augmented Reality (AR) technology and three-dimensional images are used to enhance the dining experience. Through this immersive presentation, guests can view the menu items, witness vegetables being picked from the field, watch a crab or octopus coming to the plate, observe a bee flying and landing on food, and enjoy the preparation of dishes by small, animated chefs, all accompanied by an audiovisual gastronomy show.

One of the most popular applications utilizing virtual technologies is Google Earth VR, a "three-dimensional, live computer map" of the entire world. With this application, users can virtually visit global destinations, view them from a bird's-eye perspective or through 360-degree photos, and explore streets via the Google Street View application (Roman, Kosiński, Bhatta, Niedziółka & Krasnodębski, 2022).

Emirates Airlines, a leader in the industry, provides a 360-degree, three-dimensional in-cabin Virtual Reality (VR) experience through its website and app. With this award-winning technology, passengers can explore the seating plan in three dimensions and reserve their preferred seats before online check-in (https://www.emirates.com). First Airlines, a Japanese airline, offers virtual flights departing from Tokyo. The company reports a 100% occupancy rate on its virtual flights to various destinations, including Hawaii, Rome, Paris, and New York (Gursoy, Malodia & Dhir, 2022).

Currently, 58 museums and archaeological sites can be visited virtually through the Virtual Museum website, created by the Ministry of Culture and Tourism (https://sanalmuze.gov.tr). The Ephesus Experience Museum, which was awarded the "World's Best Experience Museum" title, offers an immersive exploration of the daily life, trade, architecture, and art of Ancient Ephesus. Through cutting-edge "immerse" technologies, the museum creates a visual and sensory experience that allows visitors to virtually experience that historical period (https://basin.ktb.gov.tr). Similarly, the Hagia Sophia History and Experience Museum uses similar technologies to present the magnificent 1,700-year history of Hagia Sophia, sharing its rich stories with visitors in a virtual and digital environment (https://www.demmuseums.com).

Turkey's first virtual reality museum, MüzeVerse, offers visitors a unique opportunity to travel 4,500 years back in time through VR goggles. This immersive experience allows participants to explore the Pyramids of Cheops in a virtual world during a 45-minute journey titled "Journey to the Pyramids-Discovery of Ancient Egypt" (https://www.muzeverse.com.tr).

In 2022, Yapı Kredi Bank launched the Metaverse Museum as part of the May 19th Commemoration of Atatürk, Youth and Sports Day. The museum displayed numerous valuable items, including Atatürk's medals, amber rosary beads, wooden walking stick, gramophone, and the glasses frame belonging to Atatürk's mother, Zübeyde Hanım. These artifacts were exhibited in the Yapı Kredi building within the metaverse platform Decentraland, with the items being represented as NFTs (https://www.yapikredi.com.tr).

The Sakıp Sabancı Museum, Naval Museum, Bursa Tofaş Watch and Clock Museum, Topkapı Palace Museum, and Hatay Archaeology Museum are among the museums in Turkey that incorporate augmented reality (AR) technology (Özgüneş & Bozok, 2017).

The Louvre Museum's virtual reality project, Mona Lisa: Beyond the Glass, provides an opportunity to explore Leonardo da Vinci's Mona Lisa and its intricate details using VR glasses in a dedicated space within the museum, as well as through smart devices accessible worldwide. This eight-minute virtual experience narrates the true story of the painting, and, as the experience unfolds, the Mona Lisa begins to move, revealing how her dress was made and how her hair was designed (https://www.louvre.fr).

The works housed in the Van Gogh Museum in Amsterdam are available for viewing as an online collection through a 4K virtual tour on YouTube (https://www.vangoghmuseum.nl). Additionally, at the Van Gogh Immersive Experience Exhibition in London, visitors can immerse themselves in eight of Van Gogh's iconic paintings using VR glasses, providing a unique, interactive way to explore his art (https://vangoghexpo.com/london).

National Geographic also offers virtual trips using VR goggles, enabling users to experience activities such as canoeing in Antarctica, observing penguins, climbing glaciers, exploring the ancient Inca city of Machu Picchu in Peru, and capturing these moments through photography (https://www.meta.com/quest/fitness/).

Holodia offers a unique opportunity to transform sports activities into virtual vacations, allowing users to visit and explore destinations such as Venice, Antarctica, Cambridge, Paris, San Francisco, and many others in a virtual environment. This immersive experience is enabled when engaging in activities such as walking, running, skiing, boxing, or squatting, or by placing sensors on devices like exercise bikes, elliptical trainers, and fitness rowers, which synchronize with VR glasses (https://www.holodia.com).

Various factors, such as the desire to explore beyond Earth, scientific research, and the search for new living spaces, have contributed to the growing appeal of space travel and space tourism. This form of travel is now available as a virtual experience.

The US National Aeronautics and Space Administration (NASA)'s "Eyes on Exoplanets" application enables users to virtually fly to a planet of their choice and explore the universe in a scientifically accurate, three-dimensional environment. The "NASA Lunar Mapping and Modeling Portal" provides virtual space tours using data collected from lunar missions, offering detailed virtual maps related to elevation, inclination, risk, gravity, and mineralogy (Damjanov & Crouch, 2018). Through the "SC 360 Expedition" application at the NASA Kennedy Space Center Visitor Complex, users can virtually visit the NASA Space Center, explore Mercury, Gemini, or Apollo capsules, and view the space shuttle Atlantis. The NASA SLS VR experience takes visitors on a scientific mission to Mars, Saturn, or Jupiter while seated inside a launch rocket, simulating the experience of a professional astronaut. The "Spacewalk VR Experience" application allows users to embark on a spacewalk and repair the space shuttle, offering both day and night options. Additionally, the "Mission: Mars" VR application takes users to Mars' orbit, where they can examine satellites and vehicles (Roman, Kosiński, Bhatta, Niedziółka & Krasnodębski, 2022).

Conclusions and Implications

During the COVID-19 pandemic, the tourism sector, like many other industries, has been significantly affected by technological developments and transformations. Various factors, such as border closures, social distancing measures, and hygiene regulations, have contributed to these changes. As a result, technological advancements that might have taken years to develop and integrate into everyday life have rapidly emerged, offering unique and enhanced experiences to users. Among these advancements, Virtual Reality (VR) and Augmented Reality (AR) technologies have played a pivotal role in transforming the tourism sector. These technologies are increasingly utilized in the promotion and marketing of tourism activities, destinations, and services. By providing immersive and interactive experiences, VR and AR enable potential and existing tourists to explore destinations and tourist products in advance. This innovative approach not only enhances curiosity but also strengthens travelers' intention to visit and make purchasing decisions.

To popularize experience innovation supported by VR, AR, and other emerging technologies in the tourism sector, it is essential to provide specialized training for tourism professionals. To effectively promote and market experience innovation, tourism enterprises should first allow their employees to engage with VR and AR environments, thereby enhancing their familiarity with these technologies. Additionally, dedicated promotion and experience booths at tourism fairs can be effective in introducing virtual tourism to a broad audience of investors and consumers. Tourism destinations and attractions can also be integrated into e-sports and gaming content, which have gained significant global popularity, allowing users to experience them through VR and AR technologies. Furthermore, incorporating VR, AR, and experience innovation into the curricula of tourism faculties will ensure that future tourism professionals are well-equipped with the necessary technological competencies. In vocational training programs within tourism academies and the industry, VR and AR technologies can be utilized to simulate real-world scenarios, such as housekeeping procedures, front desk check-ins, meal preparation in professional kitchens, and restaurant service operations. Moreover, increasing the number of scientific studies based on field research and applied examples of technology-supported experience innovation will provide valuable insights for tourism professionals, academics, and other stakeholders, contributing to the sector's continuous development.

In conclusion, although experience innovation driven by VR and AR technologies significantly contributes to the tourism sector, ensuring its continuous development and sustainability requires strong collaboration among all stakeholders. By promoting interdisciplinary cooperation and integrating these technologies into long-term strategies, the tourism industry can enhance visitor experiences, drive innovation, and achieve sustainable growth.

References

Chen, J., Wu, X., &Lai, I.K.W. (2023). A systematic literature review of virtual technology in hospitality and tourism (2013–2022). *Sage Open*, 13(3), 1–18. https://doi.org/10.1177/21582440231193297.

Damjanov, K., &Crouch, D. (2018). Extra-planetary mobilities and the media prospects of virtual space tourism. *Mobilities*, 13(1), 1–13. https://doi.org/10.1080/17450101.2017.1336024.

Demirezen, B. (2019). Artırılmış gerçeklik ve sanal gerçeklik teknolojisinin turizm sektöründe kullanılabilirliği üzerine bir literatür taraması. *Uluslararası Global Turizm Araştırmaları Dergisi*, 3(1), 1–26.

Durmaz, C., Bulut, Y., &Tankuş, E. (2018). Sanal gergerçekliğin turizme entegrasyonu: Samsunda'ki 5 yıldızlı otellerde uygulama. *Turkish Journal of Marketing*, 3(1), 32–49. http://dx.doi.org/10.30685/tujom.v3i1.29.

Gursoy, D., Malodia, S., &Dhir, A. (2022). The metaverse in the hospitality and tourism industry: an overview of current trends and future research directions. *Journal of Hospitality Marketing & Management*, 31(5), 527–534. https://doi.org/10.1080/19368623.2022.2072504.

Hjalager, A.M. (2015). 100 innavtions that transformed tourism. *Journal of Travel Research*, 54(1), 3–21. https://doi.org/10.1177/0047287513516390.

Himanshu., Aggarwal, R., Gupta, S., Dhingra, S., &Kumar, A. (2025). The virtual voyage: understanding the factors affecting metaverse adoption in the tourism and hospitality industry. *Current Issues in Tourism*. https://doi.org/10.1080/13683500.2025.2475491.

Huang, Y.C., Backman, K.F., Backman, S.J., &Chang, L.L. (2016). Exploring the implications of virtual reality technology in tourism

marketing: an integrated research framework. *International Journal of Tourism Research*, 18(2), 116–128. https://doi.org/10.1002/jtr.2038.
Jinathissa, N., Manjaree, S., &Langseth, M. (2025). Implementing virtual reality to enhance guest experiences and operational practices in luxury hotels in Colombo, Sri Lanka. *Procedia Compoter Science*, 256, 594–601. https://doi.org/10.1016/j.procs.2025.02.156.
Nayyar, A., Mahapatra, B., Le, D.N., &Suseendran, G. (2018). Virtual reality (VR) & augmented reality (AR) technologies for tourism and hospitality industry. *International Journal of Engineering & Technology*, 7(2.21), 156–160. https://doi.org/10.14419/ijet.v7i2.21.11858.
Özgüneş, R.E., &Bozok, D. (2017). Turim sektörünün sanal rakibi(mi?): artırılmış gerçeklik. *Uluslararası Türk Dünyası Turizm Araştırmaları Dergisi*, 2(2), 146–160.
Pizam, A. (2010). Creating memorable experiences. *International Journal of Hospitality Management*, 29(3), 343.
Roman, M., Kosiński, R., Bhatta, K., Niedziółka, A., &Krasnodębski, A. (2022). Virtual and space tourism as new trends in travelling at the time of the covid-19 pandemic. *Sustainability*, 14(2), 628. https://doi.org/10.3390/su14020628.
Śledzik, K. (2013). Schumpeter's view on innovation and entrepreneurship. In S. Hittmar (Ed.), Management Trends in Theory and Practice (pp. 89–95). Faculty of Management Science and In-formatics, University of Zilina & Institute of Management by University of Zilina. http://dx.doi.org/10.2139/ssrn.2257783.
Soylu, A., Ceylan, S., &Özdipçiner, N.S. (2019,Kasım). Turizmde deneyim inovasyonu: Doğu Ekspresi örneği. 4th International symposium on innovative approaches in social, human and administrative sciences, Samsun, Türkiye. https://doi.org/10.36287/setsci.4.8.043.
Xi, N., &Hamari, J. (2021). Shopping in virtual reality: a literature review and future agenda. *Journal of Business Research*, 134, 37–58. https://doi.org/10.1016/j.jbusres.2021.04.075.
Yılmaz, M., &Karamustafa, K. (2022). Artırılmış gerçeklik uygulamaları ve turizm. *Seyahat ve Otel İşletmeciliği Dergisi*, 19(3), 399–422. https://doi.org/10.24010/soid.1077349.
https://basin.ktb.gov.tr
https://lepetitchef.com
https://sanalmuze.gov.tr
https://vangoghexpo.com/london
https://www.demmuseums.com

https://www.emirates.com
https://www.holodia.com
https://www.louvre.fr
https://www.meta.com/quest/fitness
https://www.muzeverse.com.tr
https://www.vangoghmuseum.nl
https://www.yapikredi.com.tr

Health Tourism as a Growing Value in the Context of Alternative Tourism

Mert CAN TEKELER[1]

Introduction

Nowadays, various touristic destinations are making intensive efforts to maximize their market share and increase their revenues from tourism. In this respect, countries are orientated to alternative tourism resources both to ensure sustainability and to diversify the touristic attractions of the destination by using various tourism resources. Health tourism, which has been very popular in recent years among alternative tourism sources and continues its upward trend day by day, promises significant potential especially for countries that have a strong medical infrastructure and make investments in the health sector.

Health tourism can be defined as a type of travel in which individuals travel from their country of permanent residence to other countries to regain their health and receive medical services in the countries they visit (Cohen, 2008). Thanks to its high-quality health facilities and the investments made in the health sector in recent years, Türkiye has become an important destination for health tourism. In this study, aiming to determine the importance of health tourism as an alternative tourism resource, the concept of health tourism will be examined comprehensively by focusing on issues such as the definition of health tourism, its history, benefits, types, the status of health tourism in Türkiye and in the world.

The Concept of Health Tourism

The concept of health can have different definitions. When the concept refers to a negative meaning, it evokes diseases. At the meantime, the concept of health is understood to mean the state of not being diseased. When the

1 Lecturer, Zonguldak Bulent Ecevit University, Devrek Vocational School, Department of Hotel, Restaurant and Catering Services, m.tekeler@beun.edu.tr, ORCID: 0000-0002-5830-3384

concept has a positive meaning, it connotes physical well-being, such as mental and physical adaptation. At present, it is more common to use the concept in the second meaning (İçöz, 2009; Jones, 2005). The concept of health can be defined as a state of social, physical and psychological well-being (Abels and Kipnis, 1998). Generally considered, health can be defined as the physical, economic and social well-being of individuals in society in addition to the prevention of diseases, illnesses and disabilities (Çam and Çılgınoğlu, 2021).

Health, as one of the most commercial, political, social and culturally emphasized concepts, has taken its place among the global markets that generate the highest income in the free market economy with the neoliberal policies implemented on a global scale (Aslanova, 2013).

Tourism is the whole of the events arising from the movement and accommodation of foreigners for a temporary period without the purpose of earning money. People's travelling away from their settlement to receive treatment has also led to the emergence of health tourism (Bulut and Şengül, 2019).

When literature has been examined, it has been observed that there are different definitions of health tourism.

Ross (2001) has defined health tourism as 'tourism generated by individuals travelling outside their permanent place of residence for health reasons' (İçöz, 2009). Boz (2004) has identified health tourism as 'the movements arising from individuals travelling outside their destination for a limited period of time for the purpose of protecting their health and healing, going to a tourism enterprise that uses natural resources and meeting their needs for cure, accommodation, nutrition and entertainment'. Cohen (2008) has also defined health tourism as 'planned travelling to a destination other than one's place of residence for the purpose of obtaining health services'.

According to the definitions provided, health tourism can be defined as tourism formed by people who travel outside their permanent settlement for a temporary period to obtain medical services.

Health tourism, which is a particular sub-branch of tourism, is differentiated from other tourism activities due to its structure. Özalp (2005) has explained the specific characteristics of health tourism as follows (Aslanova, 2013):

- Health tourism needs technical equipment and required labor force.
- It is very important that the health organization serving health tourism provides services at international standards.

- The health institution providing services for health tourism should have a staff with a command of the common foreign language.
- Health tourism is a high value-added tourism resource that supports the development of different types of tourism for both patients and their relatives.
- Promotion and marketing activities should be conducted in different languages with government support for the target market addressed. In these promotional and marketing activities, especially insurance and health institutions that will help their patients to receive healthcare services in a different country should be targeted.

The relationship between health and tourism emerges with both positive and negative dimensions. The positive aspect of the relationship between health and tourism is that tourists travel for health purposes and get treatment. The negative aspect is the risks to the health of tourists. For instance, SARS, which appeared in the Far East in 2002, caused a total loss of approximately 20 billion dollars and three million people lost their jobs in countries such as China, Hong Kong, Singapore and Vietnam (Kuo et al., 2008).

Cohen (2008) has classified tourists who participate in health tourism into five main groups:

- **Salt tourist:** Tourists who do not use any healthcare services in the host country.
- **Medical Tourist:** Tourists travelling for both tourism and medical services. Furthermore, tourists who do not go to the destination country to receive health service but decide to receive health service during their stay can be included in this group.
- **Patients on holiday:** Tourists who travel out of the destination of residence to receive health services and take advantage of holiday opportunities in the country where they are in the process of being treated.
- **Salt patient**: Tourists who travel only for the purpose of obtaining health services and do not participate in any tourism activities in the destination.

According to the classification of tourists in the health tourism market made by Cohen (2008); it can be said that the tourists in the first group are not in the target market of health care facilities, the other three groups can be selected as the target market and the most important group for health care service facilities is the tourist group, which is referred to as pure patients and travelling only to receive health services.

History of Health Tourism

The history of health tourism dates back to ancient times. Sumerians have laid the foundations of urban life and due to the development of urban life, people started to live together. This situation has created a basis for the phenomenon of travelling outside the settlement and returning to the place of residence, which is one of the requirements of tourism (Akoğlan Kozak, Evren and Çakır, 2013).

Ancient Greeks were the first civilization to establish the foundations of health tourism. 'Temples of Asclepios' is the first known health center in the world. The Romans also have contributed to the development of health tourism by treating some diseases with healing waters and establishing facilities in the regions where these healing waters were located (Bulut and Şengül, 2019).

In the Middle Ages, Rome continued to be an essential destination in terms of health tourism. During this period, hot mineral water called 'onsen' had become popular in Japan due to its healing properties. This mineral water was used for treatment purposes by soldiers who were wounded especially in wars. In this period, important developments were also experienced in Islamic countries in the name of health. Mansour Hospital, which was built in Cairo, is recognized as the largest and most advanced health institution of its period on a world scale (Demir, 2013).

In the 20th century, Europe and America became the world's most important health centers. Particularly, the 'American Board of Medical Specialists', which was established in 1993, has accelerated health tourism in the United States of America. In the 1960s, yoga centers in India became important again regarding health tourism. In the 1980s, despite the existence of high-quality health centers in the United States of America and the provision of quality services, the provision of health services became very expensive, and the costs of health care services increased on a global scale. In line with this development, people have been in search of more cost-effective treatment in different countries to receive health services. Today, these searches are continuing (Berksoy, 2022).

Types of Health Tourism

Health tourism consists of different types. Although there are different classifications regarding these types, the classification of health tourism types carried out by the Ministry of Health, which is the most competent authority in our country, is as follows (Aslanova, 2013; Canik and Özdemir, 2017):

- **Medical Tourism:** The Ministry of Health has defined medical tourism as 'planned treatment practices performed by doctors employed in 2nd and 3rd step health facilities for individuals with a specific disease'.
- **Thermal Tourism and SPA Wellness:** The Ministry of Health has identified Thermal Tourism and SPA Wellness as 'treatment supporting and rehabilitating activities carried out in hot springs and spa centers during rehabilitation periods after illness or to partially cure diseases that have caused permanent damage'.
- **Elderly and Disabled Tourism:** These are practices carried out by qualified personnel in Clinical Guesthouses, Geriatric Treatment Centers and Nursing Homes for the care and rehabilitation of elderly and disabled individuals.

Advantages of Health Tourism

It can be said that health tourism has an important role in expanding tourism all year and increasing tourism revenues. Health tourism is a large-scale industry involving many sectors. In this respect, Harrant (2003), considering the sub-sectors that have made up the health tourism industry, listed the benefits of health tourism as follows (Harrant, 2003 as cited in Aslanova, 2013):

- Health tourism enables patients to obtain quality and accurate health services from hospitals within the European Union by avoiding patients travelling long distances to access quality health services.
- Healthcare facilities benefit from the sharing of better practices, experiences and equipment to provide higher quality healthcare services with health tourism.
- Thanks to health tourism, insurance companies can provide an opportunity to provide an additional service to their customers to ensure satisfaction by reducing waiting periods.

- Health tourism provides a substantial benefit to the economy by sharing costs, resources and services.

Situation of Health Tourism in the World

Health tourism on a global scale has shown a remarkable growth in recent years. This development has taken place in parallel with the increase in demand for health services in different countries. Moreover, patients travel between countries for various reasons such as receiving quality health services, reaching these facilities in a shorter time and at the same time providing price advantage. This situation has created an important opportunity for developing countries as well as developed countries and has increased the competitiveness of developing countries in the field of health tourism (Ünver, 2024).

The number of people participating in rapidly developing health tourism has reached 800 million. Today, the leading countries that dominated this market with their technology, science and qualified personnel have been gradually losing their leadership positions. Currently, it has been observed that developing countries in the Asian continent such as Thailand, Mexico, Brazil, India and Turkey have started to come to the forefront in health tourism (Çelikkanat and Güngörmüş, 2018).

Medical tourism is an important part of health tourism. Applications such as treatment of dental and eye diseases, aesthetic applications, cardiovascular treatment, prosthesis, infertility treatment and IVF are evaluated within the scope of medical tourism. People who want to benefit from such health services turn to countries that attract attention with both success rates and favorable costs. In this regard, Southern Europe, South America and South Asian countries are particularly remarkable. Basically, this is because the prices of treatment and surgical procedures in these countries are lower than in the Nordic countries. Moreover, these countries' ease of transport, strong health infrastructure and qualified personnel constitute an important attraction factor (Aydın, 2012).

One of the European countries, Germany, has attracted considerable attention as an important center in the thermal tourism pillar of health tourism. In addition, countries such as Slovakia and Czechia, which are also among the Northern European countries, have been making significant investments in terms of thermal tourism. There are important thermal facilities in France, Spain, Italy and Russia in Europe. Japan and the United States of America are also among the important centers in terms of thermal tourism with their resources and qualified health facilities (Dikmetaş Yardan et al., 2014).

One of the Asian countries, India has recently become an international center for health tourism. In particular, the opportunity to access health services at low cost has made India a country favored by health tourists. India has many powerful features that make it favorable in health tourism. The technological infrastructure and hospital infrastructure of the health facilities at the standards of European and American countries, the fact that the doctors are trained abroad and have international experience and have a good command of western treatment methods, short waiting times and affordable health services can be considered among these strengths (Dikmetaş Yardan et al., 2014).

Situation of Health Tourism in Türkiye

Within the context of the high level of education and prosperity in developed countries, the presentation of health services is also quite costly. In these countries, social security institutions are faced with various difficulties due to rising health expenditures and costs. In this regard, social security institutions and private insurance companies in developed countries make contracts with countries that are in close proximity and provide quality medical services, thereby reducing the cost of health care services (Dikmetaş Yardan et al., 2014).

Türkiye has substantial opportunities in terms of health tourism. Besides its sea, climate, beaches, forests, plateaus, thalassotherapy facilities, factors such as having an international airline brand such as 'THY' and ease of transportation enhance Türkiye's competitiveness in the field of health tourism. Particularly in terms of medical tourism, Türkiye offers a considerable potential with its large number of qualified university hospitals, training and research hospitals and private health institutions. The most common branches receiving foreign patients are internal medicine, orthopedics, gynecology and obstetrics. Türkiye also receives a large number of foreign patients for eye surgery. Furthermore, Türkiye is in a favorable position compared to other countries in terms of hair transplantation and aesthetic applications (Health Tourism Coordination Board, 2025).

Thanks to its natural resources, Türkiye has an important opportunity in the fields of elderly and disabled tourism and thermal tourism. In order to get the expected contribution especially from elderly and disabled tourism, agreements should be made with insurance companies, the number and quality of facilities should be increased, and promotions should be made abroad. Türkiye's important opportunity in thermal tourism is that both medical tourism and thermal tourism services can be provided together in many locations. (Aydın and Aydın, 2015).

Due to its geographical location and the quality services provided through qualified health facilities, Türkiye is a country with important opportunities regarding health tourism. Moreover, another advantage of Türkiye is its expert and experienced health personnel. For instance, Türkiye is one of the few countries with 18 different disciplines in pediatrics. Likewise, in ophthalmology, Türkiye is home to top specialty disciplines. At the same time, in recent years, Türkiye's health tourism opportunities through private hospitals have been presented to participants at medical tourism fairs organized on a global scale. Also, private hospitals have established units for foreign patients, aiming to improve service quality. These units are essential for foreign patients to feel special and safe (Health Tourism Coordination Board, 2025).

Conclusion

Health is one of the most crucial components of human life. In this context, it is one of the most fundamental rights of every individual living in a country to have access to health services. In terms of access to health services, each state has an important responsibility to ensure that its citizens have equal access to health services.

Particularly in developed countries, high education and high welfare levels have been observed. Besides, the provision of health services in these countries is also costly. With the purpose of dealing with this high-cost issue, developed countries have entered into bilateral agreements with countries that are located in close proximity and offer quality health services at more affordable prices, thereby facilitating access to health services for their citizens and at the same time reducing the cost of health service provision. As a result, the emergence of a new tourism source under the name of health tourism has paved the way for the emergence of a new tourism source. Development of a new alternative tourism source under the name of health tourism has created an important opportunity for developing countries. These countries have been making substantial financial investments in the health sector to expand tourism to 12 months to increase the income from tourism and to catch the competitiveness in this field, which they have not been able to achieve in mass tourism. In recent years, Türkiye has become one of the prominent countries in the field of health tourism with factors such as the number of both public and private health facilities, increasing investments and qualified health personnel. Especially in medical tourism and thermal tourism, as well as in other areas such as thermal tourism, Türkiye stands out in health tourism.

Due to its resources, Türkiye is a country that offers a very strong potential in terms of health tourism. With each passing year, it is possible to mention that the number of tourists coming to our country in different fields of health tourism is increasing. On the other hand, Turkey is a country that should be more emphasized in health tourism with its high-quality health facilities, quality service understanding in different disciplines and well-equipped medical staff. Therefore, the Ministry of Culture and Tourism has important duties and responsibilities. Particularly in international fairs and similar organizations, the health system of our country should be promoted and marketed. The Ministry of Health and the Ministry of Culture and Tourism should increase their communication, and joint studies should be conducted to develop strategies to increase the competitiveness of Türkiye in this sector. Moreover, health tourism deficiencies should be determined, and a sustainable development policy should be implemented. It is necessary to explain the thermal tourism resources of our country to foreign tourists in a more effective manner and to develop the thermal tourism pillar of health tourism. Particular attention should be given to the foreign language competence of the employees working in health facilities and deficiencies in this field should be identified and eliminated. Policies to increase the number of accredited health facilities operating in our country should be implemented. Thanks to the research to be conducted in line with the recommendations mentioned above, Türkiye can increase its market share in the health tourism pie and generate more income from tourism all year round.

References

Abels D. J. & Kipnis V. (1998). Bioclimatology and balneology in dermatology: A dead sea perspective. *Clinics in Dermatology, 16 (6), 695–698*

Aslanova, K. (2013). Türkiye'de sağlık turizmi ve sağlık turizmi hukuku. *Avrasya Uluslararası Araştırmalar Dergisi, 3(3), 129–145*

Aydın, O. (2012). Türkiye'de alternatif bir turizm; sağlık turizmi. *Karamanoğlu Mehmetbey Üniversitesi Sosyal ve Ekonomik Araştırmalar Dergisi, 2012(2), 91–96.*

Aydın, G. & Aydın, B. K. (2015). Dünyada ve Türkiye'de sağlık turizmi pazarlama uygulamaları ve karşılaştırmalı durum analizi. *Pazarlama ve Pazarlama Araştırmaları Dergisi, 8(16), 1–22.*

Berksoy, F. (2022). *Sağlık turizminin gelişmesinde sağlık turizmi seyahat acentelerinin rolü ve önemi* (Yüksek Lisans Tezi). Aydın Adnan Menderes Üniversitesi

Boz, M. (2004). *Turizmin gelişmesinde alternatif turizm pazarlamasının önemi* (Doktora Tezi). Marmara Üniversitesi

Bulut, A. & Şengül, H. (2019). Dünyada ve Türkiye'de sağlık turizmi. *Yönetim, Ekonomi ve Pazarlama Araştırmaları Dergisi.*

Canik, S. & Güneren Özdemir, E. (2019). Sağlık turizmini konu alan lisansüstü tezlerin bibliyometrik değerlendirmesi (1988–2017). *Anatolia: Turizm Araştırmaları Dergisi, 30(2), 125–134.* https://doi.org/10.17123/atad.636870

Cohen, E. C. E. (2008). Medical tourism in Thailand. *AU-GSB e-journal, 1(1).*

Çam, O. & Çılgınoğlu, H. (2021). Sağlık turizmi kapsamında medikal turizmde öne çıkan sağlık uygulamaları. *Oğuzhan Sosyal Bilimler Dergisi, 3(2), 197–216.*

Çelikkanat, Ş. & Güngörmüş, Z. (2018). Gelişmekte olan ülkelerde sağlık turizmi ve hemşirenin sağlık turizmindeki yeri. *Social Sciences Studies Journal, 4(22), 3938–3944.*

Demir, D. (2013). *Türkiye'de medikal turizmi etkileyen faktörler üzerine bir çalışma* (Yüksek Lisans Tezi). Toros Üniversitesi

Dikmetaş Yardan, E., Dikmetaş, H., Coşkun Us, N. & Yabana, B. (2014). Türkiye ve Dünya'da sağlık turizmi. *Sağlıkta Performans ve Kalite Dergisi, 8(2), 27–42.*

Harrant, P. (2003). Hospital Cooperation in Border Regions in Europe Presantation of the Results of HOPE Study. *Free movement and cross-border cooperation in Europe: The role of hospitals&practical experiences in hospitals, Luxembourg: Entente des hospitaux Luxembourgeois.*

Hunter-Jones, P. (2005). Cancer and tourism. *Annals of Tourism Research*, *32*(1), 70–92.

İçöz, O. (2009). Sağlık Turizmi Kapsamında Medikal (Tibbi) Turizm ve Türkiye'nin Olanakları. *Yaşar Üniversitesi E-Dergisi, 4(14), 2257–2279.*

Kuo, H. I., Chen, C. C., Tseng, W. C., Ju, L. F. & Huang, B. W. (2008). Assessing impacts of SARS and Avian Flu on international tourism demand to Asia. *Tourism management, 29(5), 917–928.*

Sağlık Turizmi Koordinasyon Kurulu (2025). Türkiye'nin Sağlık Turizmindeki Önemi. http://www.saturk.gov.tr/images/pdf/tyst/06.pdf Erişim Tarihi: 03.03.2025, 12:40

Ünver, Ö. (2024). *Türkiye Sağlık Turizminin Uluslararası Platformdaki Yeri: Bibliyometrik Analiz* (Yüksek Lisans Tezi). Kütahya Dumlupınar Üniversitesi

The Effects of Healthy Nutrition and Detox Programs on Health in Detox-Themed Wellness Hotels

Havva GÖZGEÇ MUTLU[1]

Introduction

In recent years, wellness tourism has emerged as a rapidly growing sector, driven by increasing global awareness of health and well-being. Reflecting this trend, the global wellness economy reached a record $6.3 trillion in 2023 and is projected to grow to approximately $9 trillion by 2028 (Global Wellness Institute [GWI], 2024), highlighting the sector's expanding influence and economic significance. Among the various offerings in this sector, detox-themed wellness hotels have gained significant attention for their structured programs that integrate healthy nutrition and detoxification to promote both physical and mental well-being. These hotels provide guests with specialized dietary regimens, often combined with wellness activities such as yoga, meditation, hydrotherapy, and spa treatments to support overall health and rejuvenation (GWI, 2021).

Wellness tourists tend to spend more than the average traveler. In 2022, international wellness tourists spent an average of $1,764 per trip, which is 41% higher than the expenditure of other international tourists. Similarly, domestic wellness tourists spent an average of $668 per trip, representing a 175% higher expenditure compared to other domestic tourists (GWI, 2023).

Healthy nutrition and detox programs offered in wellness-themed hotels provide a wide range of health benefits. Such programs play a significant role in various physiological processes, including weight management, improvement of insulin sensitivity (Kerna et al., 2024), elimination of toxins from the body (Gibson, 1992), and reduction of oxidative stress (Panda et al., 2023). Moreover, appropriate nutrient intake and antioxidant support effectively enhance the body's detoxification mechanisms (Park, 2024).

---

[1] PhD., Dokuz Eylül University, Efes Vocational School, Hotel, Restaurant, and Catering Services Department, hgozgecm@gmail.com

This chapter explores the benefits of key nutrition programs and detox regimens featured in detox-themed wellness hotels. The discussion will cover intermittent fasting, ketogenic diets, and low-calorie dietary approaches, followed by an examination of various detox programs, including green liquid detox, water fasting detox, and green salad detox, emphasizing their physiological benefits and scientific foundations.

Healthy Nutrition Programs

The World Health Organization (WHO) defines a healthy diet as one that helps protect against malnutrition in all its forms, as well as noncommunicable disease such as diabetes, heart disease, stroke, and cancer. In wellness tourism, healthy nutrition is a fundamental component aimed at enhancing individuals' overall well-being.

Research indicates that wellness retreats often provide an organic, predominantly plant-based diet, which has been associated with significant health improvements. For instance, a study observed that participants in a week-long wellness retreat experienced substantial enhancements in various health metrics, including reductions in weight, abdominal girth, and blood pressure, following a diet rich in plant-based foods. Furthermore, wellness tourism often incorporates educational programs that encourage the adoption of healthy eating practices. These programs aim to instill long-term dietary changes by providing knowledge about nutrition and its impact on health (Cohen et al., 2017).

Another study emphasizes the importance of traditional foods, such as millets, which offer numerous health benefits. The research advocates for the revival of such nutritious foods within wellness tourism to promote healthier eating habits and lifestyles (Pagaldiviti & Dash, 2023).

Wellness hotels offer healthy nutrition programs such as intermittent fasting, the ketogenic diet, and low-calorie diets.

Intermittent Fasting

"Intermittent fasting is a broad term that encompasses a variety of programs that manipulate the timing of eating occasions by utilizing short-term fasts in order to improve body composition and overall health" (Tinsley & La Bounty, 2015, p. 661). Common IF methods include alternate-day fasting, whole-day fasting, time-restricted feeding,

and periodic fasting. In alternate-day fasting, individuals alternate between fasting days, during which minimal calories are consumed, and non-fasting days; this approach has been shown to reduce body weight, body fat, total cholesterol, and triglyceride levels. Whole-day fasting, typically involving a 24-hour fast, similarly promotes weight loss, fat reduction, and improvements in blood lipid profiles (Tinsley & La Bounty, 2015). Time-restricted feeding limits food intake to specific hours each day, while periodic fasting, such as the 5:2 diet, includes fasting on two non-consecutive days per week with unrestricted eating on the remaining days (Tinsley & La Bounty, 2015; Gonciarz et al., 2024).

By improving energy balance, IF facilitates weight and fat loss, enhances insulin sensitivity, and lowers blood glucose levels, thereby reducing the risk of diabetes (Bieńko et al., 2024). Additionally, IF contributes to better cardiovascular health by decreasing cholesterol and triglyceride levels (Gonciarz et al., 2024; Brogi et al., 2024). On a cellular level, it activates pathways that alleviate oxidative stress, regulate organelle function, and promote cellular repair, supporting longevity and protection against age-related diseases (Reddy et al., 2024). IF also improves liver health by reducing liver enzyme levels and inflammation, making it beneficial in the management of cardiovascular disease and obesity (Ranjbar et al., 2024; Diab et al., 2024).

Ketogenic Diet

The ketogenic diet (KD) is a high-fat, moderate-protein, and very low-carbohydrate dietary approach that induces ketosis by increasing the production of ketone bodies. Originally developed for therapeutic purposes, KD has demonstrated benefits in managing various conditions, including epilepsy, diabetes, obesity, and certain cancers (Filiz, Yildiz, & Gürbüz, 2022).

KD is particularly effective in treating drug-resistant epilepsy in children, with studies reporting significant seizure reduction in the majority of patients (Dressler et al., 2018). Additionally, emerging research suggests its potential in improving cognitive function and supporting the treatment of neurodegenerative diseases (Kayode et al., 2020; Rajiyah, 2020).

Beyond neurological benefits, KD contributes to metabolic health by promoting weight loss, lowering body mass index, and reducing fat mass (Pastuszak et al., 2023). It also supports glycemic control in type 2 diabetes and may serve as an adjunct therapy (Ilnicka, 2022).

Low-Calorie Diets

Low-calorie diets offer significant benefits in metabolic regulation and the management of various health conditions. Early implementation in clinical settings has been shown to reduce insulin resistance and insulin requirements, though it does not affect intensive care unit stay duration, mortality, or infection rates (Jiang & Xu, 2022).

They are also used therapeutically, such as in preparing nutritionally balanced, low-calorie meals for cancer patients to meet daily dietary needs (Rivoltini et al., 2022). In particular, positive outcomes on immune function and reduced inflammation have been reported among gastric cancer patients (Li, Han, Du & Guo, 2015).

Very low-calorie diets and ketogenic variations are effective in treating obesity and related conditions, provided they are applied under medical supervision (Gorini, Armani & Caprio, 2023). Studies indicate that these diets significantly reduce body mass index, fat mass, and waist circumference, while improving metabolic profiles by lowering cholesterol and glucose levels, thereby decreasing cardiovascular risks (Merra et al., 2017).

Detox Programs

Wellness resorts and destination spas are among the key providers of detox programs and nutritional services (Bardukova, 2024). Some of the nutritional detox programs offered by these hotels include green liquid detox, water fasting detox, and green salad detox.

Green Liquid Detox

Green detox juices have become popular among health-conscious consumers seeking to maintain a balanced, low-calorie diet (Fouchy et al., 2019). Typically prepared from a mix of fruits, vegetables, and nutrient-rich plants, these juices are valued for their antioxidant content, which helps protect the body against harmful substances. For example, green juice made with YlBio and chia gel has shown strong antioxidant properties without the need for artificial additives (Fraga et al., 2020).

One of the key ingredients in green detox juices is wheatgrass, known for its high chlorophyll, vitamin, and antioxidant levels. Often called "green blood," wheatgrass juice supports detoxification, improves digestion, and enhances blood circulation. It has also been linked to managing chronic diseases and increasing hemoglobin levels, offering benefits for conditions such as anemia and inflammatory bowel disease (Chauhan, 2014; Padalia et al., 2010).

Other variations of green detox juices include ingredients like kale, turmeric, Angelica keiskei, and rooibos tea. Regular consumption of these juices has been associated with reduced DNA damage caused by oxidative stress, particularly in smokers, suggesting a protective effect against certain cancers (Kang et al., 2004). Additionally, green grape juice (verjuice) has demonstrated antioxidant and antitumor properties, enhancing the effectiveness of some cancer treatments (Nasser, 2020).

Water Fasting

Water-only fasting (WF) involves the complete avoidance of caloric intake for a specific period and has been associated with various health benefits (Gabriel et al., 2022). Research shows that WF can lower blood pressure, reduce body weight and abdominal circumference, and decrease medication use, making it a safe option for managing hypertension (Zeiler et al., 2024).

WF may offer benefits for autoimmune conditions as well. In individuals with Hashimoto's thyroiditis, WF has been linked to improvements in symptoms by reducing thyroid-stimulating hormone (TSH) and inflammation markers (Watts et al., 2023). Additionally, WF shows positive effects on prostate health. An 8-day WF in middle-aged men improved urinary function, reduced prostate size, and led to hormonal changes without adverse effects (Letkiewicz et al., 2020). It also appears to influence gut health by lowering harmful bacteria linked to inflammation (He et al., 2019).

Green Salad Detox

Green leafy vegetables are rich in essential nutrients such as dietary fiber, vitamins (A, C, E, K), calcium, iron, magnesium, and potassium, while being naturally low in fats and calories (Randhawa et al., 2015; Rana, 2016).

Leafy greens like spinach, kale, and lettuce are particularly valued for their antioxidant content, which protects the body by neutralizing harmful free radicals. Regular consumption of these vegetables supports heart health, lowers cholesterol and blood pressure, helps regulate blood sugar, and aids in weight management. Their high-water content promotes hydration, contributing to healthy skin and hair, while calcium and vitamin K strengthen bones and support proper blood clotting (Rana, 2016).

Certain leafy vegetables, such as angelica, dandelion, and mustard leaves, are especially high in natural antioxidants and polyphenols. These compounds assist in detoxification and may protect against artery-related diseases (Park et al., 2008). Moreover, studies have shown that frequent consumption of green salads is associated with improved health outcomes, including better survival rates in oral cancer patients (Bell, 2021).

Conclusion

Wellness tourism, driven by increasing global awareness of health and well-being, has emerged as a rapidly growing industry, with detox-themed wellness hotels playing a pivotal role in promoting holistic health. These establishments integrate scientifically supported nutrition and detoxification programs, providing structured interventions that enhance both physical and mental well-being. The findings presented in this study highlight the significant benefits of such programs, emphasizing their impact on metabolic health, longevity, disease prevention, and overall wellness.

Healthy nutrition programs, such as intermittent fasting, the ketogenic diet, and low-calorie dietary approaches, have been shown to contribute to weight management, improved metabolic function, and enhanced mental clarity. Intermittent fasting, in particular, demonstrates strong potential in regulating blood sugar levels, reducing inflammation, and supporting cellular repair mechanisms, while the ketogenic diet has proven effective in managing neurological and metabolic disorders. Additionally, low-calorie diets serve as therapeutic tools, aiding in weight loss, improving cardiovascular health, and supporting individuals undergoing medical treatments.

Detox programs, including green liquid detox, water fasting detox, and green salad detox, provide complementary benefits by promoting detoxification, antioxidant support, and improved digestion. Green liquid detox programs, enriched with phytochemicals and antioxidants, help combat oxidative stress and enhance immune function. Water fasting, as an advanced detoxification strategy, has been linked to metabolic improvements, immune modulation, and reductions in inflammatory markers. Green salad detox programs, rich in fiber, essential vitamins, and minerals, support cardiovascular health, weight management, and disease prevention.

Wellness tourists typically spend significantly more than the average traveler. This trend highlights the economic importance of wellness tourism and indicates substantial growth potential. This trend underscores the growing demand for wellness-focused experiences and highlights the potential for further expansion in this sector. As wellness tourism continues to evolve, the integration of personalized nutrition plans, evidence-based detox protocols, and educational wellness programs will be crucial in enhancing guest experiences and promoting long-term health benefits.

In conclusion, detox-themed wellness hotels represent a transformative approach to health-centered tourism, providing guests with scientifically backed interventions that support sustainable well-being. Moving forward, continued research, innovation, and investment in wellness tourism infrastructure will be essential to maximizing its impact on global health, ensuring that wellness travelers receive tailored, high-quality programs that meet their evolving needs.

References

Awuchi, C. G., Igwe, V. S., & Amagwula, I. O. (2020). Nutritional diseases and nutrient toxicities: A systematic review of the diets and nutrition for prevention and treatment. *International Journal of Advanced Academic Research, 6*(1), 1–46.

Bardukova, L. (2024). Health and wellness tourism: Current trends and strategies in the Bulgarian tourism industry. *Economics and Computer Science,* (2), 22–36.

Bell, E. B., Reis, I. M., Cohen, E. R., Almuhaimid, T., Smith, D. H., Alotaibi, F., ... & Franzmann, E. J. (2021). Green salad intake is associated with improved oral cancer survival and lower soluble CD44 levels. *Nutrients, 13*(2), 372.

Bieńko, M., Rutkowska, M., Król, T., Toborek, M., Marchaj, M., Korta, K., ... & Margas, M. (2024). Sports performance and intermittent fasting: Impact on training and results, mechanisms, and health benefits in weight management, metabolic health, and chronic

diseases–A review of risks, current evidence, and future research directions. *Quality in Sport, 16,* 52500–52500.
Brogi, S., Tabanelli, R., Puca, S., & Calderone, V. (2024). Intermittent fasting: Myths, fakes and truth on this dietary regimen approach. *Foods, 13*(13), 1960.
Chauhan, M. (2014). A pilot study on wheat grass juice for its phytochemical, nutritional and therapeutic potential on chronic diseases. *International Journal of Chemical Studies, 2*(4), 27–34.
Cohen, M. M., Elliott, F., Oates, L., Schembri, A., & Mantri, N. (2017). Do wellness tourists get well? An observational study of multiple dimensions of health and well-being after a week-long retreat. *Journal of Alternative and Complementary Medicine, 23*(2), 140–148.
Diab, R., Dimachkie, L., Zein, O., Dakroub, A., & Eid, A. H. (2024). Intermittent fasting regulates metabolic homeostasis and improves cardiovascular health. *Cell Biochemistry and Biophysics,* 1–15.
Dressler, A., Haiden, N., Trimmel-Schwahofer, P., Benninger, F., Samueli, S., Gröppel, G., ... & Feucht, M. (2018). Ketogenic parenteral nutrition in 17 pediatric patients with epilepsy. *Epilepsia Open, 3*(1), 30–39.
Filiz, C., Yıldız, E., & Gürbüz, O. (2022). Ketogenic Nutrition and Health. *Izmir Democracy University Health Sciences Journal, 5*(2), 317–345.
Fouchy, M. V., Santos, I. da C., Ahn, C. S. von, Mendonça, C. R. B., & Nogueira, M. B. (2019). Detox juice development without additives and its acceptability in relation to a commercial formulation. *Brazilian Journal of Technology, 2*(2), 601–607.
Fraga, J. L., Sant'Ana, G. C. F., Silva, K. A., & Amaral, P. F. F. (2020). Green (Detox) juice physicochemical properties and stabilization effect of naturals emulsifiers. *Ciência Rural, 50* (3), 1–11.
He, Y., Yin, J., Lei, J., Liu, F., Zheng, H., Wang, S., ... & Zhou, H. (2019). Fasting challenges human gut microbiome resilience and reduces Fusobacterium. *Medicine in Microecology,* 1, 100003.
Gabriel, S., Ncube, M., Zeiler, E., Thompson, N., Karlsen, M. C., Goldman, D. M., ... & Myers, T. R. (2022). A six-week follow-up study on the sustained effects of prolonged water-only fasting and refeeding on markers of cardiometabolic risk. *Nutrients, 14*(20), 4313.
Global Wellness Institute (2021). Global wellness tourism economy report. https://globalwellnessinstitute.org/wp-content/uploads/2018/11/GWI_GlobalWellnessTourismEconomyReport.pdf (Accessed: 28.11.2024).
Global Wellness Institute (2023, November 28). Wellness tourism will cross the $1 trillion mark in 2024. https://globalwellnessinstitute.org/

global-wellness-institute-blog/2023/11/28/wellness-tourism-will-cross-the-1-trillion-mark-in-2024/ (Accessed: 28.11.2024).

Global Wellness Institute. (2024). 2024 Global Wellness Economy Monitor. https://globalwellnessinstitute.org/industry-research/2024-global-wellness-economy-monitor/ (Accessed: 28.11.2024).

Gibson, G. G. (1992). *Detoxication mechanisms and the role of nutrition. In Principles and Practice of Environmental Medicine* (pp. 139–158). Boston, MA: Springer US.

Gonciarz, M. J., Adamiec, D. K., Dąbrowska, A., Paduszyńska, N., Zaliwska, D., Czach, M., ... & Do, M. K. (2024). The impact of Intermittent Fasting on Weight Loss and Metabolic Health-a review of the latest research. *Quality in Sport, 16*, 52871–52871.

Gorini, S., Armani, A., & Caprio, M. (2023). Very low-calorie ketogenic diet: A valuable and fashionable nutritional therapy. When could it become dangerous?. *Diabetes/Metabolism Research & Reviews, 39*(8), 1.

Ilnicka, N., Abod, L., Matyja, D., Sadlik, M., & Zuziak, P. (2022). Ketogenic diet as a tool in the treatment of diseases. *Journal of Education, Health and Sport, 12*(12), 199–203.

Jiang, Q., & Xu, T. (2022). Effect of early low-calorie enteral nutrition support in critically ill patients: A systematic review and meta-analysis. *BioMed Research International, 2022*(1), 7478373.

Kang, M. H., Park, Y. K., Kim, H. Y., & Kim, T. S. (2004). Green vegetable drink consumption protects peripheral lymphocytes DNA damage in Korean smokers. *Biofactors, 22*(1–4), 245–247.

Kayode, O. T., Rotimi, D. E., Afolayan, O. A., & Kayode, A. A. A. (2020). Ketogenic diet: A nutritional remedy for some metabolic disorders. *The Journal of Education, Health and Sport, 10*(8), 180–188.

Kerna, N. A., Ngwu, D. C., Jomsky, B. M., Holets, H. M., Nnake, I., Jeremiah, S. M., ... & Adadzi, J. (2024). Impact of Detox Diets on Obesity and Metabolic Syndrome: Implications for Weight Loss, Metabolic Health, and Clinical Practice. *European Journal of Medical and Health Research, 2*(5), 135–152.

Letkiewicz, S., Pilis, K., Ślęzak, A., Pilis, A., Pilis, W., Żychowska, M., & Langfort, J. (2020). Eight days of water-only fasting promotes favorable changes in the functioning of the urogenital system of middle-aged healthy men. *Nutrients, 13*(1), 113.

Li, J. H., Han, L., Du, T. P., & Guo, M. J. (2015). The effect of low-nitrogen and low-calorie parenteral nutrition combined with enteral nutrition on inflammatory cytokines and immune functions in patients with gastric cancer: A double blind placebo trial. *European Review for Medical and Pharmacological Sciences, 19*(8), 1345–1350.

Merra, G., et al. (2017). Effects of very-low-calorie diet on body composition, metabolic state, and genes expression: a randomized double-blind placebo-controlled trial. *European Review for Medical and Pharmacological Sciences, 21*(2), 329–345.

Nasser, M., Cheikh-Ali, H., Hijazi, A., Merah, O., Al-Rekaby, A. E. A. N., & Awada, R. (2020). Phytochemical profile, antioxidant and antitumor activities of green grape juice. *Processes, 8*(5), 507.

Padalia, S., Drabu, S., Raheja, I., Gupta, A., & Dhamija, M. (2010). Multitude potential of wheatgrass juice (Green Blood): An overview. *Chronicles of young scientists, 1*(2), 23–8.

Pagaldiviti, S. R., & Dash, A. (2023). Wellness Tourism: Reviving Healthy Food and Lifestyle. *International Journal of Positivity & Well-being, 1*(1), 53–61.

Panda, C., Komarnytsky, S., Fleming, M. N., Marsh, C., Barron, K., Le Brun-Blashka, S., & Metzger, B. (2023). Guided metabolic detoxification program supports phase II detoxification enzymes and antioxidant balance in healthy participants. *Nutrients, 15*(9), 2209.

Park, C. H., Kwon, O. Y., Shim, H. J., Kim, M. H., Lee, J. H., Lee, K. J., ... & Kim, M. R. (2008). Protection of green leafy vegetable extracts against oxidation of human low-density lipoprotein. *Food Science and Biotechnology, 17*(1), 151–155.

Park, Y. C. (2024). Understanding the detoxification system and functional approach. *Journal of Korean Institute for Functional Medicine, 7*(1), 34–39. https://doi.org/10.32581/jkifm.2024.7.1.34

Pastuszak, Ż., Trawińska, L., Trawińska, J., & Stępień, J. (2023). Evaluation of effectiveness and safety of short term low fat ketogenic diet in the reduction of fat mass in patients over 40 years old, 11 May 2023, PREPRINT (Version 1) available at Research Square [https://doi.org/10.21203/rs.3.rs-2851251/v1] (Accessed: 28.10.2024).

Rajiyah, T., & Pannain, S. (2020). *The ketogenic diet in health and disease.* In Dietary Sugar, Salt and Fat in Human Health (pp. 557–619). Academic Press.

Rana, M. K. (2016). Salad Crops: Dietary Importance. *Encyclopedia of Food and Health,* 665–672.

Randhawa, M. A., Khan, A. A., Javed, M. S., & Sajid, M. W. (2015). Green leafy vegetables: a health promoting source. In *Handbook of fertility* (pp. 205–220). Academic Press.

Ranjbar, M., Shab-Bidar, S., Mohammadi, H., & Djafarian, K. (2024). Effect of Intermittent Fasting on Liver Function Tests: A Systematic Review and Meta-Analysis of Randomized Clinical Trials. *Nutrition Reviews*, *83*(3), e965-e979.

Reddy, B. L., Reddy, V. S., & Saier Jr, M. H. (2024). Health benefits of intermittent fasting. *Microbial Physiology*, *34*(1), 142–152.

Rivoltini, L., Vernieri, C., & De Braud, F. (2022). Low-calorie food preparation for the nutrition of cancer patients. https://patents.google.com/patent/US20220338522A1/en (Accessed: 29.10.2024).

Tinsley, G. M., & La Bounty, P. M. (2015). Effects of intermittent fasting on body composition and clinical health markers in humans. *Nutrition Reviews*, *73*(10), 661–674.

Watts, M., Gabriel, S., Goldhamer, A. C., & Myers, T. R. (2023). Prolonged water-only fasting in the management of hashimoto's thyroiditis: a case report. *Integrative and Complementary Therapies*, *29*(6), 309–311.

World Health Organization. (n.d.). Healthy diet. World Health Organization. Retrieved from https://www.who.int/news-room/fact-sheets/detail/healthy-diet (Accessed: 29.10.2024).

Zeiler, E., Gabriel, S., Ncube, M., Thompson, N., Scharf, E., Goldhamer, A. C., & Myers, T. R. (2024). Prolonged Water-only Fasting is a Safe and Feasible Treatment Option for Managing Stage 1 and 2 Hypertension. *medRxiv*, 2024-02.

Investigation of The Effect of Social Intelligence on Social Entrepreneurship Tendency in The Tourism Education

Olcay ÖZIŞIK YAPICI[1], *Kürşad SAYIN*[2]

Introduction

Social entrepreneurship is a subfield of entrepreneurship literature that describes organizations and initiatives that follow innovation and aim to solve various social and environmental challenges (Hietschold et.al., 2022). Social entrepreneurship contributes to social values by reducing social issues such as the needs of underprivileged children, poverty, increasing crime tendencies, and rejection of access to education. Social entrepreneurship positively affects the solution of environmental and social problems (such as equality, promotion, innovation, and employment) and contributes to society by providing business education and career counseling during the economic recession (El-Den, Adikhari & Azam, 2017). Social entrepreneurship plays an important role in the form of creating a social value and transferring this value to society (Tekin, 2019). Factors such as establishing a social enterprise, maintaining activities, providing social welfare, providing personal satisfaction, using knowledge and mastery, creating a change in society, wanting to help those who want to engage in social enterprise, being able to respond a previously insufficiently met need and ensuring sustainability have an impact on this role (Türkeş & Özgeldi, 2023).

Social entrepreneurship is one of the creative ways to cope with the complex demands of modern life. Social entrepreneurship contributes to the development of the economy, leads to innovation and gets solutions for social transformation (do Adro & Fernandes, 2022). The importance of social entrepreneurship for the economic growth of countries and the protection of economic prosperity has been increasing gradually. Institutions and economic literature have recently

---

1 Ondokuz Mayıs University, Faculty of Tourism, Department of Tourism Management, olcay.yapici@omu.edu.tr

2 Selçuk University, Silifke Taşucu Vocational School, Department of Hotel, Restaurant and Catering Services, kursadsyn@selcuk.edu.tr

been paying more attention to social entrepreneurship as the driving force of sustainable development goals (Trabskaia et.al, 2023). In addition, social entrepreneurship contributes to improving the labor market quality (Waligora, 2019), innovation and job creation (Dato-on & Kalakay, 2016) and contributes to regional development (Mozas Moral et al., 2020), as well as in crises, they create great efforts to combat serious effects (Fernandez-Guada & Martín-Lopez, 2023).

Altinay et al., (2016) also state that social entrepreneurship in the tourism sector tends to support an important development approach, especially in developing countries with excessive resource constraints and insufficient support for entrepreneurial activities in tourism by the government. For this reason, the decreasing public expenditures in recent years and the inability of the private and public sectors to meet environmental and social needs have increased the importance of social enterprises. During periods when countries are experiencing economic crises, tourism plays an important role in reducing the country's current account deficits by providing a large foreign exchange input (Ateş, Sunar & Bağcı, 2019) and supports the economic development of countries. The tourism sector has a labor-intensive feature helps to reduce unemployment in countries (Karaman & Çiçekdağ, 2023). Social Entrepreneurship Organizations have been gaining popularity worldwide as a way to solve social problems with sustainable commercial solutions in tourism. This popularity is increasing because people prefer more cost-effective, innovative and sustainable steps to overcome social issues in their region (Fatchurrohman & Khakim, 2023).

There are some research in the literature. Tu, et al. (2021) searched the intentions of graduate students toward social entrepreneurship, Garcia-Gonzalez & Ramirez-Montoya (2021) explored social entrepreneurship education through teaching and promoting. Rambe & Ndofirepi (2021) searched the intention of social entrepreneurship in Zimbabwe. The effect of social intelligence on social entrepreneurship intention has not been investigated within the scope of vocational tourism education. This study expands the perspective of social intelligence in tourism students in terms of social intelligence. In summary, we sought to answer two research questions:

Q1: How does social intelligence affect the intention of social entrepreneurship tourism students?

Q2: Which combination of sub-dimensions (social knowledge process, social skills and social awareness) of social intelligence causes an increased intention of social entrepreneurship?

The research differs from other studies examining the relationship between social intelligence and social entrepreneurship. Additionally, it can be considered as the first study examining them. Secondly, it reveals that students who get tourism education and do internships have higher social entrepreneurship intentions. This situation emphasizes the importance of internships in tourism education. In this way, students contribute to sustainability and rural development by increasing their social entrepreneurship intentions.

Literature and Hypothesis Development

Social Entrepreneurship

Social entrepreneurship is gained in the literature about thirty years ago (Newbert & Hill, 2014). Social entrepreneurship effectively creates social value and ensures social transformation (Altinay et al., 2016; Sheldon, Pollock & Daniele, 2017). It is seen that the concept of social entrepreneurship is included in the tourism literature (Kummitha et al., 2021). Social entrepreneurship has gained interest in tourism in recent years (Sheldon et al., 2017) and is of particular importance in both developing and developed countries where economic difficulties arising in obtaining financial resources by social entrepreneurs occur (Bruton, et al., 2015). Social entrepreneurship in tourism brings about social change (Sheldon et al.,. 2017). With tourism social entrepreneurship, positive results can be achieved in economic, social and environmental terms, and this situation can increase society's well-being and positive social change (Altinay et al., 2016). Tourism social entrepreneurship contribute to ensuring economic and environmental sustainability in local communities (Dahles, et al., 2019). Social entrepreneurship is considered important and necessary in tourism (Nasrolahi Vosta & Jalilvand, 2014). Therefore, when considering the students who are studying tourism as future tourists, it is important to determine the tendencies of social entrepreneurship.

Social Intelligence

Social intelligence was first defined by Edward Thorndike (1920) as the ability to understand, and manage women, men and children by acting credibly in interpersonal relationships. Among the types of intelligence in Thorndike's (1920) Three-Factor Theory of Intelligence there are; social intelligence, mechanical intelligence and abstract intelligence's definition of social intelligence is closely related to the definition of social intelligence in Howard Gardner's (1983; 1993; 2002), Theory of Multiple Intelligences. Gardner (1993) defined social intelligence as a type of multiple intelligence and proved to be different from other dimensions of intelligence.

Social intelligence has become more important due to stress, increased tensions and various complex situations. For these reasons, interest in social intelligence, defined as interpersonal competencies that facilitate effective human behavior by improving the ability to make emotionally and socially intelligent decisions, has increased again. Research on social intelligence has shown that social intelligence has two common components awareness of others and reactions and adaptation to others and other social situations (Björkqvist, Österman & Kaukiainen 2000). Social intelligence helps solve the problems of social life and overcome various social tasks. Therefore, social intelligence is an important developmental aspect of education (Weis & Sub, 2007). The development of an individual's socially intelligent behavior contributes to the development of people in reflecting on themselves and social processes, interpreting behavior, and training social skills (Frankovský, 2014). Social intelligence can be learned, developed and used as an effective life skill to manage personal and interpersonal relationships and achieve success in all areas of life (Saxena & Jain, 2013).

Social Intelligence and Social Entrepreneurship Relationship

The subjects related to the concept of social intelligence in the field of literature are; emotional intelligence (Sert & Raza, 2019), social intelligence-self-efficacy belief level (Berkant & Ekici, 2007), social intelligence- self-anxiety (Doğan et al., 2009). Although social entrepreneurship is a new concept from an academic point of view, it comes across to people as a concept whose history goes back much further. The concept of social entrepreneurship is defined as a process that emphasizes the aspect of social responsibility in addition to the economic characteristics of entrepreneurship and involves the innovative use of resources to create social value, realize social change or meet social needs (Özbilen, et al., 2020). Social entrepreneurship is a tool that involves the recognition, evaluation and use of opportunities that result in a social value for the basic social needs of society, as opposed to the well-being of personnel or stakeholders (Peredo & McLean, 2006). Hence we hypothesize that:

H1: The social information process has a significant effect on social entrepreneurship.
H_{1a}: The social information process has a significant effect on empathy.
H_{1b}: The social information process has a significant effect on moral obligation.
H_{1c}: The social information process has a significant effect on social entrepreneurial self-efficacy.
H_{1d}: The social information process has a significant effect on perceived social support.

Social entrepreneurs aim to ensure the social welfare of people in need of support and assistance in today's world and to find sustainable solutions to challenging living standards (Gencer & Aydınoğlu, 2020). Social intelligence enables an individual to communicate and cooperate with others. Therefore, the success is related to social intelligence. Social entrepreneurship requires individuals to have strong social relationships. Being able to be an entrepreneur makes it necessary for people to have strong social and human relationships (Uzun, Buran & Beydilli, 2017). Above the arguments we hypothesizes:

H_2: Social skills have a significant effect on social entrepreneurship.
H_{2a}: Social skills have a significant effect on empathy.
H_{2b}: Social skills have a significant effect on moral obligation.
H_{2c}: Social skills have a significant effect on social entrepreneurial self-efficacy.
H_{2d}: Social skills have a significant effect on perceived social support

Kömeçoğlu (2022) researched factory employees in Kastamonu, Çakanel (2018) researched Pamukkale University Faculty of Economics and Administrative Sciences students, Keser & Sari (2024) conducted a study on individuals over the age of eighteen and it was found that they reached the results that there is a positive significant relationship between intelligence and entrepreneurship. A study conducted in South Africa shows that the education of students studying in tourism and taking entrepreneurship courses affects their entrepreneurial intentions. It has been concluded that it positively affects the willingness of students getting tourism education to start tourism-related businesses and the perception of entrepreneurship (Sibusisso & Ikechukwu, 2023).

When tourism students complete their internships, confidence in their skills and abilities may increase and their self-sufficiency as future entrepreneurs may increase (Gurusamy & Ganapathy, 2023). In a study conducted in Indonesia, it was concluded that the values of students related to entrepreneurship were positively affected during internships undertaken in the scope of the start-up program (Lantu et al., 2021). Based on the above arguments, we hypothesize:

H_3: Social awareness has a significant effect on social entrepreneurship.
H_{3a}: Social awareness has a significant effect on empathy.
H_{3b}: Social awareness has a significant effect on moral obligation.
H_{3c}: Social awareness has a significant effect on social entrepreneurial self-efficacy.
H_{3d}: Social awareness has a significant effect on perceived social support.

When the relevant literature was examined, there were no studies on tourism program students on social intelligence and social entrepreneurship issues. It is believed that the research conducted by determining the relationship between the social entrepreneurship tendency and the social intelligence of students receiving tourism education will fill a gap in the literature. The importance of social entrepreneurship in tourism in solving social problems such as supporting sustainable development, creating social value, equal opportunities, unemployment and social change is increasing (McGehee et al., 2014).

Methods

Study Design

The study aims for undergraduate and postgraduate students of tourism students in a university in Türkiye. The study adopted exploratory but descriptive research design due to the nature of issues under investigation towards addressing study objectives (Iwu et al., 2019). Exploratory research design helps identify problems, generate hypotheses gain insight into a subject (Soomro et al., 2019), and formulate conclusions and recommendations.

The research population consists of students studying tourism at the higher education level in Türkiye. The research sample consists of 400 students selected randomly from the population. The survey forms were delivered to the students via the Internet and 351 questionnaires were included in the analysis. The survey form was used as a data collection tool in the research. The survey form consists of three parts. In the first section, demographic information of the participants. In the second part, the items related to the social intelligence scale and in the third part, the items related to the social entrepreneurship scale are included as a 5-point Likert type. Social Intelligence Scale, The Tromso Social Intelligence Scale, which consists of 21 items in three dimensions and whose Turkish form has been adapted by Doğan & Çetin (2009), was used. A 4-dimensional scale consisting of 17 items adapted by Akar (2020) was used for the Social Entrepreneurship scale.

Two types of variables were taken into account in the research. The Social Intelligence Scale, which has the social knowledge process, social skills, and social awareness subdimensions, was used as the independent variable, and the Social Entrepreneurship Tendency Scale, which has the empathy, social moral obligation, social entrepreneurship self-efficacy, and perceived social support subdimensions, was used as the dependent variable. The hypotheses explaining the relationships between the expressions are explained below.

Data Analyses

A computer software, Smart-Pls was used for data analysis. Validity and reliability of the variables test used. Confirmatory factor analysis (outer loadings) was performed to test construct validity. Cronbach Alpha test was applied to test the reliability of the scales. A cut-off point of between 0.5 and 0.7 is commonly used by researchers to denote the internal consistency or reliability of variables measuring a particular dimension or study construct (Collier, 2020; Iwu et al., 2019). This study considered a Cronbach's Alpha coefficient of 0.7 and above acceptable for good reliability. The average variance extracted (AVE) was calculated to measure the convergent validity of the scales. The heterotrait monotrait ratio (HTMT) and Fornell-Larcker criterion were calculated to test the discriminant validity of the scales. Variance inflation factor (InnerVIF) values were calculated to test whether the research model had a multicollinearity problem. PLS Predict analysis was applied to determine predictive power and a cross-validation predictive ability test (CVPAT) was applied to perform validation. Multiple group analysis (MGA) was conducted between those with experience in the tourism sector and those who did not.

Results

Demographic Profile

When the demographic characteristics of the individuals participating in the research are examined, 59.8% are female and 40.2% are male. When the ages of the participants are examined, 56.7% are 19–21, 25.1% are 22–24, 9.4% are 16–18, 4% are 30 and over, 3.4% are 25–27 and 1.1% are 28. -30. When the participants' education levels are examined, 67.2% have associate degree, 31.6% have a bachelor's degree and 1.1% have a postgraduate. When the participants' past experiences in the tourism sector were examined, it was determined that 52.4% of them had and 47.6% had no experience.

Table 1. Demographics features

Category		N	%
Gender	Male	141	40.2
	Female	210	59.8
Age	16–18	33	9.4
	19–21	199	56.7
	22–24	88	25.1
	25–27	12	3.4
	28–30	4	1.1
	30 and older	14	4.0
Education	Associate degree	236	67.2
	Bachelor's degree	111	31.6
	Postgraduate	4	1.1
Experience	Yes	184	52.4
	No	167	47.6

Measurement Model Assessment

In the research, the validity and reliability scores of the variables in the model were examined. Confirmatory factor analysis (outer loadings) was performed to test construct validity. Empathy2 and empathy5 items from the empathy scale, PSS4 item from the perceived social support scale, SAware2 and SAware6 items from the social awareness scale, SES4 item from the social entrepreneurship self-efficacy scale, SIP6 item from the social information process scale and, SSkills1, SSkills4 and SSkills6 items from the social skills were removed from the study because factor loadings were below 0.50 (Kaiser, 1974). Cronbach Alpha test was applied to test the reliability of the scales. When the results were examined, the reliability values of all scales were above 0.70 (Hair et al., 2019).

Table 2. Reliability and Validity Scores

Construct	λ	*OutlerVIF*	α	*rho_a*	*rho_c*	*AVE*
***Social Information Processing* (SIP)**			.847	.849	.884	.523
SIP1 <- SIP	.689	1.517				
SIP2 <- SIP	.743	1.741				
SIP3 <- SIP	.786	2.128				
SIP4 <- SIP	.755	1.795				
SIP5 <- SIP	.663	1.465				
SIP7 <- SIP	.692	1.535				
SIP8 <- SIP	.728	1.649				
***Social Skills* (SSkills)**			.801	.803	.883	.715
SSkills2 <- SSkills	.863	2.026				
SSkills3 <- SSkills	.844	1.875				
SSkills5 <- SSkills	.828	1.511				
Social Awareness (SAware)			.768	.980	.829	.499
SAware1 <- SAware	.593	1.195				
SAware3 <- SAware	.895	1.848				
SAware4 <- SAware	.776	1.833				
SAware5 <- SAware	.593	1.488				
SAware7 <- SAware	.625	1.521				
Empathy			.725	.770	.841	.638
Empathy1 <- Empathy	.797	1.468				
Empathy3 <- Empathy	.738	1.419				
Empathy4 <- Empathy	.857	1.394				
Moral Obligation (MObligation)			.866	.875	.908	.713
MObligation1 <- MObligation	.823	1.832				
MObligation2 <- MObligation	.870	2.399				
MObligation3 <- MObligation	.796	1.847				

(Continued)

Table 2. (Continued)

Construct	λ	*OutlerVIF*	α	*rho_a*	*rho_c*	*AVE*
MObligation4 <- MObligation	.885	2.453				
Social entrepreneurial self-efficacy (SES)			.788	.793	.877	.704
SES1 <- SES	.769	1.400				
SES2 <- SES	.886	2.145				
SES3 <- SES	.859	1.957				
Perceived social support (PSS)			.734	.750	.853	.662
PSS1 <- PSS	.666	1.172				
PSS2 <- PSS	.893	2.393				
PSS3 <- PSS	.863	2.265				

Note. α = Cronbach Alpha; λ = Outer loadings; OuterVIF = Outer variance inflation factor; rho_a and rho_c = Composite reliability; AVE = Averaged variance extracted.

Composite reliability was calculated to determine the "internal consistency of the scales." For this, the rho_a value of Dijkstra & Henseler (2015) was calculated and all scales gave results above .70. Secondly, rho_c was calculated for internal consistency and all scales were above 0.60 (Bagozzi & Yi, 1988). "The average variance extracted" (AVE) was calculated to measure the convergent validity of the scales. AVE values were above 0.50 (Fornell & Larcker, 1981) (See Table 2).

Table 3. Heterotrait Monotrait Ratio

No	Construct	1	2	3	4	5	6	7
1	**Empathy**							
2	**MObligation**	.815						
3	**PSS**	.515	.492					
4	**SAware**	.118	.100	.133				
5	**SES**	.608	.682	.709	.119			
6	**SIP**	.631	.580	.572	.103	.683		
7	**SSkills**	.360	.368	.678	.159	.461	.540	

Note. Numbers in bold represent the square root of AVE. MObligation= Moral obligation, PSS= Perceived social support, SAware= Social awareness, SES= Social entrepreneurial self-efficacy, SIP= Social information process, SSkills= Social skills.

To test the discriminant validity of the scales, HTMT (See Table 3) and Fornell-Larcker criterion (See Table 4) were calculated. As a result of HTMT analysis, it was determined that the values were below 0.90 (Henseler et al., 2015). In addition, the Fornell-Larcker criterion, the AVE square root values of all scales were higher than their correlation loadings with the others (Fornell & Larcker, 1981).

Table 4. Fornell Larcker Criterion

No	Construct	1	2	3	4	5	6	7
1	Empathy	**.799**						
2	MObligation	.666	**.844**					
3	PSS	.380	.387	**.814**				
4	SAware	-.024	.019	-.122	**.707**			
5	SES	.477	.561	.531	-.099	**.839**		
6	SIP	.515	.502	.445	-.051	.556	**.723**	
7	SSkills	.294	.317	.522	-.135	.367	.445	**.845**

Note. Numbers in bold represent the square root of AVE. MObligation= Moral obligation, PSS= Perceived social support, SAware= Social awareness, SES= Social entrepreneurial self-efficacy, SIP= Social information process, SSkills= Social skills.

Structural Model Assessment

InnerVIF values were calculated to test whether there was a multicollinearity problem in the research model. As a result of the analysis, it was determined that all values were below 5.00 (See Table 6). Thus, it was concluded that there was no multicollinearity problem between the scales (Becker et al., 2015). The determination coefficient of the independent variables to the dependent variables (R^2) was calculated. As a result of the analysis, the determination ratio of empathy and, social moral obligation is 0.26 and, the determination ratio of perceived social support and social entrepreneurship self-efficacy is 0.32. The effect size (f^2) of the dependent variables on the independent variables was calculated (See Table 6). Accordingly, the social knowledge process on the effect of entrepreneurial self-efficacy had the highest effect size (.29). In addition, the effect size of the social knowledge process on empathy was 0.25, the effect size of the social knowledge process on social moral obligation was 0.22 and the effect size of social skills on perceived social support was .19 (See Table 3). Since these values were above 0.15 (Cohen, 1988), they were considered moderate.

Predictive Model Assessment

PLSPredict analysis was applied to determine predictive power and CVPAT was applied to perform validation. First, the $Q^2_{predict}$ index appeared above 0. Since the estimated error coefficients are not symmetrically distributed, PLS-SEM_MAE and LM_MAE values were compared (Shmueli et al., 2019). Accordingly, the majority of LM-MAE values were higher than PLS-SEM_MAE values. Therefore, the predictive power of the model is high. CVPAT analysis was performed to test the prediction accuracy. The PLS-SEM prediction outperformed both the naive indicator average (IA) and linear model prediction (LM). Model prediction values are shown in detail in Table 5.

Table 5. PLSPredict and CVPAT Results

Construct	Items	$Q^2_{predict}$	*PLS-SEM_ MAE*	*LM_ MAE*	IA Average Loss Difference (*p*-value)	LM Average Loss Difference (*p*-value)
Empathy	Empathy1	.126	.603	.625	-.116 (.023)	-.019 (.372)
	Empathy3	.075	.718	.732		
	Empathy4	.239	.542	.534		
Moral Obligation	MObligation1	.189	.598	.619	-.151 (.003)	-.011 (.529)
	MObligation2	.171	.608	.635		
	MObligation3	.121	.632	.644		
	MObligation4	.212	.626	.634		
Perceived social support	PSS1	.157	.626	.634	-.174 (.000)	-.025 (.055)
	PSS2	.237	.681	.677		
	PSS3	.217	.647	.653		
Social entre-preneurial self-efficacy	SES1	.199	.529	.541	-.147 (.001)	-.019 (.079)
	SES2	.225	.575	.592		
	SES3	.222	.560	.553		
	SES1	.199	.529	.541		
Overall					-.147 (.000)	-.018 (.046)

Hypothesis Testing

Hypotheses were tested and presented in Figure 1 and Table 6. The social information process has a significant effect on empathy (ß = .479, t = 5.993, p < .000), on moral obligation (ß = .449, t = 6.683, p < .000), on the social entrepreneurial self-efficacy (ß = .492, t = 8.187, p < .000), and on the perceived social support (ß = .264, t = 4.285, p < .000). Therefore, hypotheses H_{1a}, H_{1b}, H_{1c} and H_{1d} are supported. Social skills have a significant effect on the moral obligation (ß = .125, t = 2.281, p < .023), on the social entrepreneurial self-efficacy (ß = .141, t = 2.529, p < .011), and on the perceived social support (ß = .398, t = 8.282, p < .000). Therefore, hypotheses H_{2b}, H_{2c} and H_{2d} are supported. However, ambiance (ß = .082, t = 1.349, p < .177) does not have a significant effect on empathy. So, H_{2a} is not supported.

Table 6. Structural Equation Model Results

Hypotheses		*ß*	*x̄*	*SD*	*t-statistic*	*p-value*	*InnerVIF*	*f*²	*Result*
H_{1a}	SIP -> Empathy	.479	.477	.080	5.993	.000***	1.247	.253	*Supported*
H_{1b}	SIP -> MObligation	.449	.448	.067	6.683	.000***	1.247	.220	*Supported*
H_{1c}	SIP -> SES	.492	.488	.060	8.187	.000***	1.247	.288	*Supported*
H_{1d}	SIP -> PSS	.264	.261	.062	4.285	.000***	1.247	.085	*Supported*
H_{2a}	SSkills -> Empathy	.082	.078	.061	1.349	.177	1.267	.007	*Not supported*
H_{2b}	SSkills -> MObligation	.125	.120	.055	2.281	.023**	1.267	.186	*Supported*
H_{2c}	SSkills -> SES	.141	.141	.056	2.529	.011**	1.267	.017	*Supported*
H_{2d}	SSkills -> PSS	.398	.398	.048	8.282	.000***	1.267	.024	*Supported*
H_{3a}	SAware -> Empathy	.013	.029	.087	.150	.880	1.019	.000	*Not supported*
H_{3b}	SAware -> MObligation	.059	.068	.077	.763	.446	1.019	.005	*Not supported*
H_{3c}	SAware -> SES	-.054	-.049	.067	.801	.423	1.019	.004	*Not supported*
H_{3d}	SAware -> PSS	-.055	-.057	.063	.866	.387	1.019	.004	*Not supported*

Social awareness doesn't have a significant effect on empathy (ß = .013, t = .150, p < .880), on moral obligation (ß = .059, t = .763, p < .446), on social entrepreneurial self-efficacy (ß = -.054, t = .801, p < .423), and on the perceived social support (ß = -.055, t = .866, p < .387). Therefore, hypotheses H_{3a}, H_{3b}, H_{3c} and H_{3d} are not supported.

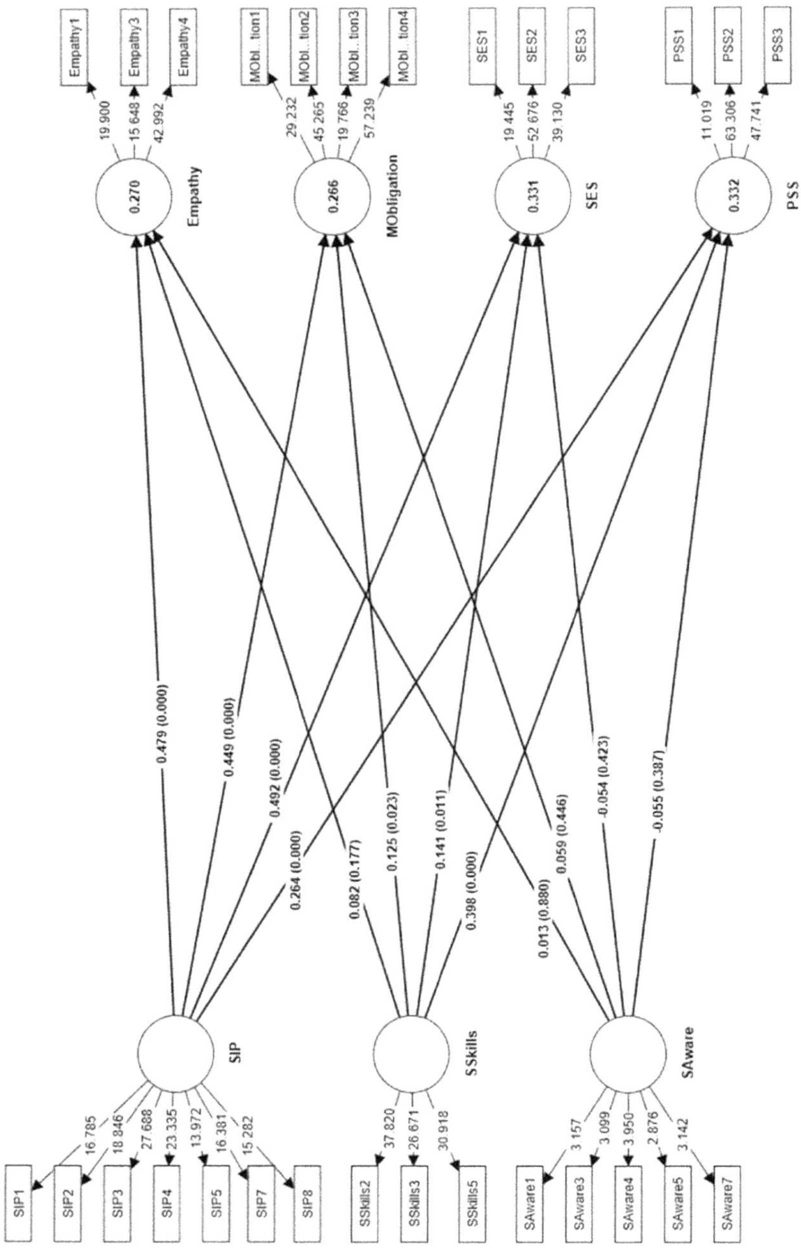

Figure 1: Research Model

Multi-Group Analysis

In the research, multiple group analysis was conducted between those who had experience in the tourism sector and those who did not. Invariance measurement assessment (Henseler et al., 2015) was applied for MGA. Configural invariance, compositional invariance, mean values, and variances of composite tests were checked in the measurement invariance for composite models (MICOM) analysis. First, it was examined whether there was a difference between the outer loadings of the two groups and no significant difference was found. In addition, validity and reliability scores were found to be good in both groups (Cheah et al., 2020). As a result of MICOM analysis, all scales in composite invariance results gave results greater than .05. In addition, the similarity of mean composite and variance values except for social entrepreneurial self-efficacy and social information process was greater than .05 and it is seen that the equality is partially achieved. In this context, the results were examined using non-parametric tests (See Table 7).

Table 7. MGA results (experienced vs inexperienced)

Hypotheses	No (β)	Yes (β)	p value (no)	p value (yes)	*Results*
SIP -> Empathy	.439	.534	.000	.000	Yes
SIP -> MObligation	.409	.483	.000	.000	Yes
SIP -> SES	.468	.488	.000	.000	Yes
SIP -> PSS	.181	.355	.039	.000	Yes
SSkills -> Empathy	.068	.092	.493	.254	No
SSkills -> MObligation	.054	.181	.576	.007	No
SSkills -> SES	.014	.260	.870	.000	No
SSkills -> PSS	.427	.342	.000	.000	Yes
SAware -> Empathy	-.028	.111	.791	.246	No
SAware -> MObligation	.046	.099	.648	.267	No
SAware -> SES	-.093	-.051	.309	.537	No
SAware -> PSS	-.079	-.059	.355	.466	No

MObligation= Moral obligation, PSS= Perceived social support, SAware= Social awareness, SES= Social entrepreneurial self-efficacy, SIP= Social information process, SSkills= Social skills.

As a result, the effect of social information processing on empathy, moral obligation, social entrepreneurial self-efficacy and perceived social support is higher for experienced people than for inexperienced ones. On the other hand, the effect of social skills on perceived social support is higher for those who are inexperienced than for those who are experienced. In other hypotheses, there is no significant difference between those who are experienced and those who are not.

Conclusions

Social entrepreneurship may be easier to support sustainable development, create social value and solve social problems such as equal opportunities, unemployment and social change. Research, the effect of the social intelligence of the students receiving tourism education on the social entrepreneurship tendency and the fact that they have completed an internship in the field has positive impact on their social entrepreneurship. Internships covering practical training on how students receiving tourism education can become social entrepreneurs are important. With the internships, tourism students' self-confidence emerges, they get to know the sector, and their social entrepreneurship and social intelligence are positively affected. It shows that those who has experience in the field of tourism education among the students may have intentions to become social entrepreneurs in tourism after graduation and may be more resistant to the difficulties encountered in this regard.

It is concluded that the social information process, one of the sub-dimensions of social intelligence, significantly affects empathy, moral obligation, social entrepreneurship self-efficacy and perceived social support, which are sub-dimensions of social entrepreneurship. At the same time, it is also concluded that social skills have a significant effect on moral obligation, social entrepreneurship self-efficacy and perceived social support in the sub-dimensions of social entrepreneurship, while they do not have a significant effect on empathy. Furthermore, according to the results of the research, while social awareness has a significant effect on empathy, it does not have a significant effect on moral obligation, social entrepreneurship self-efficacy and perceived social support.

In addition, another result of the research is that the effect of the social information process of tourism students on empathy, moral obligation, social entrepreneurship self-efficacy and perceived social support is higher for students with industry experience than for those without experience.

Moreover, the effect of social skills on perceived social support is higher for the inexperienced than for the experienced. In other dimensions, there is no significant difference between those who are experienced and those who are not. As a result, the importance of sector experience (internship) in tourism education is emerging.

Theoretical Implications

This study has contributed to the existing literature on social intelligence, sector experience and social entrepreneurship tendency by investigating the effect of social intelligence on social entrepreneurship tendency in the context of sector experience. The research findings show some theoretical results:

(1) The effects of social intelligence on social entrepreneurship tendency: It reveals the effect of social intelligence on social entrepreneurship tendency in the context of students receiving tourism education. This finding is in line with previous studies emphasizing the effects of social intelligence on social entrepreneurship (Çakanel, 2018; Keser & Sarı, 2024).

(2) Understanding the role of the social information process on the social entrepreneurship trend in the context of students receiving vocational education (internship): According to Babson (2012), younger generations with work experience show an effort to maintain the entrepreneurship and leadership of people in the business environments where they work. From this point of view, the experiences of the social group in which individuals are involved in entrepreneurial processes provide important contributions to the future social entrepreneurial characteristics of individuals. According to Stewart, Lumpkin & Katz (2010), the previous social environment and work experiences of individuals and their families can significantly contribute to their social entrepreneurship potential (Aydoğmuş, 2019). The research shows that the effects of the social information process on the tendency of social entrepreneurship are higher in students who have received tourism education and have industry experience.

Practical Implications

Social entrepreneurship is important to create innovative solutions to the increasing social problems (Akkan & Süygün, 2016). According to the findings obtained from this research, it can be said that individuals must have high social intelligence for social entrepreneurship to be carried out effectively by individuals. Educational activities and studies that will increase the social intelligence of students should be carried out by educational institutions and public and non-governmental organizations aimed at students who is going to work in the tourism sector after their education. Educational activities and studies will be useful for students to develop empathy abilities in their educational and educational life, to communicate correctly with individuals and to support their emotional and social intelligence. At the same time, it is also considered important to ensure that students participate in social life by actively participating in sector practices during their education, communicating with people from different cultures, participating in teamwork with non-governmental organizations, visiting tourism enterprises within the framework of university-sector cooperation in the implementation of some of their courses.

Recommendations

According to the findings obtained from the research, suggestions have been made to support the authorities working at universities in Turkey to enable a larger number of students to become social entrepreneurs in the field of tourism. In many of the universities that provide tourism education, entrepreneurship education is provided. It may be useful to design the subject of social entrepreneurship in such a way that it will be included within the scope of the entrepreneurship course. It is recommended to follow the student's development related to entrepreneurship during the education period when the entrepreneurship course is taught at the associate degree and undergraduate levels.

It is recommended that students be made aware of social entrepreneurship and that studies be carried out aimed at being effective in many areas together with social entrepreneurship. Emphasizing that social entrepreneurship on a global scale is related to the Development Goals of the United Nations can encourage and support students.

It is proposed to increase the practices for university students who provide tourism education to become more intertwined with the sector and gain experience. Experienced students who know their sector are more willing to engage in social entrepreneurship. Therefore, it may be useful to review the adequacy of the application hours of the courses in the curriculum once again.

Limitations and Further Research

The fact that the research was conducted with 351 students who are studying tourism in Turkey is a limitation. Research covering different analyses on social entrepreneurship can be carried out by reaching a larger number of students getting a tourism education.

References

Akar, H. (2020). Sosyal girişimcilik öncülleri ölçeğinin Türkçe'ye uyarlanması: geçerlik ve güvenirlik çalışması, *Turkish Studies- Education*, 15(2), 579–593.

Akkan, E., & Süygün, M. S. (2016). Lisans öğrencilerinin özyeterlik algıları ve başarma güdülerinin sosyal girişimcilik eğilimlerine etkilerinin incelenmesi. *Journal of Economics and Administrative Sciences*, (XVIII)2, 35–63.

Altinay, L., Sigala, M., & Waligo, V. (2016). Social value creation through tourism enterprise. *Tourism Management*, 54, 404–417. https://doi.org/10.1016/j.tourman.2015.12.011.

Ateş, A., Sunar, H. & Bağcı, T. (2023). An implementation proposal for the reintegration of casinos into Turkish Tourism. *Akşehir Meslek Yüksekokulu Sosyal Bilimler Dergisi*, 16, 78–88.

Aydoğmuş, M. (2019). Öğretmen adaylarının sosyal girişimcilik özelliklerinin farklı değişkenler açısından incelenmesi. *EKEV Akademi Dergisi*, (23)80, 339–352.

Babson, C.B. (2012). How can family business be more entrepreneurial?https://www.forbes.com/sites/babson/2012/05/11/how-can-family-business-be-more-entrepreneurial/?sh=456985f3270a

Bagozzi, R. P., & Yi, Y. (1988). On the evaluation of structural equation models. *Journal of the Academy of Marketing Science*, 16, 74–94.

Becker, J. M., Ringle, C. M., Sarstedt, M., & Völckner, F. (2015). How collinearity affects mixture regression results. *Marketing Letters*, 26, 643–659. https://doi.org/10.1007/s11002-014-9299-9

Berkant, H.G., & Ekici, G. (2007). Sınıf öğretmeni adaylarının fen öğretiminde öğretmen özyeterlik inanç düzeyleri ile zeka türleri arasındaki ilişkinin değerlendirilmesi. *Ç.Ü. Sosyal Bilimler Enstitüsü Dergisi*, 16(1), 113–132.

Björkqvist, K., Österman, K., & Kaukiainen, A. (2000). Social intelligence - empathy= aggression? *Aggression and violent behavior*, 5(2), 191–200.

Bruton, G. D., Khavul, S., Siegel, D., & Wright, M. (2015). New financial alternatives in seeding entrepreneurship: Microfinance, crowdfunding, and peer-to-peer innovations. *Entrepreneurship: Theory & Practice*, 39(1), 9–26.

Cheah, J. H., Thurasamy, R., Memon, M. A., Chuah, F., & Ting, H. (2020). Multigroup analysis using SmartPLS: Step-by-step guidelines for business research. *Asian Journal of Business Research*, 10(3), I-XIX.

Cohen, J. (1988). *Statistical Power Analysis For The Behavioral Sciences* (2nd ed.). Lawrence Erlbaum Associates.

Collier, J. E. (2020). Applied *Structural Equation Modelling Using AMOS*: Basic to advanced techniques. Routledge.

Çakanel, Z. (2018). Duygusal Zekâ, Sosyal Zekâ ve Sosyal Girişimcilik Arasındaki İlişkinin İncelenmesi, Pamukkale Üniversitesi Yüksek lisans tezi. Sosyal Bilimler Enstitüsü, Pamukkale Üniversitesi.

Dahles, H., Khieng, S., Verver, M. & Manders, I. (2019). Social entrepreneurship and tourism in Cambodia: Advancing community engagement. *Journal of Sustainable Tourism*, 28, 816–833.

Dato-on, M. C., & Kalakay, J. (2016). The winding road of social entrepreneurship definitions: a systematic literature review. *Social Enterprise Journal*, (12)2, 131–160.

Dijkstra, T. K., & Henseler, J. (2015). Consistent and asymptotically normal PLS estimators for linear structural equations. *Computational Statistics & Data* Analysis, 81, 10–23.

do Adro, F., & Fernandes, C. (2022). Social entrepreneurship and social ınnovation: Looking inside the box and moving out of it. Innovation. *The European Journal of Social Science Research* 35: 704–730.

Doğan, T., & Çetin, B. (2009). Tromso sosyal zekâ ölçeği Türkçe formunun faktör yapısı, geçerlik ve güvenirlik çalışması. *Kuram ve Uygulamada Eğitim Bilimleri / Educational Sciences: Theory & Practice* 9 (2), Bahar, 691–720.

Doğan, T., Totan, T., & Sapmaz, F. (2009). Üniversite öğrencilerinde benlik saygısı ve sosyal zeka. *Sakarya Üniversitesi Eğitim Fakültesi Dergisi*, 17, 235–247.

El-Den, J., Adikhari, P. & Azam, S. (2017). Social media in the service of social entrepreneurship: Identifying factors for better services. *Journal of Advances in Humanities and Social Sciences*, 3(2), 105–104.

Fatchurrohman, M., & Khakim, M.A. (2023). Development of social entrepreneurship organizations based on social bricolage and diamond models to support village tourism ınnovation. *Journal of Economics, Finance and Management Studies*, (6) 11, 5435–5440.

Fernandez-Guada, J., & Martín-Lopez, S. (2023). Gender differences in social entrepreneurship: Evidence from Spain. *Women's Studies International Forum* 96, 1–3. https://www.sciencedirect.com/science/article/pii/S0277539522001042.

Fornell, C., & Larcker, D. F. (1981). Evaluating structural equation models with unobservable variables and measurement error. *Journal of Marketing Research*, 18(1), 39–50.

Frankovský, M., & Birknerová, Z. (2014). Measuring social intelligence-The MESI Methodology. *Asian Social Science*. 10(6). 90–97. ISSN 1911-2017.

Garcia-Gonzalez, A., & Ramirez-Montoya, M. S. (2021). Social entrepreneurship education: changemaker training at the university. *Higher Education, Skills and Work-Based Learning*, (11) 5, 1236–1251.

Gencer, Z.T. & Aydınlıoğlu, Ö. (2020). Sosyal sorumlu mu sosyal girişimci mi? Kavramsal boyutuyla sosyal girişimciliğin dönüşümü. *Gümüşhane Üniversitesi İletişim Elektronik Fakültesi* Dergisi, 8(2), 1057–1080.

Gurusamy, L. & Ganapathy, S. (2023). Entrepreneurıal skılls through ınternshıp traınıng: Focus on tourısm and hotel management. Journal of Research Administration, 5(2), 1630–1642.

Hair, J. F., Risher, J. J., Sarstedt, M., and Ringle, C. M. (2019). When to use and how to report the results of PLS-SEM. *European Business Review*, 31(1), 2–24.

Henseler, J., Ringle, C.M. & Sarstedt, M. (2015). A new criterion for assessing discriminant validity in variance-based structural equation modeling, *Journal of the Academy of Marketing Science*, 43(19, 115–135.

Hietschold, N., Voegtlin, C., Scherer, A.G., & Gehman, J. (2022). Pathways to social value and social change: An integrative review of the social entrepreneurship literature. *International Journal of Management Reviews*, 25, 564–586.

Iwu, C. G., Opute, P. A., Nchu, R., Eresia-Eke, C., Tengeh, R. K., Jaiyeoba, O., & Aliyu, O. A. (2019). Entrepreneurship education, curriculum and lecturer-competency as antecedents of student entrepreneurial intention. *The International Journal of Management Education*, 17(1), 1–13.

Kaiser, H. F. (1974). An index of factorial simplicity. *Psychometrika*, 39(1), 31–36.

Karaman, A., & Çiçekdağı, M. (2023). Genç neslin turizm etkinliklerini değerlendirmesi: Şeb-i Arus törenleri örneği. *KMÜ Sosyal ve Ekonomik Araştırmalar Dergisi*, 25(45), 1225–1243.

Keser, E., & Sarı, H. (2024). Sosyal girişimciliğin sosyal zekâ kapsamında incelenmesi: Ağrı İlinde bir araştırma. İşletme Ekonomi ve Yönetim Araştırmaları Dergisi, (7)1, 37–54.

Kömeçoğlu, F. (2022). Sosyal Zekâ ve Sosyal Girişimcilik İlişkisi: Kastamonu İli Örneği. Yüksek Lisans Tezi. Danışman: Şükran Sırkıntıoğlu Yıldırım, Kastamonu Üniversitesi Sosyal Bilimler Enstitüsü İşletme Ana Bilim Dalı. Kastamonu

Kummitha, H. R., Kolloju, N., Jancsik, A., & Szalok, Z. C. (2021). Can tourism social entrepreneurship organizations contribute to the development of ecotourism and local communities: Understanding the perception of local communities. *Sustainability*, 13, 11031, 1–15.

Lantu, D. C., Suharto, Y., Fachira, I., Permatasari, A. & Anggadwita, G. (2022). Experiential learning model: improving entrepreneurial values through an internship program at start-ups. *Higher Education, Skills and Work-Based Learning*, 12(1), 107–125.

McGehee, N.G., Kline, C. & Knollenberg, W. (2014), Social movements and tourism-related local action. *Annals of Tourism Research,* (48)1, 140–155.

Nasrolahi Vosta, L. & Jalilvand, M.R. (2014), Examining the influence of social capital on rural women entrepreneurship: an empirical study in Iran. *World Journal of Entrepreneurship, Management and Sustainable Development*, (10)3, 209–227.

Newbert, S. L., and Hill, R. P. (2014). Setting the stage for paradigm development: A 'small-tent' approach to social entrepreneurship. *Journal of Social Entrepreneurship*, 5(3), 243–269. https://doi.org/10.1080/19420676.2014.889738.

Özbilen, F.M., Canbulat, T., & Çekiç, O. (2020). Evaluation of preservice classroom teachers' self-efficacy and social entrepreneurship levels. *The Journal of Buca Faculty of Education*, 50, 274–297.

Rambe, P., & Ndofirepi, T. M. (2021). Explaining Social Entrepreneurial Intentions among college students in Zimbabwe, *Journal Of Socıal Entrepreneurship*, (12) 2, 175–196.

Peredo, A. M., & McLean, M. (2006). Social entrepreneurship: A critical review of the concept. *Journal of World Busin*ess 41, 56–65

Saxena, S., & Jain, R, K. (2013). Social ıntelligence of undergraduate students ın relation to their gender and subject stream. *IOSR Journal of Research & Method in Education (IOSR-JRME*), 1(1), 1–4.

Sert, M., & Traş, Z. (2019). Ergenlerin Duygusal Zekâ Düzeyleri ile Özyeterlik Algıları Arasındaki İlişkinin İncelenmesi. *Manas Sosyal Araştırmalar Dergisi*, 8(1), 1205–1220.

Sheldon, P. J., Pollock, A., & Daniele, R. (2017). *Social Entrepreneurship And Tourism: Setting The Stage*. In P. J. Sheldon, & D. R. Esemaier (Eds.),

Social entrepreneurship and tourism: Philosophy and Practice, 1–18. Springer.

Shmueli, G., Sarstedt, M., Hair, J. F., Cheah, J. H., Ting, H., Vaithilingam, S., & Ringle, C. M. (2019). Predictive model assessment in PLS-SEM: Guidelines for using PLSpredict. *European Journal of Marketing*, 53(11), 2322–2347.

Soomro, B. A., Abdelwahed, N. A. A., & Shah, N. (2019). The influence of demographic factors on the business success of entrepreneurs: An empirical study from the small and medium-sized enterprises context of Pakistan. *International Journal of Entrepreneurship*, 23(2), 1–12.

Tekin, E. (2019). *Girişimcilik ve Sosyal Sermaye*. Konya: Eğitim Yayınevi.

Thorndike, E. L. (1920). Intelligence and its use. *Harper's Magazine*, 140, 227–235. https://pdfs.semanticscholar.org/c671/1105ef4b89f302c3188c99e066d440e9b640.pdf?_ga=2.263938572.1425426494.1675003802-964435170.1569196919.

Trabskaia, I., Gorgadze, A., Raudsaar, M., & Myyrylainen, H. (2023). A bibliometric analysis of social entrepreneurship and entrepreneurial ecosystems. *Administrative Sciences* (13) 75, 1–17.

Tu, B., Bhowmik, R., Hasan, M. K., Asheq, A. A., Rahaman, M.A., & Chen, X. (2021). Graduate students' behavioral ıntention towards social entrepreneurship: Role of social vision, ınnovativeness, social proactiveness, and risk taking. *Sustainability*, 13, 6386, 1–23. https://www.mdpi.com/2071-1050/13/11/6386.

Türkeş, N., & Özgeldi, M. (2023). Sosyal girişimcilerin motivasyon faktörleri nelerdir? Farklı ülkelerde yapılan çalışmaların incelenmesi. *DOU Dergi*, 24(2), 435–456.

Uzun, H., Buran, A.Ç., & Beydilli, E.T. (2017). The effect of social ıntelligence on entrepreneurship ıntention: A Research on vocational college students. *Research Journal Of Polıtıcs, Economıcs And Management*, 5(5), 1–13.

Waligora, A. (2019). Social entrepreneurship vs free market. Typology (of lack) of cooperation between social economy entities and enterprises. Tyumen State University Herald. *Social, Economic and Law Research*, 5(1), 23–35.

Weis, S., & Süb, H.-M. (2007). Reviving the search for social intelligence-A multitrait-multimethod study of its structure and construct validity. *Personality and Individual Differences, 42*(1), 3–14.

Quantum Leadership: A Bibliometric Analysis of Emerging Research Field

Zehra Binnur AVUNDUK[1], *F. Füsun İSTANBULLU DİNÇER*[2], *Mithat Zeki DİNÇER*[3]

Introduction

By the 1920s, physicists had established a ground breaking discipline known as quantum mechanics. Etymologically, the term "quantum" denotes "a discrete quantity of something," while "mechanics" refers to "the study of motion." Consequently, quantum mechanics is defined as the study of the behaviour and interactions of subatomic particles. Beyond its foundational role in physics, the principles of quantum theory have been extrapolated to other disciplines, including organizational and leadership studies (Dargahi, 2013; Shelton, 1999). Because in all fields, the interaction between components is a central consideration and it cannot be conceptualized independently of physics. Particularly in an organizational context, within leader-member interactions, there exists a systematic and harmonious relationship between each subordinate component and the overarching structure.

In the contemporary era, society is experiencing a profound economic and technological revolution. However, many organizational leaders face significant challenges in steering their institutions through this period of dynamic change. To effectively navigate this transformative landscape, leaders must develop a comprehensive understanding of the evolving paradigms of leadership in the 21st century and recognize the imperative for a paradigm shift in organizational thinking. Visionary leaders must prioritize the cultivation and utilization of the intellectual capital inherent within their teams. In this context, the primary responsibility of leaders is to harness their strategic

[1] Asst. Prof. Dr., Corresponding author, Istanbul University-Cerrahpasa, Vocational School of Social Sciences, Istanbul, Türkiye, zehrabinnur.avunduk@iuc.edu.tr; ID: 0000-0001-6627-6454

[2] Prof. Dr., Istanbul University, Faculty of Economics, Department of Tourism Management, Istanbul, Türkiye, istanbul@istanbul.edu.tr, ID: 0000-0003-2338-2462

[3] Prof. Dr., Istanbul Nişantaşı University, Faculty of Economics, Administrative and Social Sciences, Istanbul, Türkiye, mithatzeki.dincer@nisantasi.edu.tr, ID: 0000-0002-4928-8303

capabilities to convert information into actionable knowledge, thereby serving as the catalyst for innovation (Harsono et al., 2025).

Nowadays, numerous determinants shape the conditions of working life. Globalization, technological advancements, shifting workforce expectations, and unforeseen disruptions and crises collectively necessitate organizational adaptation and innovation. These dynamic factors further complicate the governance of workplace conditions. To navigate such challenges effectively, organizations must not only leverage the opportunities presented by these evolving circumstances but also mitigate associated risks. Success in this regard is contingent upon organizational and employee receptiveness to innovation, the cultivation of high-level organizational intelligence, and the capacity to demonstrate adaptive leadership amid turbulent and crisis-laden environments (Bilgen & Elçi, 2022). However, innovations carry the inherent risk of disrupting existing systems and practices (Ford, 1996). Consequently, organizational managers play a pivotal role in fostering and strategically managing employee innovation (Kremer et al., 2019). The contemporary work environment, characterized by complexity, volatility, uncertainty, and rapid change, presents significant challenges for leadership (Groves & Feyerherm, 2022; Ochiaka et al., 2024). This necessitates the adoption of modern leadership paradigms, such as QL, which is better suited to navigate today's dynamic conditions. Quantum theory provides a foundational framework for rethinking management strategies, emphasizing adaptability and nonlinear organizational dynamics.

This study asserts that QL, with its emphasis on nonlinear thinking, relational entanglement, and adaptive flexibility (Shelton & Darling, 2004), is uniquely positioned to bridge the gap between digital transformation imperatives and the human dimensions of organizational change. Despite its potential, QL remains marginalized in mainstream leadership discourse. A critical review reveals two lacunae: (1) a paucity of empirical studies validating its efficacy in digital-cultural transitions (Kupers, 2020), and (2) fragmented theoretical engagement, with limited synthesis of its interdisciplinary foundations. This deficit impedes both scholarly consensus and practical implementation, particularly in globalized contexts where leaders must harmonize diverse perspectives while sustaining innovation (Nielsen & Thomsen, 2018). To address these gaps, this study conducts a systematic bibliometric analysis of QL literature, leveraging scientometric tools (Donthu et al., 2021) to map the intellectual evolution of QL, identifying seminal works and emerging thematic clusters, and explore interdisciplinary linkages between QL and adjacent fields (e.g. technology management, organizational psychology).

The main aim of this study is to facilitate researchers and scholars in comprehending the existing knowledge base within the research domain and its current global interconnections. Additionally, this investigation seeks to enhance the coordination of research networks among countries, authors, and affiliated academic institutions. Moreover, this research proposes a research agenda to transition from theoretical abstraction to empirically validated frameworks. By employing bibliometric analysis across Web of Science and Scopus datasets, this study provides a foundational scaffold for future inquiry. Its contributions are theoretical-consolidates dispersed literature into a coherent taxonomy, methodological-demonstrates bibliometrics' utility in evaluating nascent leadership theories, and practical-identifies implementable strategies for cultivating quantum-ready cultures in digitally transforming organizations. The findings aim to catalyze scholarly dialogue while equipping practitioners with evidence-based insights to harness quantum principles in mitigating digital-age leadership deficits.

This chapter proceeds in the following order. First, a literature review is conducted, followed by the methodology of the study, the data collection method, the analysis of the collected data, and finally, the discussion and conclusion sections, which include the interpretation of the findings.

Literature Review

Albert Einstein famously characterized quantum entanglement as "spooky action at a distance" referring to the phenomenon whereby two particles maintain a correlated physical state despite spatial separation. This quantum principle suggests that the dissociation of entangled entities results in systemic instability and state degradation. When properly aligned, quantum particles exhibit interdependent properties regardless of physical separation—a phenomenon with profound implications for leadership theory. In the context of QL, entanglement manifests as a persistent relational connection between leaders and followers, mediated through shared cognitive frameworks (knowledge, skills, attitudes), spiritual alignment, and collective consciousness. Quantum leaders demonstrate complete state correlation with their followers across multiple dimensions: vision, mission, value systems, strategic objectives, epistemic foundations, and affective states. This entanglement dynamic necessitates leadership practices that foster universal participation in organizational planning and vision articulation. The QL paradigm promotes a holographic organizational model, where the collective operates with unitary coherence—analogous to a single living organism at scale (Xiao et al., 2023).

The essence of modern leadership lies in the ability to foster competitive advantage by embracing emerging leadership roles and acquiring the requisite skills to address contemporary challenges (Snyman, 2008). Shelton and Darling (2003) conceptualized quantum organizations as entities in which all stakeholders possess the capacity to access the infinite potential of the quantum field. Such organizations are characterized as learning-oriented entities, where continuous learning is institutionalized as a cultural norm. The QL model acknowledges the fluidity of organizational structures in terms of time, space, and human interactions, thereby facilitating the development of diverse information, communication, and collaborative networks (Chadwick, 2010; Curtin, 2011). This approach yields several benefits, including the cultivation of multifaceted relationships and the empowerment of individuals as leaders through enhanced communication, cooperation, and teamwork (Cai et al., 2024; Değirmenci & Utku, 2000). Therefore, employees are more likely to recognize their potential, expand their capabilities, and achieve superior outcomes through skill development rather than passive adherence to hierarchical directives (Keskinkılıç Kara, 2013). Traditional organizations typically operate at the edge of equilibrium, whereas quantum organizations thrive at the edge of chaos. This fundamental distinction underscores the necessity for traditional organizations to transition toward a model that prioritizes creativity and innovation.

The digital era has precipitated unprecedented disruptions in organizational dynamics, demanding leadership paradigms capable of reconciling technological acceleration with human-centric adaptability (Schwab & Davis, 2018). As traditional leadership models struggle to address the volatility of modern business ecosystems—marked by rapid technological obsolescence, cross-cultural workforce integration, and shifting employee motivation frameworks—QL emerges as a theoretically grounded yet underexplored alternative (Malloch, 2017).

Within social sciences disciplines, management, and organizational studies, the quantum approach has received limited but noteworthy attention. For instance, Lord et al. (2015) employed it to investigate organizational change, Hahn & Knight (2021) utilized it to analyze organizational paradoxes, and Dyck & Greidanus (2017) leveraged it to deepen insights into sustainable organizing theory (Leminen et al., 2023). While conventional economic paradigms remain grounded in Newtonian mechanistic principles—emphasizing predictability, causality, rationality, and objectivity—the intricate realities of modern human systems may be more aptly understood through a non-Newtonian quantum framework. This alternative approach accounts for relativity, uncertainty, indeterminism, complexity and interconnectedness (Leong, 2022; Murphy, 2021; Dyck & Greidanus, 2017).

The Newtonian paradigm, being inherently mechanistic, is limited to explaining linear, predictable, controllable, and isolated systems (Bhattacharjee, 2017; Hahn & Knight, 2021). In contrast, the quantum approach proves particularly suitable for examining dynamic and non-linear systems, such as innovation processes (Leong, 2022; Ottosson, 2003; Ottosson & Björk, 2004; Zhao et al., 2022). Given its foundational principles, quantum theory facilitates the exploration of novel research trajectories in organizational studies, offering a more nuanced lens for understanding complex phenomena.

Existing research has extensively examined the impact of various leadership styles on employees' constructive deviance, emphasizing the mediating role of psychological and motivational mechanisms. For instance, studies on authentic leadership (Wu&Du, 2021) and differential leadership (Zhao, 2019) have demonstrated that shifts in employees' psychological states and intrinsic motivations serve as critical pathways through which leadership influences constructive deviant behaviours. Furthermore, investigations into QL have explored its effects on employees' psychological and emotional well-being, particularly in relation to resource depletion (Xin et al., 2021a) and self-determination (Xin et al., 2021b). Characterized by its holistic, inclusive, interactive, and adaptive nature, QL facilitates the effective mobilization of both task-related resources (e.g., work experience, expertise, and knowledge) and affective resources (e.g., psychological safety and trust). By leveraging these multidimensional resources, QL fosters employee engagement through diverse channels. When employees perceive their resource pool as abundant, this perception cultivates a positive cognitive-affective state, thereby increasing their propensity for constructive deviance (Zhang et al., 2024).

Empirical evidence supports the notion that leaders who provide adequate resources during leader-employee interactions yield favourable outcomes (Tepper et al., 2018). Consistent with Conservation of Resources Theory (Hobfoll, 2001), employees who possess sufficient resources are less likely to interpret workplace demands as threatening, which subsequently promotes adaptive and proactive behaviours. Zhang et al. (2024) examine the mechanisms through which quantum leaders influence employee work behaviour via task-related processes. Employing an interval data collection method, the study gathered 516 questionnaires to analyse the antecedent mechanisms and pathways by which emerging QL fosters constructive deviance. The results demonstrated that QL enhances constructive deviance through three distinct pathways: (1) the affective pathway mediated by recovery experiences, (2) the task-based pathway mediated by job design, and (3) the sequential mediation pathway incorporating both mechanisms. Furthermore, openness to experience was found to play a significant moderating role, strengthening the relationships between QL and recovery experiences, as well as between QL and job design.

From a different standpoint, QL not only fosters an environment where employees are encouraged to openly demonstrate their professional competencies but also emphasizes their personal expression and holistic development. This approach places significant emphasis on nurturing employees' profound value experiences, as well as their spiritual growth and self-actualization (Xin et al., 2021a). Moreover, innovative thinking serves as a pivotal driver of success, uniting employees in a constructive manner. By eliminating rigid structures, excessive control, and communication barriers, quantum leadership cultivates a dynamic workplace that stimulates continuous innovation. Consequently, this enhances employees' capacity for self-driven innovation, thereby improving their overall innovative performance (Hanine & Nita, 2019). In this regard, Cai et al. (2024) examined the impact of QL on employee innovation performance, analyzing the sequential mediating roles of organizational intelligence and knowledge sharing using data collected from 345 employees in Chinese SMEs during the Fourth Industrial Revolution. They found that organizational intelligence and knowledge sharing have multiple serial mediating effects on QL and innovation performance. Furthermore, the interaction between innovative culture and knowledge sharing strengthens employees' innovation performance.

Methodology and Data Collection

Bibliometric Analysis

Bibliometric analysis provides valuable, practical, and chronological statistics that researchers or practitioners can further utilize in a field to assess systematic progress (Duque Oliva et al., 2006; Katoch, 2022). It quantitatively investigates bibliographic material (Broadus, 1987) and offers an overview of a research area based on various indicators. As delineated by Donthu et al. (2021), the methodological framework for conducting bibliometric analysis comprises the following sequential steps shown in Figure 1.

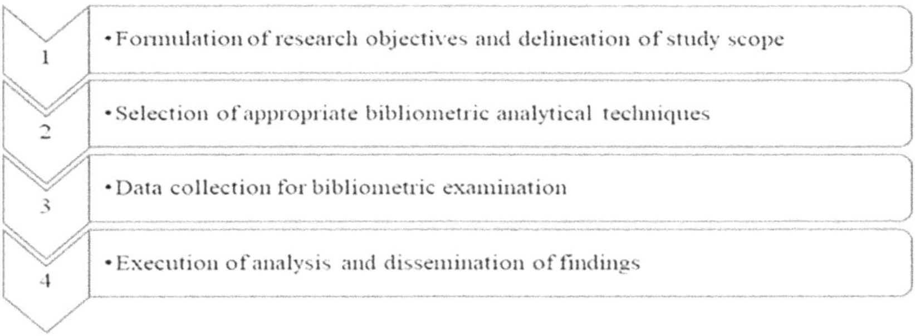

Figure 1. Methodological framework of bibliometric analysis

Source: Adapted from Donthu (2021)

Science mapping is characterized as a complex and intricate process due to its multi-stage nature and the frequent necessity for specialized software tools, many of which are proprietary. Addressing this challenge, Aria and Cuccurullo (2017) developed Bibliometrix, an open-source software tool specifically designed to facilitate comprehensive scientific mapping analysis. This study provides a comprehensive methodological guide for conducting bibliometric analysis utilizing the Bibliometrix package in R, implemented through RStudio with the Biblioshiny graphical interface.

Data Source

The initial phase of bibliometric research entails determining the database to be used for gathering the relevant materials-publications (Avunduk, 2023). Web of Science (WOS), established as one of the oldest scholarly databases, is a proprietary resource under Clarivate Analytics. Historically, WOS served as the primary source for bibliometric data until the introduction of Scopus by Elsevier in 2004, which rapidly emerged as its key competitor, challenging WOS's dominant position in academic indexing. Scopus operates as a multidisciplinary bibliographic database and is recognized for its extensive journal coverage, prioritizing the breadth of scholarly content, whereas WOS emphasizes selectivity in its indexed publications (Visser et al., 2021).

Both databases are widely regarded as the most prominent data sources for academic and scientific research (Zhu & Liu, 2020), particularly in bibliometric analyses (Ellegaard, 2018; Visser et al., 2021). Given their complementary strengths—WOS offering greater historical coverage and Scopus encompassing a broader range of academic journals—researchers such as Sánchez et al. (2017) advocate for the concurrent use of both databases to ensure a more comprehensive retrieval of scholarly literature (Caputo & Kargina, 2022). Therefore, this research has used the merged datasets from both databases to accomplish thorough and valid results.

Analysis and Results

Searching on the databases conducted on February 2, 2025, in the Web of Science (WOS) and Scopus, was restricted exclusively to the term "quantum leadership." Given the limited number of studies on this topic, the research encompassed all scientific disciplines and publication years. To ensure the retrieval of all potential publications, the term "quantum leadership" was searched across all categories—title, abstract, author keywords, and topic—using the "OR" operator. The initial search yielded 23 publications in WOS and 32 publications in Scopus. From the combined total of 55 publications, 16 duplicates (indexed in both databases) were removed, resulting in a final dataset of 39 unique publications for analysis.

According to the research findings, the publication years cover the years 1999–2024. The number of single-authored publications is 19, while internationally co-authored publications account for 2.564%. The number of publications per co-author is 1.95, and the total number of author-provided keywords is 108. The mean publication age is 6.69 years, and the average number of citations per publication is 2.974. The institutional affiliations that have had the greatest relevance in the field of QL are based on the number of documents associated with each affiliation. According to this category, Hacettepe University in Türkiye holds the first position, indicating that this institution has had a significant influence and contributions in the field of QL. While the list features universities from various countries around the world, it is particularly noteworthy that two institutions from Türkiye - including Gebze Technical University - are represented, which is a source of academic pride.

According to the distribution of publication sources by year, the data reveals that Springer Proceedings in Complexity and Nursing Administration Quarterly have emerged as the most prominent sources. Moreover, the distribution of publication sources is based on the number of topic-related publications, the data indicate that Springer Proceedings in Complexity, Nursing Administration Quarterly, and Nursing Science Quarterly have shown particularly high publication frequency in recent years. According to the thematic map analysis, the ascending themes are quantum leadership, human, organization, and nursing education article. Conversely, transformational leadership demonstrates a diminishing tendency.

Figure 2. Co-occurrence Analysis

Source: Authors' production by using the Bibliometrix – R program

Figure 2 illustrates a co-occurrence network that provides valuable insights into the relationships between keywords and associated subdomains (Nica, 2024) within the field of QL. According to the co-occurrence analysis, two main themes represented by red and blue colors are observed. While the density in the red group is higher, QL is discussed in this cluster alongside topics such as humans, leadership, management and organization, nursing education, and healthcare systems. In the blue theme, QL appears to be addressed under subject headings such as adult, skill, and nurse.

Figure 3. Word Cloud

As shown in Figure 3, the word cloud analysis highlights prominent terms related to QL such as leadership, human, adult, organization and management, skill, nurse attitude, nursing, female-male, health care delivery and system, education, professional competence, clinical competence, psychological aspect, and organizational innovation. The most frequently used words in publications related to QL are "leadership," "human," "nursing," "article," "adult," and "humans."

According to the most globally cited documents data, the works by Kennerly (2000) and Watson et al. (2018) in the nursing discipline have received the highest number of citations. In addition, the USA, Iran, Canada, and Türkiye emerge as the most frequently cited countries. The USA, followed by Türkiye and Iran, have consistently dominated QL research output across the examined timeframe. According to research productivity across nations, the USA, Türkiye, Russia, and China as the most active contributors in this field.

The temporal distribution of average annual citations, with peak values recorded in 2000, 2018, and 2021. The temporal pattern of scholarly output indicates a general upward trajectory in QL-related publications beginning in 2017. While a brief downturn occurred in 2021, the trend quickly resumed its positive growth pattern.

Conclusion

This chapter presented a systematic bibliometric analysis of QL, mapping its intellectual evolution, thematic clusters, and interdisciplinary linkages. The findings reveal that QL research has predominantly flourished in nursing and health sciences, with limited penetration into management and organizational studies. Despite its theoretical promise in addressing the complexities of the digital era—such as adaptive flexibility, relational entanglement, and innovation—QL remains underrepresented in business and social science literature. The bibliometric data underscore a critical gap: while QL principles align with contemporary organizational needs (e.g., nonlinear dynamics, globalization, and digital transformation), empirical validation and theoretical integration in management contexts are scarce.

This study has explored the concept of QL, emphasizing its growing prominence in nursing and health sciences while highlighting its underdeveloped potential in management and organizational studies. Bibliometric analysis indicates that while QL is gaining traction in healthcare contexts, its theoretical and practical applications in business management remain limited. The findings suggest that QL, with its emphasis on adaptability, relational dynamics, and systemic thinking, could offer valuable insights into contemporary organizational challenges, particularly in volatile and complex environments.

QL represents a paradigm poised to reconcile the human and technological dimensions of 21st-century organizations. Its bibliometric "shadow" in healthcare signals the untapped potential for business management. Although QL has been predominantly addressed in health sciences management and particularly in the nursing discipline, it is evident that there are very few studies on this subject. There is a need for this topic to be examined across all disciplines through theoretical, empirical, and exploratory research. Particularly, conducting various studies and developing measurement scales by researchers in the field of management and organization could facilitate the application of these outputs and scales by other disciplines as well.

Discussion, Implications, and Future Research Directions

The bibliometric analysis reveals a significant gap in the application of QL within social sciences, particularly in management and cross-sectoral organizational studies. Unlike traditional leadership models, QL emphasizes nonlinearity, interconnectedness, and holistic decision-making—elements that align well with today's rapidly changing business landscapes. However, the lack of empirical studies and theoretical frameworks in business contexts limits its broader adoption. The bibliometric analysis highlights three key insights:

Firstly, QL's stronghold in healthcare reflects its applicability to human-centric, adaptive environments (e.g., nursing education, and patient care systems). However, its potential in corporate, entrepreneurial, and cross-sectoral management remains underexplored. Secondly, co-occurrence and thematic maps identify clusters around "organizational innovation," "psychological safety," and "knowledge sharing," yet lack robust connections to mainstream leadership theories (e.g., transformational, complexity leadership). Finally, the USA, Türkiye, and Iran lead QL research output. This suggests that cultural or institutional influences play a role in QL adoption. Greater collaboration between Western and non-Western scholars could enrich the field's diversity.

A critical discussion point is whether QL can transcend its current healthcare-centric applications to address leadership challenges in corporate, entrepreneurial, and public-sector environments. Future research should investigate and develop metrics to quantify QL's impact on innovation performance and employee resilience. Future research should explore QL's role in hybrid work models and artificial intelligence-driven organizations and comparative studies between QL and established leadership theories (e.g., transformational, situational, or complexity leadership). Moreover, this study presents to future studies following research questions:

- How QL principles can be operationalized in business settings?
- How does QL impact organizational resilience and innovation?

To remain competitive, organizations must adapt to function effectively at the edge of chaos. This transformation presents novel challenges for leaders, compelling them to transcend conventional management practices and adopt innovative approaches. The next generation of quantum leaders must reimagine the organizational environment to facilitate self-organization and foster adaptive, resilient systems.

Bibliography

Aria, M., & Cuccurullo, C. (2017). Bibliometrix: An R-tool for comprehensive science mapping analysis. *Journal of Informetrics*, 11(4), 959–975.

Avunduk, Z. B. (2023). Endüstri 4.0 ve İnovasyon Etkileşiminin Uluslararası Literatürdeki Yansımaları: Bibliyometrik Bir Analiz (Reflections of Industry 4.0 and Innovation Interaction in International Literature: A Bibliometric Analysis). *Journal of Entrepreneurship and Innovation Management*, 12(2), 59–80.

Bhattacharjee, A. (2017). Ancient philosophy, quantum reality and management. *Purushartha*, 10(1), 43–52.

Bibliometrix (http://www.bibliometrix.org), accessed: February 2, 2025.

Bilgen, A. and Elçi, M. (2022). The mediating role of organizational intelligence in the relationship between quantum leadership and innovative behavior. *Frontiers in Psychology*, 13:1051028.

Broadus, R. (1987). Toward a definition of "bibliometrics." *Scientometrics*, 12(5–6), 373–379.

Cai, H., Zhu, L., & Jin, X. (2024). Validating the Causal Relationship between Quantum Leadership and Employee Innovation Performance from the Perspective of Organizational Sustainability. *Sustainability*, 16(18), 7884.

Caputo, A., & Kargina, M. (2022). A user-friendly method to merge Scopus and Web of Science data during bibliometric analysis. *Journal of Marketing Analytics*, 10(1), 82–88.

Chadwick, M. M. (2010). Creating order out of chaos: a leadership approach. *Association of Operating Room Nurses Journal*, 91, 154–170.

Curtin, L. (2011). Quantum leadership: succeeding in interesting times. Nurse Leader, 9, 35–38.

Dargahi, H. (2013). Quantum leadership: the implication for Iranian nursing leaders. *Acta Medica Iranica*, 411–417.

Değirmenci, M., Utku, Ş. (2000). Yönetim ve Örgüt Yapısına Kuantum Mekaniği Açısından Bir Bakış. *Doğuş Üniversitesi Dergisi*, 2, 76–83.

Donthu, N., Kumar, S., Mukherjee, D., Pandey, N., & Lim, W. M. (2021). How to conduct a bibliometric analysis: An overview and guidelines. *Journal of Business Research*, 133, 285–296.

Duque Oliva, E. J., Cervera Taulet, A., & Rodríguez Romero, C. (2006). A bibliometric analysis of models measuring the concept of perceived quality inproviding internet service. *Innovar*, 16(28), 223–243.

Dyck, B., & Greidanus, N.S. (2017). Quantum Sustainable Organizing Theory: A Study of Organization Theory as if Matter Mattered. *Journal of Management Inquiry*, 26(1), 32–46.

Ellegaard, O. (2018). The application of bibliometric analysis: Disciplinary and user aspects. *Scientometrics*, 116 (1), 181–202.

Ford, C. M. (1996). A theory of individual creative action in multiple social domains. *Academy of Management Review*, 21(4), 1112–1142.

Groves, K. S., & Feyerherm, A. E. (2022). Developing a leadership potential model for the new era of work and organizations. *Leadership & Organization Development Journal*, 43(6), 978–998.

Hahn, T., & Knight, E. (2021). The Ontology of Organizational Paradox: A Quantum Approach, *The Academy of Management Review*, 46(2), 362–384.

Hanine, S., & Nita, M. A. (2019). The paradigm of quantum leadership: ontology, praxis and application to management. *Revue Internationale des Sciences de Gestion*, 3(2), 837–858.

Harsono, T. W., Hidayat, K., Iqbal, M., & Abdillah, Y. (2025). Exploring the effect of transformational leadership and knowledge management in enhancing innovative performance: a mediating role of innovation capability. *Journal of Manufacturing Technology Management*, 36(1), 227–250.

Hobfoll, S.E. (2001), "The influence of culture, community, and the nested-self in the stress process: advancing conservation of resources theory," *Applied Psychology*, 50(3), 337–421.

Katoch, R. (2022). IoT research in supply chain management and logistics: A bibliometric analysis using vosviewer software. *Materials Today: Proceedings*, 56, 2505–2515.

Kennerly, S. (2000). Perceived worker autonomy: the foundation for shared governance. JONA: *The Journal of Nursing Administration*, 30(12), 611–617.

Keskinkılıç Kara,S.B. (2013).Yeni bilim ve liderlik, *Akademik Bakış Dergisi*,34,1–13.

Kremer, H., Villamor, I., & Aguinis, H. (2019). Innovation leadership: Best-practice recommendations for promoting employee creativity, voice, and knowledge sharing. *Business Horizons*, 62(1), 65–74.

Leong, D. (2022). Probabilistic Interpretation of Observer Effect on Entrepreneurial Opportunity, *Organizacijac*, 55(4), 243–258.

Leminen, S., Rajahonka, M., Westerlund, M., (2023). Innovation in Living Labs: A Quantum Approach,*Journal of Innovation Management*, 11(4), 1–22.

Lord, R.G., Dinh, J.E., & Hoffman, E.L. (2015). A quantum approach to time and organizational change. *Academy of Management Review*, 40(2), 263–290.

Malloch, K. (2017). Quantum leadership: Creating sustainable value in health care. Jones & Bartlett Learning.

Murphy, M. (2021). Markets are constantly collapsing: Reconceptualizing 'the market' as a quantum social wavefunction. *Competition and Change*, 25(5), 561–579.

Nica, I. (2024). Bibliometric mapping in the landscape of cybernetics: insights into global research networks. *Kybernetes*.

Ochiaka, D.I., Virginia, U.N., Eneh, N.A. (2024). Quantum Leadership: Navigating Complexity and Uncertainty in the Modern Business Landscape, *ICPA Journal of Business Administration and Economics*, 1(2), 27–31.

Ottosson, S. (2003). Participation action research - A key to improved knowledge of management, *Technovation*, 23(2), 87–94.

Ottosson, S., & Björk, E. (2004). Research on dynamic systems - Some considerations. *Technovation*, 24(11), 863–869.

Sánchez, A.D., M.D. Del Río, and J.Á. García. 2017. Bibliometric analysis of publications on wine tourism in the databases Scopus and WoS. *European Research on Management and Business Economics*. 23 (1): 8–15.

Schwab, K., & Davis, N. (2018). Shaping the future of the fourth industrial revolution. Crown Currency.

Shelton C. (1999). Quantum leaps. Boston, MA: Butterworth-Heinemann.

Shelton, C.D., Darling,JR.(2003).From theory to practice: using new science concepts to create learning organizations.*The Learning Organizations*, 10(6):353–60.
Shelton, C. D., & Darling, J. R. (2004). From chaos to order: Exploring new frontiers in conflict management. *Organization Development Journal*, 22, 22–41.
Snyman, L. (2008). South African executive leadership an art form: skills requirements. Master's dissertation, Gordon Institute of Business Science, University of Pretoria.
Tepper, B.J., Dimotakis, N., Lambert, L.S., Koopman, J., Matta, F.K., Park, H.M. and Goo, W. (2018), "Examining follower responses to transformational leadership from a dynamic, personenvironment fit perspective," *Academy of Management Journal*, 61(4), 1343–1369.
Visser, M., Van Eck, N.J. and Waltman, L. (2021), "Large-scale comparison of bibliographic data sources: Scopus, Web of Science, Dimensions, Crossref, and Microsoft Academic," *Quantitative Science Studies*, 2(1), 20–41.
Watson J, Porter-O'Grady T, Horton-Deutsch S, Malloch K. (2018). "Quantum Caring Leadership: Integrating Quantum Leadership with Caring Science," *Nursing Science Quarterly*; 31(3):253–258.
Wu, S.J. and Du, M.Z. (2021), "The influence of authentic leadership on employees' constructive deviance: a chain mediating model," *Soft Science*, 35(3), 101–107.
Xiao, Y., Yang, Y., Wang, X., Liu, Q., & Gu, M. (2023). "Quantum uncertainty principles for measurements with interventions," *Physical Review Letters*, 130(24), 240201.
Xin, J., Han, Y., Liu, S.J. and Sun, W.M. (2021a), "The influence of quantum leadership on employees' work engagement in uncertain situations," *Science Research Management*, 6, 176–183.
Xin, J., Tu, Y.F. and Jiao, W.H. (2021b), "Will quantum leadership improve employee workplace wellbeing in the context of epidemic uncertainty? The role of person-organization fit and emotional exhaustion," *Journal of Shanghai University of Finance and Economics*, 23(1), 15–26.
Zhang, Y., Chen, X., & Tan, C. (2024). A mechanistic study of the influence of quantum leadership on employees' constructive deviance. *Chinese Management Studies*.
Zhao, J.J. (2019), "Research on the influence mechanism of differential leadership on knowledge workers' constructive performance deviance," *Journal of Business Economics*, 11, 42–54.

Zhao, J., Li, S., Xi, X., & Gong, C. (2022). A quantum mechanics-based framework for knowledge-based innovation. *Journal of Knowledge Management*, 26(3), 642–680.

Zhu, J., and W. Liu. 2020. A tale of two databases: The use of Web of Science and Scopus in academic papers. *Scientometrics*, 123: 321–335.

Perceived Supervisor Support, Perceived Co-Worker Support and Occupational Commitment: Testing Mediation Effect of Subjective Well-Being

Efe Kaan ULU[1]

Introduction

In the food businesses, chefs are a crucial component of the service chain. In this sector, where even a smile can be impactful, having satisfied employees is essential for businesses. Therefore, studies on chefs' psychological well-being are essential for both academic knowledge and industry practices. Employee turnover in the service sector has been a topic of academic research for many years, as the high rate of labor turnover remains one of the main issues. (Ghiselli et al., 2001). The financial impact of labor turnover is a major reason for academic research on this topic. The expenses associated with of rectuiting and training an employee are estimated to reach up to 200% of the employee's annual salary (Griffeth & Hom, 2001).

Particularly in food and beverage establishments, the labor turnover rate is considerably high and exceeds the sector's average (Holmes et al., 2022; Met & Sarioğlan, 2010), and this is recognized as a global issue (Mac Con Iomaire et al., 2021). Consequently, understanding the reasons behind chefs' turnover in human resource-focused studies is critical for businesses to reduce costs associated with workforce loss and enhance employee qualifications. Students in fields such as gastronomy and culinary arts, culinary management, and tourism—who represent the future workforce of the sector—also constitute an important group for further research. According to Akoğlu et al. (2017), only 40.4% of students believe they can work in the gastronomy sector for many years.

The tourism sector is known for certain disadvantages concerning working conditions. These include unfavorable working environments, inadequate human resource management practices, long working hours compared to

1 Istanbul Bilgi University, Faculty of Applied Sciences, Department of Gastronomy and Culinary Arts, uefekaan@gmail.com

other sectors, low wages, limited investment in training, weak professional career prospects, and low social prestige (Ariza-Montes et al., 2018). Chefs, in particular, are among the occupational groups facing the most challenging conditions within the service sector. Chefs frequently compromise their physical and mental well-being in order to succeed in their careers, despite the fact that the field has become more well-known and appealing as a result of the popularity of television shows in recent years (Bloisi & Hoel, 2008; Murray-Gibbons & Gibbons, 2007).

This situation ultimately imposes significant pressure on their personal and family lives. Indeed, Harris and (Harris & Giuffre, 2010) indicate that chefs frequently experience feelings of guilt and work-family conflicts due to their professional careers. However, empirical studies investigating this profession remain limited (Robinson et al., 2014). These unfavorable circumstances could make the service sector less appealing and deter people from pursuing jobs in it. Therefore, research focusing on the well-being of chefs (Ulu & Vatan, 2023) and students training in this field, along with the implementation of appropriate human resource practices, is of great importance.

Employees' work lives are shaped by various internal and external organizational factors. Internal organizational factors include the nature of the job, compensation, working conditions, relationships with colleagues, and communication with supervisors. During their tenure within an organization, employees spend the most time with their colleagues and supervisors. Therefore, it can be argued that employees' positive or negative perceptions of their jobs and organizations are likely shaped by their interactions with those they are in constant close contact with (Tozkoparan & Tenteriz, 2022). In the context of chefs, working environments characterized by hierarchical structures and chains of command suggest that interactions with executive chefs and coworkers are likely to influence their attitudes toward their jobs. This research aims to examine the relationship between perceived supervisor support, perceived coworker support, subjective well-being, and occupational commitment. Furthermore, this relationship is analyzed through the lens of social exchange theory. The reciprocity principle in social exchange theory suggests that each party in a relationship holds valuable resources that can be exchanged for mutual benefit (Rhoades & Eisenberger, 2002). When individuals feel supported and valued in their workplaces, they are more inclined to put in greater effort to achieve organizational goals. Consequently, both the staff and the organization benefit from this relationship. This dynamic becomes particularly significant in labor-intensive sectors such as tourism, where human interactions are paramount (Tuna Arslan & Aydin, 2023).

Identifying the factors that enhance the well-being of chefs, who experience high labor turnover rates, as well as students training in this field, is crucial for reducing turnover rates and increasing professional commitment. This study investigates the impact of subjective well-being (SWB) on perceived supervisor support, perceived coworker support, and chefs' professional commitment in professional kitchens, known for their challenging working conditions and high stress levels. Given the absence of research on this specific topic, this study is expected to contribute to the existing literature. The study is designed with the hypothesis that perceived supervisor support (PSS) and perceived coworker support positively influence professional commitment. It is also hypothesized that SWB mediates the relationship between perceived supervisor support, perceived coworker support, and professional commitment. Understanding the relationship between these concepts is expected to benefit the culinary profession, which is known for its high labor turnover rate. By implementing appropriate workplace policies and fostering supervisor and coworker support, restaurant and hotel managers, as well as human resource professionals, can potentially enhance chefs' commitment to their profession.

Literature

Subjective Well-Being

SWB refers to the value individuals place on themselves based on their self-evaluation (Rosenberg, 1989). This evaluation relies on subjective judgments in physical, mental, and social domains of life (Dlugonski & Motl, 2012). In other words, it is related to how much individuals like themselves and how much they value themselves (Blascovich et al., 1991). SWB enables individuals to maintain feelings of happiness and competence when facing challenges (Pierce & Gardner, 2004). Personal SWB also enhances individuals' awareness of discriminatory treatment (Seery & Quinton, 2015). High SWB can reduce employees' likelihood of leaving their jobs while increasing organizational commitment when encountering interpersonal discrimination. Perceived discrimination refers to negative attitudes, judgments, or unfair treatment directed toward individuals (Jackson & Jackson, 2019).

Perceived Supervisor Support

Perceived supervisor support (PSS) refers to the level of support employees receive from their supervisors (Maertz et al., 2007). It is defined as employees' perceptions of supervisors who value their contributions and care about their well-being (Eisenberger et al., 2002). In other words, it involves employees feeling valued, appreciated, and supported by their supervisors, including a sense that their happiness and well-being are important and that their concerns are addressed (Tuna Arslan & Aydin, 2023).

Since supervisors are responsible for rewarding, protecting, motivating, and inspiring their subordinates, they play a vital role in organizations (Phungsoonthorn & Charoensukmongkol, 2020). Supervisor support positively impacts employee performance and reduces uncertainty (Gordon et al., 2019; Skiba & Wildman, 2019; Tummers et al., 2018). Because of the F&B sector's fragile structure, supervisor support is crucial for lowering kitchen staff tension during uncertain times and preserving a healthy work environment.

Workers may be reluctant to voice their opinions and are more likely to quit if they don't feel that their supervisors are supporting them enough (İnce, 2016). On the other hand, workers who feel that their supervisors are supporting them can recuperate from job difficulties that cause them to become unhappy and physically and emotionally exhausted (Ertürk et al., 2016). Research on hotel workers has demonstrated that supervisor assistance improves employee happiness (Bardakoğlu et al., 2017), lowers workplace stress, and boosts job engagement (Karatepe et al., 2020). Additionally, it significantly improves both general life happiness and job satisfaction (Cole et al., 2006; Kale, 2015).

Perceived Co-Worker Support

Perceived coworker support (PCS) refers to the level of support employees receive from their colleagues in workplace interactions (H. J. Kim et al., 2017). In other words, coworker support can be defined as the extent to which employees believe that their colleagues will assist them with work-related issues or provide task-related information (Chung et al., 2021). (Hochschild, 2012) found that employees often experience emotional dissonance when they are required to express emotions different from what they truly feel. As a result, employees tend to form a community with other coworkers to share their genuine feelings as a way to alleviate stress. The presence of empathetic individuals who can genuinely understand and relate to workplace experiences and the associated emotions can mitigate the impact of stressors during both positive and challenging moments. Coworker support has been found to have an positive impact on organizational commitment, job performance, and job satisfaction (Guchait et al., 2014; H. J. Kim et al., 2017; Limpanitgul et al., 2014). Attitudes of employees' are shaped not only by workplace policies but also by their relationships with coworkers and supervisors (Aydın & Kalemci Tüzün, 2019).

Occupational Commitment

Occupational commitment can be defined as the psychological relationship of individuals with their occupation and their emotional responses related to it (Lee et al., 2000; Meyer et al., 1993). It also can be described as the willingness to remain and work in a occupation (Carson & Bedeian, 1994). Occupational commitment has three dimensions as continuance, normative, and affective commitment (Meyer et al., 1993).

The people's attachment to their jobs because of the expenses of quitting the job is explained as continuance commitment (Meyer et al., 1993). The urge to stay in a profession because of perceived obligations is explained by the normative commitment dimension. For instance, normative occupational commitment may arise as a result of family members working in the same sector (Meyer et al., 1993).Conversely, affective commitment is the psychological and emotional attachment to the occupation and the willingness to work without obligations (Mercurio, 2015).

While continuance and normative commitment dimensions explain why individuals stay in their occupations due of obligation, affective commitment reflects their voluntary choice to stay. People who have deep emotional attachments to their occupation priotitize their personel development and actively pursue the educational opportunities for their careers (Meyer et al., 1993).

Hypotheses

Employees who perceive that their supervisors trust them and provide autonomy exhibit increased commitment to their current jobs (Kidd & Smewing, 2001). This proves the significance of positive feedback from supervisors and the importance of delegating responsibility with trust. According to studies, perceived supervisor support increase employees' commitment to their jobs and reduces turnover intentions (Aydın & Kalemci Tüzün, 2019; Chen et al., 2002; Rathi & Lee, 2017; Rousseau & Aubé, 2010). Support from supervisors enhances the quality of work life and lowers turnover intentions by creating a happy work environment. Sungu et al. (2019) explored the connection between occupational commitment and organizational commitment, highlighting that individuals with strong organizational commitment tend to also exhibit high levels of occupational commitment. Based on these relationships, the following hypothesis is proposed:

Hypothesis 1: PSS has a positive effect on occupational commitment.

There is limited research in the literature regarding support from coworkers, but the relationship between coworkers and individuals can be evaluated similarly to the relationship between organizations, managers, and individuals due to the social exchange between them (Aydın & Kalemci Tüzün, 2019). It is acknowledged that coworkers can influence employees in the workplace (Charoensukmongkol et al., 2016; Kram, 1983; Raabe & Beehr, 2003), and the social exchange between them can affect employee performance (Seers, 1989; Sherony & Green, 2002). Based on these relationships, it is hypothesized that coworker support may affect professional commitment, leading to the following hypothesis:

Hypothesis 2: PCS has a positive effect on occupational commitment.

Individuals with high SWB are more ambitious and determined in achieving success. These individuals are emotionally resilient, difficult to persuade, optimistic, and possess a holistic self-awareness (Sevim & Artan, 2021). In contrast, individuals with low SWB are emotionally less stable and more easily influenced by daily events. They constantly seek acceptance and approval from others, but are also persistently uncertain about achieving this (Hogg & Vaughan, 2008). While people with high SWB have a strong sense of self and are less impacted by external factors, those with low SWB are more sensitive and susceptible to them (Sevim & Artan, 2021). According to the study by (Nahum-Shani et al., 2014), perceived managerial support in people with high SWB reduces the negative effects of managerial undermining on job stress. It has also been found that high SWB can reduce employees' intentions to leave their jos in negative situations (Jackson & Jackson, 2019). Even when perceived support in the work environment is low, individuals with high SWB may not experience an impact on their occupational commitment. Also, coworker support can protect people from the negative effects of stressfull enviroments, enhance positive emotions and increase SWB. It is hypothesized that SWB may play a mediating role in the relationship between perceived supervisor support, perceived coworker support, and occupational commitment.

Hypothesis 3: SWB mediates the relationship between PSS and occupational commitment.Hypothesis 4: SWB mediates the relationship between PCS and occupational commitment.
Hypothesis 5: PSS has a positive effect on SWB.
Hypothesis 6: PCS has a positive effect on SWB.

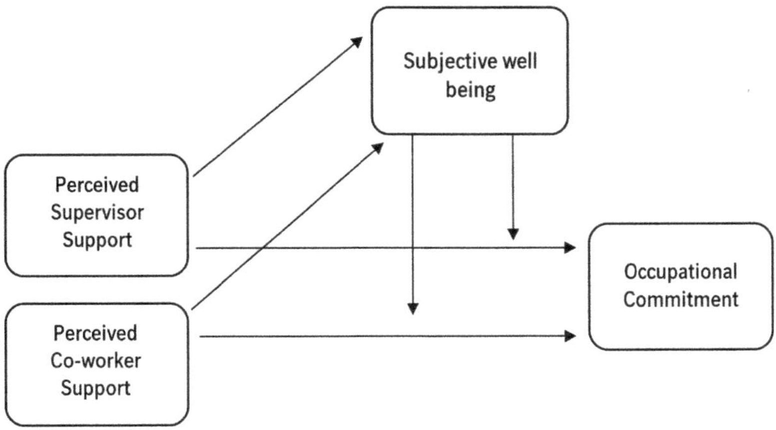

Figure 1. Research Model

Method

The aim of this empirical study is to analyze the mediating role of SWB in the relationship between perceived supervisor support, perceived coworker support, and occupational commitment. Additionally, this relationship is tested using a sample of undergraduate students enrolled in gastronomy and culinary arts programs. The details of the variables, measurement scales, sample, data collection, and data analysis are presented below.

Scales

The survey instrument utilized in this study comprises four sections, each designed to capture specific aspects relevant to the research objectives. All survey items were evaluated using a 5-point Likert-type scale. The first section contains demographic questions aimed at gathering background information about the participants. The second section includes the occupational commitment scale, which measures the level of occupational commitment. The third section incorporates scales that measure perceived supervisor support and perceived coworker support,. Finally, the fourth section consists of the SWB scale. To measure SWB, Rosenberg's 10-item SWB scale was used (Rosenberg, 1965). Occupational commitment was assessed using the 11-item occupational commitment scale developed by Blau et al. (Blau et al., 1993). Additionally, perceived supervisor support and perceived coworker support were measured using Turkish-adapted scales compiled by Giray and Şahin (Giray & Sahin, 2012), ensuring cultural relevance and applicability within the study context.

The occupational commitment scale was subjected to a translation-back translation process to ensure accuracy and conceptual equivalence across languages (Brislin, 1970). Two English instructors and a tourism scholar first translated the scale from English into Turkish. Then, the other two experts retranslated the scale into English to reduce meaning loss. This process aimed to preserve the integrity of the scale.

Sample

This study employed a disproportionate quota sampling method. The research was conducted in Istanbul, Turkey, the country's most developed city in terms of gastronomy, and included Gastronomy and Culinary Arts Departments of universities within the scope of the study. The participants consisted of third- and fourth-year undergraduate students majoring in Gastronomy and Culinary Arts who had completed at least three months of internship or had prior work experience in the sector. Existing research suggests that students who have undergone internships or gained practical work experience develop a better understanding of the service sector and hold more realistic attitudes toward their careers, including career goals and expectations (H. Kim & Park, 2013; Robinson et al., 2016).

In line with the relevant literature, it can be argued that students in their third and fourth years, particularly those with internship or work experience in restaurants or tourism enterprises, exhibit stronger and more realistic professional attitudes (Kusluvan et al., 2022).

Data Collection

The data for this study were collected through an online survey questionnaire, which included demographic questions and measurement scales designed to assess the structural relationships within the research model. To reach a broader participant base, faculty members from various universities were asked to distribute the online survey to their students. As a result, data collection took place between September 28, 2024, and November 3, 2024. A total of 300 potential respondents were contacted during between September 28, 2024, and January 07, 2025 and 196 responses received. Our final sample consisted of 196 students with ages ranging from 18 to 34 years. The data obtained through the survey were analyzed using descriptive statistical analysis methods. IBM SPSS 27.0 (Statistical Package for the Social Sciences) program was used for statistical analysis of the research.

Analysis

Dimensionality, Reliability and Validity of the Scales

As two of the scales were relatively untested for reliability and validity and two scales have undergone cross-cultural translation, we used exploratory factor analysis, reliability analysis and confirmatory factor analysis to establish the dimensionality, reliability and validity of the scales. Before the exploratory factor analysis, the dataset was checked for suitability and normality for factor and regression analysis. As shown in Table 1, all values fell within the acceptable range of -1 to +1, indicating that the dependent and independent variables exhibited a normal distribution (Hair et al., 2019). The principle component method of extraction is used for exploratory factor analysis. Factor loadings of scale items, removed items, number of eigenvalues greater than 1 and variances explained are presented in Appendix 1.

The PSSe, consisting of 11 items, demonstrated a KMO value of 0.859 and Bartlett's Test of Sphericity result of $\chi^2 = 1025.627$, $p < 0.001$. The Cronbach's Alpha value was calculated as 0.920, indicating excellent internal reliability. The total variance explained by the single factor structure was 59.040%. The PCS Scale initially presented a two-factor solution, with the second factor comprising items 7, 8, and 9. However, the Cronbach's Alpha value for this factor was 0.613, which was below the acceptable threshold. After removing item 7, the factor and reliability analysis were repeated, resulting in a single-factor structure with 8 items. The KMO value was 0.838, and Bartlett's Test of Sphericity yielded $\chi^2 = 772.339$, $p < 0.001$. The Cronbach's Alpha value increased to 0.900, and the total variance explained was 59.756%. The Occupational Commitment Scale consisted of 9 items after the exclusion of item 8 due to an insufficient factor loading. The KMO value was 0.879, and Bartlett's Test of Sphericity yielded $\chi^2 = 1011.740$, $p < 0.001$. The Cronbach's Alpha value was calculated as 0.923, indicating excellent internal reliability. The total variance explained was 63.358%. The SWB Scale demonstrated a single-factor structure with 5 items after the removal of items 1 and 7 due to low factor loadings. The KMO value was 0.862, and Bartlett's Test of Sphericity resulted in $\chi^2 = 330.544$, $p < 0.001$. The Cronbach's Alpha value was 0.866, indicating good internal reliability. The total variance explained was 65.385%.

Following the exploratory factor analysis, a confirmatory factor analysis was conducted. All indicator variables significantly loaded onto their corresponding latent variables, with standardized regression weights exceeding 0.50. To assess the reliability and convergent validity of the scales, composite reliability (CR) and average variance extracted (AVE) were calculated for each scale (Table 2). CR scores and, AVE values were satisfactory (Table 2). However, this did not compromise the scales' convergent validity, as CR values exceeded the acceptable threshold of 0.60 (Table 2) (Malhotra, 2020).

Table 1: Tests of normality for research variables.

	N	Skewness	Standard Error	Kurtosis	Standard Error
Perceived supervisor support	148	-,581	,199	,057	,396
Perceived co-worker support	148	-,547	,199	,245	,396

	N	Skewness	Standard Error	Kurtosis	Standard Error
Occupational commitment	148	-480	,199	-,306	,396
Subjective well being	148	-,250	,199	-,421	,396

Table 2: Reliability and Convergent Validity of Research Constructs.

Scales	Number of Items	Cronbach Alpha (α)	Composite Reliability (CR)	Average Variance Extracted (AVE)
Perceived supervisor support	11	,920	0.964	0.713
Perceived co-worker support	8	,900	0.947	0.697
Occupational commitment	9	,923	0.965	0.758
Subjective well being	10	,866	0.867	0.568

Results

Descriptive Statistics about Research Participants and Variables

60% of the participant in the study were female, and the rest were male. 90% of the students were in the 3rd, and 4th grade. Almost all of the students (90%) were between 19–23 years old. The departments where most internship experiences took place was restaurant kitchens (76%), and pastry (%10). Further descriptive statistics are given about study variables in Table 3. The grand mean scores for the primary research variables are presented in Table 3. The results indicate a relatively positive perception of supervisor support (M = 3.51, SD = 0.777) and co-worker support (M = 3.66, SD = 0.738). Occupational commitment was evaluated at a moderate level (M = 3.27, SD = 0.945), while SWB received the lowest score (M = 3.02, SD = 0.901).

These findings suggest that while students generally perceive adequate support from their supervisors and colleagues, their occupational commitment and SWB remain moderate, with SWB being notably lower. Further descriptive statistics provide detailed insights into the study variables.

Table 3: Mean Scores for Research Variables

Scales	Grand mean[a]	Standart deviation
Perceived supervisor support	3,51	,77748
Perceived co-worker support	3,66	,73800
Occupational commitment	3,27	,94569
Subjective well being	3,02	90094

a Higher grand mean reflects a higher perception of PSS, PCS, occuational commitment, and SWB (1=Strongly disagree, 5=Strongly agree)

Table 4: Correlations

Variables	PSS	PCS	Occupational commitment	SWB
PSS	1	0.533**	0.312**	0.353**
PCS	0.533**	1	0.309**	0.377**
Occupational commitment	0.312**	0.309**	1	0.577**
SWB	0.353**	0.377**	0.577**	1

Based on the correlation results presented in Table 4, perceived supervisor support is significantly and positively correlated with perceived co-worker support ($r = 0.533$, $p < 0.01$), occupational commitment ($r = 0.312$, $p < 0.01$), and SWB ($r = 0.353$, $p < 0.01$). Similarly, perceived co-worker support is strongly and positively associated with occupational commitment ($r = 0.309$, $p < 0.01$) and SWB ($r = 0.377$, $p < 0.01$).

A notable finding is the significant positive relationship between occupational commitment and SWB ($r = 0.577$, $p < 0.01$), suggesting that individuals who report higher levels of SWB are more likely to exhibit a strong occupational commitment. These findings highlight the interconnectedness between support systems in the workplace, self-perception, and occupational attitudes. These findings suggest that individuals with higher levels of perceived support of supervisors and co-workers tend to report stronger occupational commitment, and SWB.

Hypothesis Testing

In the study, multiple linear regression analysis was applied to measure the effect of perceived co-worker support and perceived supervisor support on occupational commitment. The analysis results demonstrated that the model was statistically significant ($F = 10.417$, $p < 0.001$, $R^2 = 0.126$), indicating that occupational commitment can be statistically predicted based on these two support factors.

Regarding individual predictors, perceived supervisor support was found to have a statistically significant positive effect on occupational commitment (Beta = 0.206, $t = 2.240$, $p = 0.027$). Similarly, perceived co-worker support also had a statistically significant positive effect on occupational commitment (Beta = 0.199, $t = 2.169$, $p = 0.032$). The explanatory power of the model ($R^2 = 0.126$) was relatively low, indicating that while the predictors were statistically significant, their contribution to explaining occupational commitment was limited.

The significance values presented (<0.05) highlight the positive and meaningful impact of both support types on occupational commitment. Notably, the Beta value for perceived supervisor support (0.206) suggests a slightly stronger standardized effect compared to perceived co-worker support (0.199), though the difference is minimal.

Table 5: Regression analysis results

	B	Std. Error	Beta	t	Sig.
PSS	0.250	0.112	0.206	2.240	0.027
PCS	0.255	0.118	0.199	2.169	0.032

a. Dependent variable: Occupational commitment

Mediation Analysis Findings

The mediation analyses reveal that SWB fully mediates the relationship between perceived co-worker and supervisor support and occupational commitment. Initially, both perceived supervisor support ($B = 0.379$, $p < 0.001$) and perceived co-worker support ($B = 0.396$, $p < 0.001$) had significant direct effects on occupational commitment. Additionally, both types of support significantly predicted SWB (B = 0.409, p < 0.001 for supervisor support; B = 0.461, p < 0.001 for co-worker support). However, when SWB was included in the models, the direct effects of both support types on occupational commitment became non-significant (B = 0.150, p = 0.087 and B = 0.136, p = 0.146, respectively), while SWB remained a strong and significant predictor in both models (B = 0.560, p < 0.001 for supervisor support; B = 0.564, p < 0.001 for co-worker support). Thus, the study's third, fourth, fifth, and sixth hypotheses are supported.

Table 6. Mediation Analysis Results: The Mediating Role of SWB Between PSS and Occupational Commitment

Model	Independent Variable	B	Std. Error	Beta	t	Sig.
PSS >Occupational Commitment	PSS	0.379	0.096	0.312	3.966	0.000
PSS > SWB	PSS	0.409	0.090	0.353	4.555	0.000
PSS + SWB > Occupational Commitment	PSS	0.150	0.087	0.124	1.724	0.087
	SWB	0.560	0.075	0.533	7.436	0.000

Table 7. Mediation Analysis Results: The Mediating Role of SWB Between PCS and Occupational Commitment

Model	Independent Variable	B	Std. Error	Beta	t	Sig.
PCS >Occupational Commitment	PCS	0.396	0.101	0.309	3.923	0.000
Perceived Co-worker Support > SWB	PCS	0.461	0.094	0.377	4.922	0.000
PCS + SWB > Occupational Commitment	PCS	0.136	0.093	0.106	1.461	0. 146
	SWB	0.564	0.076	0.537	7.387	0.000

Discussion and Conlusion

The findings of this study suggest several key insights into the factors affecting occupational commitment among undergraduate gastronomy and culinary arts department students, with a particular focus on the roles of perceived co-worker and supervisor support, and SWB.

The correlation analysis highlights significant positive relationships between both PCS and, PSS, occupational commitment, and SWB. Notably, occupational commitment and SWB were strongly correlated ($r = 0.577$, $p < 0.01$), which implies that gastronomy and culinary arts students with higher well-being are likely to exhibit stronger occupational commitment. Moreover, the correlation between supervisor and co-worker support ($r = 0.533$, $p < 0.01$) is significant, underscoring the importance of a supportive work environment.

The regression analysis results show that both PSS and PCS significantly predict occupational commitment, with supervisor support having a slightly stronger effect. However, the model's relatively low explanatory power ($R^2 = 0.126$) suggests that other factors may also contribute to occupational commitment beyond these two forms of support. Despite this, the statistically significant relationships affirm the importance of workplace support in fostering students' commitment to their future careers and results are consistent with previous researches (Aydın & Kalemci Tüzün, 2019; Fernet et al., 2021).

Mediation analysis revealed that SWB fully mediates the relationship between both supervisor and co-worker support and occupational commitment, consistent with previous studies in the literature (Ilyas et al., 2023; Singhal & Rastogi, 2018). The direct effects of supervisor and co-worker support on occupational commitment became non-significant once SWB was included in the model. Locke and Latham, (2006) suggested that employees commit to their career goals through the experience of SWB, aiming to enhance their performance in achieving those goals. This indicates that enhancing people's SWB through workplace support could be a crucial pathway to strengthening occupational commitment and reducing turnover intentions (Skakon et al., 2010).

The results of this study provide valuable insights for managers of food service businesses. Experience of intern cooks during internships or early career stages shapes their future career choices and it is essential to prioritize the well-being and development of interns. Positive internship experiences can foster strong professional commitment and enthusiasm for the occupation, whereas a negative experience may lead to dissatisfaction and even career abandonment. Since well-educated individuals hold strong potential for the sector, strategically planned and well-supported internship programs are essential.

Within the framework of social exchange theory, support from colleagues triggers positive feelings and effects satisfaction and commitment.While colleague support can promote collabration and knowledge sharing managerial support can also create perception that their efforts are valued. Conversely, a lack of peer support or insufficient managerial recognition may lead employees to perceive an imbalance between their efforts and rewards, resulting in decreased job performance or a tendency to search alternative jobs. Thus, fostering a balanced social exchange in kitchen environments requires supports from both colleagues and management.

In conclusion, results offer valuable insights about the relationship between perceived support from supervisors and co-workers, occupational commitment, and SWB. Additionally, the mediating role of SWB suggests that improving gastronomy and culinary arts students' SWB can have a positive impact on their commitment. While the study focuses on key factors influencing occupational commitment, further researches can focus on other potential variables to explore how these relationships might vary across different occupations.

References

Ariza-Montes, A., Arjona-Fuentes, J. M., Han, H., & Law, R. (2018). The price of success: A study on chefs' subjective well-being, job satisfaction, and human values. *International Journal of Hospitality Management*, *69*, 84–93. https://doi.org/10.1016/j.ijhm.2017.10.006

Aydın, E., & Kalemci Tüzün, I. (2019). Organizational support sources and job performance relations: What about occupational commitment? *Anatolia*, *30*(3), 379–389. https://doi.org/10.1080/13032917.2019.1597740

Bardakoğlu, Ö., Akgündüz, Y., Kızılcalıoğlu, G., & Yeşilyurt, H. (2017). Otel işletmelerinde algılanan yönetici desteğinin çalışanların iş stresi ve mutluluk düzeylerine etkisi. 8. *Uluslararası Girişimcilik Kongresi*, 103–116.

Blascovich, J., Tomaka, J., Robinson, J. P., Shaver, P. R., & Wrightsman, L. S. (1991). Measures of self-esteem. *Measures of Personality and Social Psychological Attitudes*, *1*(2), 115–160.

Blau, G., Paul, A., & John, N. S. (1993). On developing a general index of work commitment. *Journal of Vocational Behavior*, *42*(3), 298–314.

Bloisi, W., & Hoel, H. (2008). Abusive work practices and bullying among chefs: A review of the literature. *International Journal of Hospitality Management*, *27*(4), 649–656. https://doi.org/10.1016/j.ijhm.2007.09.001

Brislin, R. W. (1970). Back-Translation for Cross-Cultural Research. *Journal of Cross-Cultural Psychology*, *1*(3), 185–216. https://doi.org/10.1177/135910457000100301

Carson, K. D., & Bedeian, A. G. (1994). Career commitment: Construction of a measure and examination of its psychometric properties. *Journal of Vocational Behavior*, *44*(3), 237–262.

Charoensukmongkol, P., Moqbel, M., & Gutierrez-Wirsching, S. (2016). The role of coworker and supervisor support on job burnout and job satisfaction. *Journal of Advances in Management Research*, *13*(1). https://doi.org/10.1108/JAMR-06-2014-0037

Chen, Z. X., Tsui, A. S., & Farh, J. (2002). Loyalty to supervisor vs. organizational commitment: Relationships to employee performance in China. *Journal of Occupational and Organizational Psychology*, *75*(3), 339–356. https://doi.org/10.1348/096317902320369749

Chung, H., Quan, W., Koo, B., Ariza-Montes, A., Vega-Muñoz, A., Giorgi, G., & Han, H. (2021). A Threat of Customer Incivility and Job Stress to Hotel Employee Retention: Do Supervisor and Co-Worker Supports Reduce Turnover Rates? *International Journal of Environmental Research and Public Health*, *18*(12), Article 12. https://doi.org/10.3390/ijerph18126616

Cole, M. S., Bruch, H., & Vogel, B. (2006). Emotion as mediators of the relations between perceived supervisor support and psychological hardiness on employee cynicism. *Journal of Organizational Behavior*, *27*(4), 463–484. https://doi.org/10.1002/job.381

Dlugonski, D., & Motl, R. W. (2012). Possible antecedents and consequences of self-esteem in persons with multiple sclerosis: Preliminary evidence from a cross-sectional analysis. *Rehabilitation Psychology*, *57*(1), 35–42. https://doi.org/10.1037/a0027176

Eisenberger, R., Stinglhamber, F., Vandenberghe, C., Sucharski, I. L., & Rhoades, L. (2002). Perceived supervisor support: Contributions to perceived organizational support and employee retention. *Journal of Applied Psychology*, *87*(3), 565.

Ertürk, A., Kara, S. B. K., & Güneş, D. Z. (2016). Duygusal Emek ve Psikolojik İyi Oluş: Bir Yordayıcı Olarak Yönetsel Destek Algısı. *Abant İzzet Baysal Üniversitesi Eğitim Fakültesi Dergisi*, *16*(4), Article 4.

Fernet, C., Gillet, N., Austin, S., Trépanier, S., & Drouin-Rousseau, S. (2021). Predicting nurses' occupational commitment and turnover intention: The role of autonomous motivation and supervisor and coworker behaviours. *Journal of Nursing Management*, *29*(8), 2611–2619. https://doi.org/10.1111/jonm.13433

Giray, M. D., & Sahin, D. N. (2012). Algilanan örgütsel, yönetici ve çalisma arkadaslari destegi ölçekleri: Geçerlik ve güvenirlik çalismasi. *Turk Psikoloji Yazilari*, *15*(30), 1.

Gordon, S., Tang, C.-H., Day, J., & Adler, H. (2019). Supervisor support and turnover in hotels: Does subjective well-being mediate the relationship? *International Journal of Contemporary Hospitality Management*, *31*(1), 496–512.

Griffeth, R. W., & Hom, P. W. (2001). *Retaining valued employees*. sage. https://books.google.com/books?hl=en&lr=&id=8JLeS54I184C&oi=fnd&pg=PP9&dq=Griffeth,+Rodger+W.,+and+Peter+W.+Hom+(2001),+Retaining+ValuedEmployees,+Thousand+Oaks,+CA:+Sage&ots=zndfETb8kt&sig=9IPoWn5oaadVz9OWa9PEYI0gF9w

Guchait, P., Paşamehmetoğlu, A., & Dawson, M. (2014). Perceived supervisor and co-worker support for error management: Impact on perceived psychological safety and service recovery performance. *International Journal of Hospitality Management*, *41*, 28–37.

Harris, D. A., & Giuffre, P. (2010). "The Price You Pay": How Female Professional Chefs Negotiate Work and Family. *Gender Issues*, *27*(1), 27–52. https://doi.org/10.1007/s12147-010-9086-8

Hochschild, A. R. (2012). *The Managed Heart: Commercialization of Human Feeling*. University of California Press. https://doi.org/10.1525/9780520951853

Hogg, M. A., & Vaughan, G. M. (2008). *Social psychology*. Pearson Education. https://scholar.google.com/scholar?cluster=8650055098386546870&hl=en&oi=scholarr

Holmes, M. R., McAdams, B., Gibbs, C., & D'Angelo, A. (2022). Is the foodservice industry perceived as being palatable by those looking to enter it? *Journal of Foodservice Business Research*, *25*(4), 414–433. https://doi.org/10.1080/15378020.2021.1950507

Ilyas, S., Abid, G., & Ashfaq, F. (2023). The impact of perceived organizational support on professional commitment: A moderation of burnout and mediation of well-being. *International Journal of Sociology and Social Policy*, *43*(7/8), 710–726. https://doi.org/10.1108/IJSSP-06-2022-0170

İnce, A. R. (2016). Algilanan örgütsel desteğİn İşe adanmişlik üzerİndekİ etkİsİnde yönetİcİ desteğİnİn aracilik rolü. *Elektronik Sosyal Bilimler Dergisi*, *15*(57). https://dergipark.org.tr/en/pub/esosder/article/263211
Jackson, S., & Jackson, L. T. (2019). Self-esteem: Its mediating effects on the relationship between discrimination at work and employee organisation commitment and turn-over intention. *Journal of Psychology in Africa*, *29*(1), 13–21. https://doi.org/10.1080/14330237.2019.1568065
Johns, N., & Menzel, P. J. (1999). "If you can't stand the heat!" ... Kitchen violence and culinary art. *International Journal of Hospitality Management*, *18*, 99–109. https://doi.org/10.1016/S0278-4319(99)00013-4
Kale, E. (2015). Lider Desteği ve İş Arkadaşlari Desteğinin İş Performansi Üzerine Etkileri: İş Tatmini ve Yaşam Tatmininin Araci Rolü. *Uluslararası İktisadi ve İdari İncelemeler Dergisi*, *14*. https://dergipark.org.tr/en/pub/ulikidince/article/232160
Karatepe, O. M., Aboramadan, M., & Dahleez, K. A. (2020). Does climate for creativity mediate the impact of servant leadership on management innovation and innovative behavior in the hotel industry? *International Journal of Contemporary Hospitality Management*, *32*(8), 2497–2517. https://doi.org/10.1108/IJCHM-03-2020-0219
Kidd, J. M., & Smewing, C. (2001). The role of the supervisor in career and organizational commitment. *European Journal of Work and Organizational Psychology*, *10*(1), 25–40. https://doi.org/10.1080/13594320042000016
Kim, H. J., Hur, W.-M., Moon, T.-W., & Jun, J.-K. (2017). Is all support equal? The moderating effects of supervisor, coworker, and organizational support on the link between emotional labor and job performance. *BRQ Business Research Quarterly*, *20*(2), 124–136. https://doi.org/10.1016/j.brq.2016.11.002
Kim, H., & Park, E. J. (2013). The role of social experience in undergraduates' career perceptions through internships. *Journal of Hospitality, Leisure, Sport & Tourism Education*, *12*(1), 70–78. https://doi.org/10.1016/j.jhlste.2012.11.003
Kram, K. E. (1983). Phases of the Mentor Relationship. *Academy of Management Journal*, *26*(4), 608–625. https://doi.org/10.2307/255910
Kusluvan, H., Akova, O., & Kusluvan, S. (2022). Occupational stigma and career commitment: Testing mediation and moderation effects of occupational self-esteem. *International Journal of Hospitality Management*, *102*, 103149. https://doi.org/10.1016/j.ijhm.2022.103149

Lee, K., Carswell, J. J., & Allen, N. J. (2000). A meta-analytic review of occupational commitment: Relations with person-and work-related variables. *Journal of Applied Psychology*, *85*(5), 799.

Limpanitgul, T., Boonchoo, P., & Photiyarach, S. (2014). Coworker support and organisational commitment: A comparative study of Thai employees working in Thai and American airlines. *Journal of Hospitality and Tourism Management*, *21*, 100–107.

Locke, E. A., & Latham, G. P. (2006). New Directions in Goal-Setting Theory. *Current Directions in Psychological Science*, *15*(5), 265–268. https://doi.org/10.1111/j.1467-8721.2006.00449.x

Mac Con Iomaire, M., Afifi, M. F., & Healy, J. (2021). Chefs' perspectives of failures in foodservice kitchens, part 1: A phenomenological exploration of the concepts, types, and causes of food production failure. *Journal of Foodservice Business Research*, *24*(2), 177–214. https://doi.org/10.1080/15378020.2020.1842955

Maertz, C. P., Griffeth, R. W., Campbell, N. S., & Allen, D. G. (2007). The effects of perceived organizational support and perceived supervisor support on employee turnover. *Journal of Organizational Behavior*, *28*(8), 1059–1075. https://doi.org/10.1002/job.472

Malhotra, N. K. (2020). *Marketing research: An applied orientation* (Seventh edition, global edition). Pearson.

Mercurio, Z. A. (2015). Affective Commitment as a Core Essence of Organizational Commitment: An Integrative Literature Review. *Human Resource Development Review*, *14*(4), 389–414. https://doi.org/10.1177/1534484315603612

Met, Ö. L., & Sarioğlan, M. (2010). Otel İşletmeleri Yiyecek İçecek Üniterlerinde İşgören Devri Üzerine Görgül Bir Araştırma. *Adıyaman Üniversitesi Sosyal Bilimler Enstitüsü Dergisi*.

Meyer, J. P., Allen, N. J., & Smith, C. A. (1993). Commitment to organizations and occupations: Extension and test of a three-component conceptualization. *Journal of Applied Psychology*, *78*(4), 538.

Murray-Gibbons, R., & Gibbons, C. (2007). Occupational stress in the chef profession. *International Journal of Contemporary Hospitality Management*, *19*(1), 32–42. https://doi.org/10.1108/09596110710724143

Nahum-Shani, I., Henderson, M. M., Lim, S., & Vinokur, A. D. (2014). Supervisor support: Does supervisor support buffer or exacerbate the adverse effects of supervisor undermining? *The Journal of Applied Psychology*, *99*(3), 484–503. https://doi.org/10.1037/a0035313

Phungsoonthorn, T., & Charoensukmongkol, P. (2020). Antecedents and outcomes associated with a sense of place toward the organization of Myanmar migrant workers in Thailand. *Equality, Diversity and Inclusion: An International Journal*, *39*(2), 195–218.
Pierce, J. L., & Gardner, D. G. (2004). Self-Esteem Within the Work and Organizational Context: A Review of the Organization-Based Self-Esteem Literature. *Journal of Management*, *30*(5), 591–622. https://doi.org/10.1016/j.jm.2003.10.001
Raabe, B., & Beehr, T. A. (2003). Formal mentoring versus supervisor and coworker relationships: Differences in perceptions and impact. *Journal of Organizational Behavior*, *24*(3), 271–293. https://doi.org/10.1002/job.193
Rathi, N., & Lee, K. (2017). Understanding the role of supervisor support in retaining employees and enhancing their satisfaction with life. *Personnel Review*, *46*(8), 1605–1619. https://doi.org/10.1108/PR-11-2015-0287
Rhoades, L., & Eisenberger, R. (2002). Perceived organizational support: A review of the literature. *Journal of Applied Psychology*, *87*(4), 698.
Robinson, R. N. S., Ruhanen, L., & Breakey, N. M. (2016). Tourism and hospitality internships: Influences on student career aspirations. *Current Issues in Tourism*, *19*(6), 513–527. https://doi.org/10.1080/13683500.2015.1020772
Robinson, R. N. S., Solnet, D. J., & Breakey, N. (2014). A phenomenological approach to hospitality management research: Chefs' occupational commitment. *International Journal of Hospitality Management*, *43*, 65–75. https://doi.org/10.1016/j.ijhm.2014.08.004
Rosenberg, M. (1965). Society and Adolescent Self-image. *Princeton, NJ: Princeton University Press.*
Rosenberg, M. (1989). *Society and the adolescent self-image. Revised edition.* Middletown, CT: Wesleyan University Press.
Rousseau, V., & Aubé, C. (2010). Social Support at Work and Affective Commitment to the Organization: The Moderating Effect of Job Resource Adequacy and Ambient Conditions. *The Journal of Social Psychology*, *150*(4), 321–340. https://doi.org/10.1080/00224540903365380
Seers, A. (1989). Team-member exchange quality: A new construct for role-making research. *Organizational Behavior and Human Decision Processes*, *43*(1), 118–135.

Seery, M. D., & Quinton, W. J. (2015). Targeting Prejudice: Personal Self-Esteem as a Resource for Asians' Attributions to Racial Discrimination. *Social Psychological and Personality Science*, *6*(6), 677–684. https://doi.org/10.1177/1948550615575756

Sevim, K., & Artan, T. (2021). Yüksek ve düşük benlik saygısını etkileyen faktörler. *Toplumsal Politika Dergisi*, *2*(2), 109–121.

Sherony, K. M., & Green, S. G. (2002). Coworker exchange: Relationships between coworkers, leader-member exchange, and work attitudes. *Journal of Applied Psychology*, *87*(3), 542.

Singhal, H., & Rastogi, R. (2018). Psychological capital and career commitment: The mediating effect of subjective well-being. *Management Decision*, *56*(2), 458–473. https://doi.org/10.1108/MD-06-2017-0579

Skakon, J., Nielsen, K., Borg, V., & Guzman, J. (2010). Are leaders' well-being, behaviours and style associated with the affective well-being of their employees? A systematic review of three decades of research. *Work & Stress*, *24*(2), 107–139. https://doi.org/10.1080/02678373.2010.495262

Skiba, T., & Wildman, J. L. (2019). Uncertainty Reducer, Exchange Deepener, or Self-Determination Enhancer? Feeling Trust Versus Feeling Trusted in Supervisor-Subordinate Relationships. *Journal of Business and Psychology*, *34*(2), 219–235. https://doi.org/10.1007/s10869-018-9537-x

Sungu, L. J., Weng, Q. (Derek), & Xu, X. (2019). Organizational commitment and job performance: Examining the moderating roles of occupational commitment and transformational leadership. *International Journal of Selection and Assessment*, *27*(3), 280–290. https://doi.org/10.1111/ijsa.12256

Tozkoparan, G., & Tenteriz, Y. (2022). Algılanan Yönetici Desteği ve Algılanan Çalışma Arkadaşları Desteğinin Örgütsel Vatandaşlık Davranışı Üzerine Etkisi. *Business and Economics Research Journal*, *13*(1), 89–107. https://doi.org/10.20409/berj.2022.363

Tummers, L., Steijn, B., Nevicka, B., & Heerema, M. (2018). The Effects of Leadership and Job Autonomy on Vitality: Survey and Experimental Evidence. *Review of Public Personnel Administration*, *38*(3), 355–377. https://doi.org/10.1177/0734371X16671980

Tuna Arslan, P., & Aydin, E. (2023). Otel İşletmelerinde Yenilikçi İş Davranışlarının Örgütsel Destek, Yönetici Desteği ve Çalışma Arkadaşı

Desteği Üzerine Etkileri: İstanbul İli Örneği. *Seyahat ve Otel İşletmeciliği Dergisi*, *20*(1), 73–94. https://doi.org/10.24010/soid.1144883

Ulu, E. K., & Vatan, A. (2023). The relationship between cooks' flow experience, job satisfaction and subjective well-being. *Journal of Foodservice Business Research*, *0*(0), 1–25. https://doi.org/10.1080/15378020.2023.2251371

The Value of Leisure and Volunteering

Akyay UYGUR[1], Halise Dilek SEVİN[2], Orhun TOPAÇOĞLU[3]

Introduction

Leisure time is defined as a period that individuals spend according to their own preferences, free from work, responsibilities, and other obligations (Haywood, 2019). Throughout history, the meaning of leisure time has varied in line with the cultural structures, economic conditions, and social frameworks of societies. In Ancient Greece, leisure time was considered not only a period of rest but also a valuable opportunity for individuals to enhance their cultural development and fulfill their social responsibilities. During this period, leisure time was seen as the opposite of work and was closely associated with the concept of "freedom" (Balme, 1984). Aristotle emphasized the spiritual and intellectual development of individuals by advocating for engagement in activities that are "valuable in themselves" (Telfer, 1987).

Today, the effective utilization of leisure time has become a crucial aspect of both personal development and social responsibility. In this context, volunteer services play a significant role. Volunteering is an activity in which individuals use their free time to help others, address social problems, and contribute to societal transformation. Moreover, volunteer activities provide individuals with opportunities for personal growth, improving their social skills, creating career opportunities, and gaining diverse cultural experiences (Wilson, 2000; Stukas et al., 2016). Volunteering not only provides social benefits but also positively affects individuals' psychological well-being, enhances their ability to empathize, and strengthens their social interactions and personal resilience

1 Prof. Dr. Ankara Hacı Bayram Veli University, Tourism Faculty, Department of Recreation Management, akyay.uygur@hbv.edu.tr

2 Assoc. Prof. Dr. Ankara Hacı Bayram Veli University, Tourism Faculty, Department of Recreation Management, halise.sevin@hbv.edu.tr

3 Std. Ankara Hacı Bayram Veli University, Tourism Faculty, Department of Recreation Management, topacogluorhun@gmail.com

(Ali et al., 2016; Chung et al., 2021). Furthermore, volunteering fosters a sense of belonging, encouraging individuals to contribute to society and supporting social integration (Sveen, 2022). Research indicates that individuals who participate in volunteer activities report higher levels of life satisfaction and better coping mechanisms for psychological challenges (Binder, 2015). Additionally, volunteering increases empathy, strengthens interactions with diverse social groups, and promotes a more inclusive perspective on social diversity (Musick & Wilson, 2007). From the perspective of social capital, volunteering expands individuals' social networks and enhances social trust. Thus, engaging in leisure activities that contribute to social welfare, rather than solely for recreation, offers substantial benefits to both individuals and society (Putnam, 2000).

The Concept and Importance of Leisure Time

Leisure time is defined as the period individuals can spend voluntarily and arbitrarily after fulfilling their obligatory responsibilities and basic needs (Lundberg et al., 1934; Karaküçük, 2014). Before the Industrial Revolution, leisure time was largely perceived as non-working time. However, with industrialization, it evolved into a broader social concept as the working class began to gain more free time (Freudenberger & Cummins, 1976). Particularly in the 20th century, leisure time increased due to the reduction in working hours and the influence of labor unions (Veal, 2017). In the 21st century, digitalization and the spread of flexible working models have significantly transformed both the form and utilization of leisure time. Today, individuals engage in leisure activities not only for passive rest or entertainment but also for social interaction, virtual socialization, and productive hobbies through digital platforms (Topaçoğlu & Kılavuz, 2022).

Stebbins (2013) highlights that leisure time plays a crucial role in individuals' physical and mental renewal processes. He categorizes leisure into three types: serious leisure, casual leisure, and project-based leisure. Serious leisure includes activities that provide long-term satisfaction, such as artistic endeavors, academic studies, or creative pursuits. Casual leisure comprises short-term activities primarily focused on relaxation or entertainment, while project-based leisure refers to temporary yet relatively complex activities undertaken occasionally (Stebbins, 2013).

Among these leisure types, the extent to which an individual benefits from leisure time largely depends on how it is utilized. Individuals who practice effective time management and engage in leisure activities that support personal development, social interaction, and health-oriented pursuits report higher life satisfaction (Topaçoğlu & Uygur, 2024). Notably, participation in serious leisure activities fosters meaningful and sustainable interests, contributing to long-term psychological and emotional well-being (Stebbins & Graham, 2004).

The significance of leisure time extends beyond relaxation; it also plays a vital role in determining life satisfaction and happiness levels (Sabırlı et al., 2019). Studies indicate that regular participation in leisure activities reduces stress levels and helps individuals cope with psychological distress (Iwasaki & Mannell, 2000). At the same time, leisure activities strengthen individuals' social ties and prevent social isolation by increasing community belonging (Toepell, 2013).

The positive effects of leisure are not limited to physical and mental well-being but also encompass psychosocial development. When individuals utilize their free time effectively, they acquire experiences that boost self-confidence, enhance self-esteem, and develop personal skills (Mannell & Kleiber, 1997). In particular, volunteering activities provide opportunities for self-actualization while fostering a sense of belonging and promoting social integration (Hustinx et al., 2010; AL-shahrani, et al., 2019).

Beyond individual benefits, leisure time also plays a crucial role in society. The engagement of individuals in productive and socially interactive activities strengthens social cohesion. Voluntary activities, in particular, contribute to social integration by fostering stronger interpersonal ties (Vannier et al., 2021). Engaging in collective activities promotes a shared sense of belonging, enhances social harmony, and allows individuals to perceive themselves as integral members of society (Luque-Suárez et al., 2021). Therefore, the societal impact of leisure time not only supports individual psychosocial development but also facilitates the construction of a more cohesive and resilient social structure (Topaçoğlu & Kılavuz, 2023).

Leisure time contributes to personal development and recreation while generating significant economic and psychosocial effects. From an economic perspective, leisure activities influence numerous sectors, both directly and indirectly. Recreation, private-public sector initiatives, and volunteer work are among the key areas that generate economic value (Arslan, 2021; Birol & Aydın, 2020). Volunteer activities, while strengthening civil society organizations, also support individuals' integration into the labor market by enhancing their economic conditions and career prospects (Fényes et al., 2021). From a psychosocial standpoint, participation in volunteer work fosters self-esteem, self-efficacy, and social solidarity (Wilson, 2000). Research indicates that volunteering enhances psychological well-being, alleviates depression and stress, prevents social withdrawal, and improves overall life satisfaction (Wilson, 2000; Mojza et al., 2010; Jiang, et al., 2018). Moreover, effective use of leisure time strengthens social relationships and facilitates individuals' integration into society (Stebbins & Graham, 2004).

Volunteering

The term "volunteering," derived from the French word voluntaire, was initially used to refer to voluntary participation in military service. Its Latin root, voluntarius, signifies the free will of an individual. Over time, the meaning of the term has evolved, and it is now commonly associated with non-military activities (Afşar, 2019). Volunteering can be defined as a voluntary activity carried out without financial gain, aimed at benefiting others, and often conducted within an institutional framework. This concept is fundamentally based on the principles of free participation and the absence of economic incentives. However, the perception and role of volunteering in society vary across countries (Çakı, 2014). These variations stem not only from religious and cultural heritage but also from historical backgrounds, political dynamics, and geographical conditions (Agerhem & Cross, 2004).

Volunteering is generally considered an altruistic act, as it involves selfless actions intended to help others or support a cause without expecting direct benefits. However, purely altruistic volunteering is relatively rare in practice. In contemporary discussions, volunteering is increasingly seen as a form of "investment" rather than solely a selfless act, as it is often motivated by factors such as career advancement and personal development. Particularly for disadvantaged youth, volunteering serves as a valuable tool for career exploration and employment opportunities. Studies on volunteering motivations frequently highlight reasons such as helping others, experiencing new cultures, expanding social networks, breaking away from routine activities, testing personal limits, acquiring new skills, making career decisions, and gaining independence from family (Çakı, 2014).

Volunteering is not merely an individual endeavor; it is also an expression of active citizenship and social responsibility. Rooted in principles of solidarity, commitment, and participation, volunteering contributes significantly to the well-being of both individuals and society (Afşar, 2019). Unlike amateur activities and hobbies, volunteering is distinguished by two fundamental characteristics. The first is altruism—while volunteers may also have personal motivations, their primary goal is to contribute to society meaningfully. This contrasts with amateur or hobbyist pursuits, which typically operate within commercial, artistic, scientific, or recreational domains. Secondly, volunteers dedicate their time and effort based on a sense of social responsibility, believing that addressing societal issues and assisting others is a fundamental duty (Parker, 2000).

Intersection of Leisure and Volunteering

Volunteering covers activities that individuals engage in without expecting any financial reward, aiming to contribute to society (Güder, 2006). The relationship between leisure time and volunteering forms an essential intersection in terms of personal development and societal contribution. Specifically, it has been shown that voluntary activities have a positive effect on individuals' psychological well-being (Mojza et al., 2010).

Volunteering activities are inherently linked to serious leisure, which is defined as an amateur, hobby, or volunteer activity that is of sufficient significance and interest for the participant to develop a career by acquiring and expressing specialized skills, knowledge, and experience (Stebbins, 2007; Stebbins, 1996). In this context, volunteering is viewed as a serious leisure activity where individuals contribute to society by using their free time in a structured and purposeful manner. Altruism and the desire to help others are key motivators for engaging in such activities, as outlined in leisure behavior. Volunteering is an activity that an individual performs voluntarily, without monetary reward, in a formal setting for the benefit of strangers (Lee & Kim, 2017). When volunteering is considered within the scope of serious leisure, it transcends individual preferences and encompasses long-term commitment and a sense of social belonging (Özmete, 2010; Carlton, 2015). The varying levels of engagement in volunteering are crucial for understanding the extent to which individuals participate and the motivations that drive these activities (Clary & Snyder, 1999).

Volunteering as a serious leisure activity is classified into six distinct levels, each reflecting different stages of engagement and commitment. The first level involves volunteering activities that challenge participants to cope with difficulties, including financial and time constraints, requiring perseverance and persistence. The second level includes volunteering activities that individuals undertake with the aim of achieving a "career" marked by significant commitment and accomplishment. The third level involves specialized activities that require knowledge, talent, and skills, such as scientific, sporting, or artistic expertise, where only individuals with such expertise can participate. At the fourth level, volunteering produces lasting benefits such as self-actualization, personal growth, sense of achievement, and social belonging. The fifth level involves activities where individuals define and maintain a worldview that supports their unique identity, and the sixth level is characterized by volunteering activities that help individuals define their identity in opposition to temporary and widespread leisure activities (Stebbins & Graham, 2004; Stebbins, 2007).

These six levels provide a framework for understanding how individuals structure their leisure time and approach volunteering activities. Volunteering is more than just a leisure pursuit; it allows individuals to develop their identity, hone their skills, and integrate more fully into their social roles (Grönlund, 2011). Viewing volunteering through the lens of serious leisure is essential for understanding the participants' commitment to the activity and their underlying motivations. While the initial stages of volunteering demand perseverance in overcoming challenges, the advanced stages enable individuals to specialize, contribute to their personal development, and establish a unique identity within society. This hierarchical approach offers a systematic perspective on how individuals interpret their leisure time and the impact of volunteering on both individual development and social dynamics. Volunteering, therefore, is a process that unfolds over time, offering individuals opportunities for both self-actualization and societal contribution (Stebbins & Graham, 2004; Stebbins, 2007).

Given that volunteering develops at various stages on both individual and societal levels, it becomes evident that the process is structured in diverse ways. While volunteering within serious leisure time is characterized by long-term commitment and integration into one's personal identity, it also exhibits a more flexible and activity-oriented structure. Volunteering activities within certain organizations and events present an area of participation shaped by different motivations (Noordegraaf, 2016; Pi, et al., 2014; Lu & Schuett, 2014).

Volunteering activities within leisure time are particularly significant as they provide meaningful engagement not only for the individuals involved but also for society at large. For instance, nature conservation projects, cultural events, sports organizations, and the tourism sector are key areas where individuals can contribute to society while enjoying their leisure (Noordegraaf, 2016; Bang & Ross, 2019). Volunteer participation plays a crucial role in organizing recreational activities, ensuring their sustainability and allowing individuals to gain social capital (Angosto et al., 2021). Volunteers are particularly essential in the success of sports events, where they participate not only in the events but also in the organizational processes (Bang & Ross, 2019; Angosto et al., 2021). Moreover, the experiences gained through volunteering can contribute significantly to career development, providing individuals with an opportunity to use their free time for both social and personal benefits (Nichols et al., 2016; Fényes, 2021).

Volunteering plays an especially critical role in large-scale organizations and international events. Major sporting events such as the Olympics and the World Cup would struggle to ensure their sustainability without volunteer participation. Volunteers contribute to the success of these events through a wide range of activities, including logistical support, participant guidance, media assistance, and health and security services (Kemp, 2022; Fairley et al., 2007). Volunteer participation in these events strengthens social connections and enables individuals to make meaningful use of their leisure time, while also contributing to the economic sustainability of the event by reducing organizational costs (Fairley et al., 2007). In large-scale events like the Olympics, the training and motivation of volunteers are key factors determining the success of the event, encouraging both the event's success and the global adoption of the volunteering concept (Khoo & Engelhorn, 2011; Karlis, 2013).

Results

Leisure time is a vital concept in which individuals develop themselves, rest, and strengthen their social ties by stepping out of their daily routines. In addition to its positive effects on psychological and physical health, it also serves as a means of searching for personal meaning and fulfillment (Stebbins, 2007). Moreover, leisure time holds significant potential for making social contributions. Volunteering enhances the meaningfulness of leisure time and enables individuals to reinforce their sense of responsibility, empathy, and solidarity (Wilson, 2000). Volunteering activities offer long-term psychological and social benefits not only to individuals but also to society at large (Musick & Wilson, 2007).

Volunteering activities emerge as one of the most socially efficient uses of free time. Utilizing free time through the framework of "serious leisure" represents a process where volunteer work contributes to the development of personal identity and overall life satisfaction. The literature indicates that associating volunteering with leisure time is a phenomenon that significantly enhances the well-being of societies. Volunteering provides valuable benefits, including personal development, a sense of social belonging, trust, and the strengthening of social connections (Stebbins, 1996). In a study conducted by Luque-Suarez et al. (2021), it was concluded that volunteering activities fostered a sense of belonging and social welfare among university students. Volunteering activities not only reinforce individuals' sense of social belonging but also contribute to the enhancement of social values such as mutual trust, solidarity, and cooperation within society. Kristiansen et al. (2015) examined the effects of long-term participation in volunteer communities on social processes, finding high levels of social integration, collective identity formation, and the strengthening of sustainable social solidarity.

Volunteering during leisure time provides significant benefits for both individuals and societies. It enhances individuals' quality of life and promotes sustainable social development by strengthening social ties. Therefore, engaging in voluntary activities during leisure time not only fosters individual fulfillment but also contributes to the improvement of social welfare (Hustinx et al., 2010).

References

Afşar, S. T. (2019). Gönüllülük olgusunu sosyolojik teorilerle anlama denemesi. Gençlik Araştırmaları Dergisi, 7(Gençlik ve Gönüllülük), 27–58.
Agerhem, S., & Cross, S. R. (2004). Volunteering in South Asia and South East Asia. International Federation of Red Cross and Red Crescent Societies.
Ali, S. B., Khan, N. A., & Zehra, A. (2016). Effect of volunteerism on mental health and happiness. International Journal of Humanities and Social Sciences, 5(2), 123–130.
AL-shahrani, H. F., & Hammad, M. A. (2019). The Role of Voluntary Work in the Improvement of The Social Capital and Sustainable Development in Saudi Society. Humanities & Social Sciences Reviews, 7(5), 1104–1120.
Angosto, S., Bang, H., Bravo, G. A., Diaz-Suarez, A., & López-Gullón, J. M. (2021). Motivations and future intentions in sport event volunteering: a systematic review. Sustainability, 13(22), 12454.
Arslan, H. M. (2021) Ekonomik İstihdam Aracı Olarak Rekreasyon. Nobel.
Balme, M. (1984). Antik Yunan'da Boş Zaman Anlayışı. İstanbul: Yunan Filolojisi Yayınları
Bang, H., & Ross, S. D. (2009). Volunteer motivation and satisfaction. Journal of venue and Event Management, 1(1), 61–77.
Binder, M. (2015). Volunteering and life satisfaction: a closer look at the hypothesis that volunteering more strongly benefits the unhappy. Applied Economics Letters, 22(11), 874–885.
Birol, S. Ş., & Aydın, E. (2020). Türkiye'de rekreasyon birimi yapılanması için sektör değerlendirmesi. Gençlik Araştırmaları Dergisi, 8(22), 136–146.
Çakı, F. (2014). Türk sosyolojisinde yeni bir alan: Gönüllülük araştırmaları. İstanbul University Journal of Sociology, 3(29), 185–209.
Carlton, S. (2015). Connecting, belonging: Volunteering, wellbeing and leadership among refugee youth. International journal of disaster risk reduction, 13, 342–349.
Chung, Y. W., Im, S., & Kim, J. E. (2021). Can empathy help individuals and society? Through the lens of volunteering and mental health. In Healthcare (Vol. 9, No. 11, p. 1406). MDPI.
Clary, E. G., & Snyder, M. (1999). The motivations to volunteer: Theoretical and practical considerations. Current directions in psychological science, 8(5), 156–159.

Fairley, S., Kellett, P., & Green, B. C. (2007). Volunteering abroad: Motives for travel to volunteer at the Athens Olympic Games. Journal of sport management, 21(1), 41–57.
Fényes, H., Markos, V., & Mohácsi, M. (2021). Volunteering among higher education students as part of individual career management. Corvinus Journal of Sociology and Social Policy, 12(2), 3–22.
Freudenberger, H., & Cummins, G. (1976). Health, work and leisure before the industrial revolution. Explorations in Economic History, 13(1), 1.
Grönlund, H. (2011). Identity and volunteering intertwined: Reflections on the values of young adults. VOLUNTAS: International journal of voluntary and nonprofit organizations, 22, 852–874.
Güder, N (2006) STK'lar İçin Gönüllülük ve Gönüllü Yönetimi Rehberi, Ankara: Sivil Toplum Geliştirme Merkezi, 2006, s. 4.
Haywood, C. (2019). The Concept of Leisure in Modern Society. London: Routledge.
Hustinx, L., Cnaan, R. A., & Handy, F. (2010). Navigating theories of volunteering: A hybrid map for a complex phenomenon. Journal for the Theory of Social Behaviour, 40(4), 410–434.
Iwasaki, Y., & Mannell, R. C. (2000). Hierarchical dimensions of leisure stress coping. Leisure sciences, 22(3), 163–181.
Jiang, D., Hosking, D., Burns, R., & Anstey, K. J. (2019). Volunteering benefits life satisfaction over 4 years: The moderating role of social network size. Australian Journal of Psychology, 71(2), 183–192.
Karaküçük, S. (2014). Rekreasyon: boş zamanları değerlendirme.
Karlis, G. (2013). The Olympic Volunteer and Olympic Education: Suggestions for Administrators of Olympic Games. Voluntaris, 1(2), 116–126.
Kemp, S. (2002). The hidden workforce: Volunteers' learning in the Olympics. Journal of European Industrial Training, 26(2/3/4), 109–116.
Khoo, S., & Engelhorn, R. (2011). Volunteer motivations at a national Special Olympics event. Adapted physical activity quarterly, 28(1), 27–39.
Kristiansen, E., Skirstad, B., Parent, M. M., & Waddington, I. (2015). 'We can do it': Community, resistance, social solidarity, and long-term volunteering at a sport event. Sport management review, 18(2), 256–267.
Lee, Y., & Kim, M. (2018). Serious leisure characteristics of older adult volunteers: The case of an international sporting event. World Leisure Journal, 60(1), 45–57.

Lu, J., & Schuett, M. A. (2014). Examining the relationship between motivation, enduring involvement and volunteer experience: The case of outdoor recreation voluntary associations. Leisure Sciences, 36(1), 68–87.
Lundberg, G. A., Komarovsky, M., & McInerny, M. A. (1934). Leisure a suburban study. Columbia University Press.
Luque-Suárez, M., Olmos-Gomez, M. D. C., Castán-García, M., & Portillo-Sánchez, R. (2021). Promoting emotional and social well-being and a sense of belonging in adolescents through participation in volunteering. In Healthcare (Vol. 9, No. 3, p. 359). MDPI.
Mannell, R. C. (1997). A social psychology of leisure. Cato Avenue State College.
Mojza, E. J., Lorenz, C., Sonnentag, S., & Binnewies, C. (2010). Daily recovery experiences: the role of volunteer work during leisure time. Journal of Occupational Health Psychology, 15(1), 60–74. (s. 62–68).
Musick, M. A., & Wilson, J. (2007). Volunteers: A social profile. Indiana University Press.
Nichols, G., Hogg, E., Knight, C., Mirfin-Boukouris, H., & Storr, R. (2016). Motivations of sport volunteers in England: A review for Sport England.
Noordegraaf, M. A. (2016). Volunteering: Is it a waste of time or best experience ever. The Sport Journal, 19, 1–16.
Özmete, E. (2010). Sosyal kapitali yapılandırmak: Sosyal iletişim ağı ve yardımseverlik-gönüllülük. Sosyal Politika Çalışmaları Dergisi, 21(21), 83–104.
Parker, S. (2000). Volunteering as serious leisure. Leisure/Loisir, 25(3–4) 147–155.
Pi, L. L., Lin, Y. H., Chen, C. Y., Chiu, J. C., & Chen, Y. M. (2014). Serious leisure, motivation to volunteer and subjective well-being of volunteers in recreational events. Social indicators research, 119, 1485–1494.
Putnam, R. D. (2000). Bowling alone: The collapse and revival of American community. Simon and schuster.
Sabırlı, T.N., Argan, M., Yetim, G. ve Hızal, A. (2019). Boş zaman tatmini, yaşam tatmini ve mutlulukta sosyal medya bağlılığının rolü: 50 yaş ve üzeri bireyler üzerinde bir araştırma. CBÜ Beden Eğitimi ve Spor Bilimleri Dergisi,14(2), 263–279.
Stebbins, R. (2013). Unpaid work of love: Defining the work–leisure axis of volunteering. Leisure Studies, 32(3), 339–345.

Stebbins, R. A. (1996). Volunteering: A serious leisure perspective. Nonprofit and voluntary sector quarterly, 25(2), 211–224.
Stebbins, R. A. (2007). Serious leisure: A social psychological perspective. Springer Science & Business Media.
Stebbins, R. A., & Graham, M. (Eds.). (2004). Volunteering as leisure/leisure as volunteering: An international assessment. Cabi.
Stukas, A. A., Snyder, M., & Clary, E. G. (2016). Understanding and encouraging volunteerism and community involvement. The Journal of social psychology, 156(3), 243–255.
Sveen, S., Anthun, K. S., Batt-Rawden, K. B., & Tingvold, L. (2023). Immigrants' experiences of volunteering; A meta-ethnography. Nonprofit and voluntary sector quarterly, 52(3), 569–588.
Telfer, E. (1987). Aristotle on Leisure and Happiness. Oxford: Oxford University Press
Toepell, K. (2013). Stress Management and the Role of Leisure. Journal of Social Psychology
Topaçoğlu, O., & Kılavuz, M. E. S. (2022). Rekreasyon ve sosyalleşme açısından metaverse incelemesi. Journal of Recreation and Tourism Research, 9(3), 18–37.
Topaçoğlu, O., & Kılavuz, M. E. S. (2023) Rekreasyon ve Sanal Sosyalleşme. Özgün Zaman Etkinliklerine Rekreatif Katılım Ve Kampüs Rekreasyonu.
Topaçoğlu, O., & Uygur, A. (2024). A Conceptual Perspective on Unmanageable Leisure Time. Journal of Tourism Intelligence and Smartness, 7(1), 25–35.
Vannier, C., Mulligan, H., Wilkinson, A., Elder, S., Malik, A., Morrish, D., ... & Epton, M. (2021). Strengthening community connection and personal well-being through volunteering in New Zealand. Health & Social Care in the Community, 29(6), 1971–1979.
Veal, A. J. (2017). Leisure, sport and tourism, politics, policy and planning (4th ed.). CABI Publishing.
Wilson, J. (2000). Volunteering. Annual Review of Sociology, 26, 215–240.

Artificial Intelligence Literacy: A Study on Tourism (Undergraduate) Students

Barış TURAN[1], *Özgür ÇALKIN*[2]

Introduction

Individuals interact with AI applications regardless of their level of awareness. Particularly, as AI becomes increasingly integrated into personal, professional, and educational aspects of life, AI literacy has become a crucial necessity (Laupichler et al., 2023). Indeed, many researchers emphasize the importance of AI literacy (Lo; 2023; Farrelly & Baker, 2023). There are various studies on the use of artificial intelligence (AI) in education, its potential effects on education, and the roles of AI in learning processes (Farrelly and Baker, 2023; Yetişensoy and Rapoport, 2023; Akinwalere and Inanov, 2022). Assessing the knowledge of new generation students in understanding, using, and interacting with AI applications can provide important insights into the development and use of AI applications. Understanding the impact of AI technologies in the field of education requires examining the level of knowledge and usage habits of individuals regarding these technologies (Elçiçek, 2024). Researchers have emphasized that artificial intelligence literacy has basic processes such as "understanding and knowing" (Kong, Cheung & Zhang, 2021; Ng, et al., 2021). Thus, in order for AI to be effectively utilized, individuals must understand and be familiar with this technology. This highlights the critical importance of AI literacy. This study aims to examine the artificial intelligence literacy levels of tourism students in terms of different variables (gender, department, etc.) and to determine the differences in awareness, usage, evaluation and ethical dimensions of artificial intelligence.

1 Ass.Prof., Kafkas University, Sarıkamış Tourism Faculty, Department of Gastronomy and Culinary Arts, baris.turan@kafkas.edu.tr

2 Ass.Prof., Kafkas University, Sarıkamış Tourism Faculty, Department of Tourism Guidance, ozgur.calkin@kafkas.edu.tr

Artificial Intelligence Literacy

Research on AI literacy has been increasing over the years. A review of various definitions reveals common features, suggesting that AI literacy encompasses multiple dimensions such as knowledge, skills, tools, attitudes, methods, applications, and ethics (Zhao, Wu & Luo, 2022; Ng et al., 2022; B. Wang et al., 2023; Lo, 2023). AI literacy is defined as the ability to understand fundamental AI functions, accurately identify AI-related products, and effectively use and evaluate them (Wang et al., 2023; Lin et al., 2021; Ng et al., 2021). It encompasses a broad spectrum of competencies, including comprehending, evaluating, ethically using, and developing AI technologies (Bozkurt, 2023). A strong AI literacy level enables individuals to grasp fundamental AI concepts, critically assess its advantages and disadvantages, and adopt a well-informed perspective (Kong et al., 2022). Ng et al. (2021) emphasize that AI literacy is not merely about knowing how to use AI systems but also requires understanding the fundamental principles, operational mechanisms, and ethical concerns underlying these systems. In this regard, fostering AI literacy among individuals is crucial for ensuring the ethical and conscious use of this technology (Polat, 2024).

Artificial intelligence is used in education as well as in many other areas. However, how consciously it is used is debatable. Research indicates that both theoretical and practical education are necessary to enhance AI literacy among university students (Cerny, 2024). Laupichler et al. (2022) argue that AI skills should be considered not merely as a subset of digital literacy but as a critical competency that directly impacts individuals' daily lives. Although it is known that research on artificial intelligence (AI) applications in Turkey has been predominantly conducted in technology-related fields, in recent years, various studies incorporating AI have been carried out in different disciplines as the scope of the topic has expanded. On the other hand, studies on the use of AI in education (Coşkun & Gülleroğlu, 2021; Arslan, 2020) and AI literacy (Banaz & Demirel, 2024; Türel et al., 2024; Elçiçek, 2024) have also begun to emerge. However, no research has been found on AI literacy among undergraduate students in the field of social sciences.

The Impact of AI on Undergraduate Student Academic Performance

Artificial Intelligence (AI) is increasingly being integrated into higher education, transforming learning environments and academic performance. The rapid development of AI technologies has led to significant advances in education. AI applications such as intelligent tutoring systems, automated grading, and personalized learning platforms have reshaped the traditional learning experience (Lo, 2023). As universities adopt AI-enabled tools, it is crucial to understand their impact on student achievement and academic performance.

Note that students who use AI-enabled study aids perform better on exams and assignments than those who rely solely on traditional learning methods. AI tools have been particularly effective in personalized learning by providing students with specific guidance and adaptive assessments (Elçiçek, 2024). Farrelly and Baker (2023) found that AI-enabled learning platforms improve student comprehension and retention. Similarly, Yetişensoy and Rapoport (2023) conducted a systematic review highlighting how AI tools provide real-time feedback, allowing students to adjust their learning strategies accordingly. Additionally, Akinwalere and Inanov (2022) compared different disciplines and found that AI-enabled education had a more significant impact on STEM students due to the nature of its automated problem-solving features.

However, some concerns have been raised about the potential disadvantages of AI. Arslan (2020) emphasized that over-reliance on AI tools could reduce critical thinking skills and reduce students' ability to work independently. Furthermore, Panigrahi (2020) discussed ethical issues such as data privacy and biases in AI algorithms that could affect learning outcomes. Studies suggest that AI has the potential to significantly improve the academic performance of undergraduate students by providing personalized learning experiences and real-time feedback. However, challenges such as ethical concerns and over-reliance on AI need to be addressed.

The Importance of AI Literacy for Undergraduate Students in Social Sciences

Today, AI is widely utilized by university students in their academic endeavors. AI-based applications are commonly employed for assignments, exam preparation, problem-solving, and answering questions. However, AI is not limited to these academic tasks; it also includes advanced concepts such as deep learning and data-driven machine learning (Türel et al., 2024). AI tools aim to increase students' learning speed, enhance their individual learning experiences, and provide instructors with a more efficient assessment process (Akinwalere & Inanov, 2022). AI literacy, on the other hand, helps individuals not only acquire technical knowledge but also understand ethical and social dimensions (Türel et al., 2024).

AI literacy is not solely relevant to technical disciplines but is also of paramount importance for undergraduate students in social sciences. Fields such as economics, sociology, psychology, political science, and communication increasingly incorporate AI applications like data analysis, text mining, sentiment analysis, and predictive modeling. These technologies provide social science students with significant opportunities to analyze large datasets, interpret social behavior patterns, and support decision-making processes (Karaoğlan Yılmaz & Yılmaz, 2023).

For social scientists, AI literacy is also essential for developing ethical awareness and social responsibility. It is crucial to critically evaluate the neutrality of AI-driven decision-making processes, detect algorithmic biases, and assess AI applications within ethical frameworks (Laupichler et al., 2022). For example, students studying law or public administration should be knowledgeable about topics such as the integration of AI into the justice system, the protection of personal data, and algorithmic discrimination. On the other hand, for media and communication students, AI literacy enables them to act consciously in areas such as digital content creation, fact-checking, and combating disinformation. Similarly, gastronomy students can benefit from AI as an effective tool for creating new recipes, selecting food visuals, and easily implementing fusion and molecular gastronomy applications. In this context, it is crucial for students in the field of social sciences to develop the ability to use AI technologies consciously and ethically (Erdoğan & Ekşioğlu, 2024).

The use of AI applications by students in tasks such as data collection, data generation, and verifying data accuracy is significant in terms of the socialization of these technologies, understanding their working principles, and evaluating their positive and negative impacts on information gathering and creation. Therefore, it is essential to develop a critical AI literacy approach that questions how AI technologies function in knowledge construction, data collection, and truth verification. Ultimately, having skills such as tracing the source of information, cross-examining data from various platforms, being aware of potential algorithmic biases, organizing and verifying AI-generated outputs is vital in AI literacy (Uçar, 2024; Popenici & Kerr, 2017).

There are various studies on the use and development of artificial intelligence (AI) in education, its potential impacts on education, future scenarios, and the roles of AI in learning processes (Lo, 2023; Farrelly & Baker, 2023; Yetişensoy & Rapoport, 2023; Akinwalere & Inanov, 2022; Arslan, 2020; Panigrahi, 2020). Assessing the knowledge of next-generation students who will shape the future regarding understanding, using, and interacting with AI applications can provide important insights into the development and utilization of AI applications. Understanding the impact of AI technologies in the field of education requires examining individuals' knowledge levels and usage habits related to these technologies (Elçiçek, 2024).

As a result, AI literacy helps social science students enhance their professional skills, develop critical and data-driven solutions to social issues, and fulfill their ethical responsibilities. Therefore, it is necessary to develop educational programs aimed at increasing AI literacy in the field of social sciences and to integrate this subject into curriculum content.Formun Altı

Methodology

Research Aim

Individuals interact with AI applications regardless of their level of awareness. Particularly, as AI becomes increasingly integrated into personal, professional, and educational aspects of life, AI literacy has become a crucial necessity (Laupichler et al., 2023). Indeed, many researchers emphasize the importance of AI literacy. Kong, Cheung & Zhang (2021) propose that AI literacy consists of three main components: understanding, usage, and evaluation, with "understanding" being the fundamental aspect. On the other hand, Ng et al., (2021) developed a taxonomy for AI literacy aligned with Bloom's Taxonomy of cognitive levels. In this taxonomy, the two lowest levels are defined as "Knowing and Understanding AI." Thus, in order for AI to be effectively utilized, individuals must understand and be familiar with this technology. This highlights the critical importance of AI literacy. This study aims to examine the AI literacy levels of students based on various factors (e.g., gender, department) and assess the impact of AI literacy on their academic and everyday lives. In line with the purpose of the research, the following hypotheses were developed: **H_1:** There is a significant difference between the AI literacy of university students and gender. **H_2:** There is a significant difference between the AI literacy of university students and age. **H_3:** There is a significant difference between the AI literacy of university students and the department they study

Research Method

In the study, the relational survey model, one of the general survey models based on the quantitative research method, was preferred. The survey model refers to an approach that aims to describe the characteristics of a group of individuals within a specific time frame (Creswell, 2011). The relational survey model, on the other hand, seeks to examine the relationship between two or more variables (Karasar, 2016). The study population comprises undergraduate students enrolled in social sciences at a state university in Turkey during the 2024/25 academic year. The sample consists of students who voluntarily participated in the study, selected using a non-probability convenience sampling method.

Data Collection Tool and Data Anlysis

In the study, the survey method was used as the data collection tool. The survey consists of two sections. In the first section, the "Artificial Intelligence Literacy Scale" was used to assess students' AI literacy levels. This scale was developed by Wang et al. (2023) and adapted into Turkish by Erdoğan & Ekşioğlu (2024). The scale consists of 12 items grouped under four factors (sub-dimensions): Awareness (4 items), Usage (3 items), Evaluation (3 items) and Ethics (3 items). The scale follows a 5-point Likert-type rating system, where higher scores indicate a higher level of AI literacy. In the second section, demographic variables were included. Data were analyzed using the SPSS package program. Reliability, descriptive and relational tests were used in the interpretation of the findings.

Findings

The sample consists of 342 respondents. 61.4% of the respondents are female and 38.6% are male. In terms of age groups, 268 people (%78.4) are between the ages of 18–24; 62 people (18.1) are between the ages of 25–34; 11 people (%3.2) are between the ages of 35–44 and 1 person (%.3) is between the ages of 45–54. 192 (%56.1) of the respondents are undergraduate students in the gastronomy and culinary arts department; 96 (%28.1) are undergraduate students in the tourism guiding department; and 54 (%15.8) are undergraduate students in the tourism management department. The distribution of the students respondents in the research by class is close to each other. The respondents of the study were students of gastronomy and culinary arts, tourism guidance and tourism management at 7 different universities. 32.2% of the participants were from university 1; 15.5% from university 2; 14.6% from university 6; 11.4% from university 4; 10.2% from university 7; 8.5% from university 3 and 7.6% from university 5.

The arithmetic mean values of the respondents' answers to the AI literacy scale are presented in *Table 1*. In the AI literacy awareness dimension, participants perceive themselves to be highly able to distinguish between smart devices and non-smart devices. In the Usage dimension, they perceive themselves to be able to use AI applications in their daily work and to increase their efficiency in their work. In the evaluation dimension, respondents have a high perception of recognizing AI applications. In the ethics dimension, respondents perceive that they act in accordance with ethical principles when using AI applications.

Table 1. Mean and Standard Deviation Values

Dimensions of Artificial Intelligence Literacy	Mean	Std.Devi.
Awareness		
I can distinguish between smart devices and non-smart devices.	4,36	,86
I do not know how AI technology can help me.	3,08	1,2
I can identify the AI technology employed in the applications and products I use.	3,63	,93

(Continued)

Table 1. (Continued)

Dimensions of Artificial Intelligence Literacy	Mean	Std.Devi.
Usage		
I can skilfully use AI applications or products to help me with my daily work.	3,69	,89
It is usually hard for me to learn to use a new AI application or product.	3,3	1,21
I can use AI applications or products to improve my work efficiency.	4,19	,8
Evaluation		
I can evaluate the capabilities and limitations of an AI application or product after using it for a while.	4,04	,87
I can choose a proper solution from various solutions provided by a smartagent.	3,97	,9
I can choose the most appropriate AI application or product from a variety for a particular task.	3,85	,94
Ethics		
I always comply with ethical principles when using AI applications or products.	3,95	,95
I am never alert to privacy and information security issues when using AI applications or products.	4	,89
I am always alert to the abuse of AI technology.	4,14	,96

According to the T-test results, there is no significant difference between the gender variable and artificial intelligence literacy. $\mathbf{H_1}$ hypothesis is rejected (***Sig. >0.05***). The results of the one-way ANOVA test conducted to determine the relationship between age and department variables and artificial intelligence literacy are presented in *Table 2*. Artificial intelligence literacy differs significantly according to the age variable. $\mathbf{H_2}$ hypothesis is accepted. Younger students (18–24 years old) demonstrated higher AI literacy levels compared to older students, suggesting that exposure to technology from an early age plays a crucial role in AI proficiency. No significant difference was found between artificial intelligence literacy and department. $\mathbf{H_3}$ hypothesis is rejected.

Table 2. One-Way ANOVA Test Results According to Age and Department Variables

Variable	N	Mean	Std. Deviation	F	Sig.	Variable		Mean Difference (I-J)	Std. Error	Sig.
Age										
18–24	241	47,71	5,014	5,941	,003	**18–24**	25–34	2,386*	,588	,000
25–34	58	45,33	3,739				35–44	,350	,751	,888
35–44	11	47,36	2,248			**25–34**	18–24	-2,386*	,588	,000
							35–44	-2,036	,837	,059
						35–44	18–24	-,350	,751	,888
							25–34	2,036	,837	,059
Department										
Gastronomy and Culinary Arts	169	46,98	4,799	1,069	,345	**Gastronomy and Culinary Arts**	Tourism Guidance	-,896	,647	,352
Tourism Guidance	90	47,88	5,043				Tourism Management	-,077	,717	,994
Tourism Management	51	47,06	4,393			**Tourism Guidance**	Gastronomy and Culinary Arts	,896	,647	,352
							Tourism Management	,819	,813	,574
						Tourism Management	Gastronomy and Culinary Arts	,077	,717	,994
							Tourism Guidance	-,819	,813	,574

Sig. >0.05 significance level

Results

This study aimed to assess the artificial intelligence (AI) literacy levels of undergraduate tourism students, focusing on different variables such as gender, age, and department. The findings indicate that AI literacy is an essential competency that enables students to interact with AI technologies effectively, evaluate their capabilities, and apply them ethically in academic and professional settings.

The study's results show that while students generally have a good level of AI literacy, significant variations exist based on age groups. Younger students (18–24 years old) demonstrated higher AI literacy levels compared to older students, suggesting that exposure to technology from an early age plays a crucial role in AI proficiency. However, no statistically significant differences were found between AI literacy and gender or department, implying that AI literacy development is not necessarily influenced by these factors. This suggests that AI literacy training should be accessible and beneficial to all students, regardless of their demographic characteristics.

The analysis of AI literacy dimensions revealed that students scored highly in awareness and ethics, indicating that they are conscious of AI's role and ethical considerations. However, some students struggled with AI usage and evaluation, suggesting the need for more hands-on experience with AI applications. This finding aligns with previous studies emphasizing the necessity of both theoretical knowledge and practical application to enhance AI literacy (Cerny, 2024; Laupichler et al., 2023).

The implications of these findings highlight the importance of integrating AI literacy into educational curricula, particularly for students in social sciences and tourism-related disciplines. Given the increasing reliance on AI in various sectors, universities should develop AI-focused educational programs that provide students with critical skills to navigate the digital landscape effectively. Furthermore, AI literacy should not be limited to technical fields but should also be tailored to address ethical, societal, and application-based aspects relevant to non-technical disciplines.

Overall, this study underscores the growing significance of AI literacy in higher education and the necessity of structured educational strategies to enhance students' AI competencies. Future research could explore additional variables, such as students' prior experiences with AI technologies and their perceptions of AI's role in their future careers, to gain a deeper understanding of how AI literacy can be further developed.

References

Akinwalere, S. N., & Ivanov, V. (2022). Artificial intelligence in higher education: Challenges and opportunities. *Border Crossing, 12*(1), 1–15. https://doi.org/10.33182/bc.v12i1.2015.

Arslan, K. (2020). Artificial Intelligence and Applicatıons in Education. *Western Anatolia Journal of Educational Sciences, 11*(1), 71–88.

Banaz, E., & Demirel, O. (2024). Investigation of artificial intelligence literacy of prospective Turkish teachers according to different variables. *The Journal of Buca Faculty of Education, 60*, 1516–1529.

Bozkurt, A. (2023). ChatGPT, generative artificial intelligence and algorithmic paradigm shift. *CRES Journal: Critical Reviews in Educational Sciences, 4*(1), 63–72.

Černý, M. (2024). University students' conceptualisation of AI literacy: Theory and empirical evidence. *Social Sciences, 13*(3), 129. https://doi.org/10.3390/socsci13030129.

Coşkun, F., & Gülleroğlu, H. D. (2021). The historical development of artificial intelligence and its use in education. *Ankara University Faculty of Educational Sciences Journal, 54*(3), 947–966.

Creswell, J. W. (2011). *Educational research: planning, conducting, and evaluating quantitative and qualitative research (4th Ed.).* London: Pearson.

Çelebi, C., Demir, U., & Karakuş, F. (2023). Examining studies on artificial ıntelligence literacy using the systematic review method. *Necmettin Erbakan Üniversitesi Ereğli Eğitim Fakültesi Dergisi, 5*(2), 535–560.

Elçiçek, M. (2024). A study on students' artificial ıntelligence (aı) literacy. *Journal of Information and Communication Technologies, 6*(1), 24–35. https://doi.org/10.53694/bited.1460106.

Eniş-Erdoğan, T., & Ekşioğlu, S. (2024). Adaptation of artificial ıntelligence literacy scale into Turkish. *TEBD, 22*(2), 1196–1211. https://doi.org/10.37217/tebd.1496716.

Farrelly, T., & Baker, N. (2023). Generative artificial intelligence: Implications and considerations for higher education practice. *Education Sciences, 13*(11), 1109.

Karakuş, A., Geçgel, Ş., & Çetin, M. (2024). An emerging paradigm: artificial ıntelligence literacy. *International Journal of Active Learning (IJAL), 8*(1), 50–63.

Karaoğlan Yılmaz, F. G., & Yılmaz, R. (2023). Adaptation of artificial ıntelligence literacy scale into Turkish. *Journal of Information and Communication Technologies, 5*(2), 172–190. https://doi.org/10.53694/bited.1376831.

Karasar, N. (2016). *Scientific research method: Concepts, principles, techniques, second edition 31st edition.* Ankara: Nobel Academic Publishing.

Kong, S. C., Cheung, W. Y., & Zhang, G. (2021). Evaluation of an artificial intelligence literacy course for university students with diverse study backgrounds. *Computers and Education: Artificial Intelligence, 2*, 1–12.

Kong, S. C., Cheung, W. Y., & Zhang, G. (2022). Evaluating artificial intelligence literacy courses for fostering conceptual learning, literacy and empowerment in university students: Refocusing to conceptual building. *Computers in Human Behavior Reports, 7*, 100223.

Laupichler, M. C., Aster, A., Haverkamp, N., & Raupach, T. (2023). Development of the "scale for the assessment of non-experts' AI literacy"–An exploratory factor analysis. *Computers in Human Behavior Reports, 12*, 100338.

Laupichler, M. C., Aster, A., Schirch, J., & Raupach, T. (2022). Artificial intelligence literacy in higher and adult education: A scoping literature review. *Computers and Education: Artificial Intelligence*, 100101.

Lin, P. Y., Chai, C. S., Jong, M. Y., Dai, Y., Guo, Y., & Qin, J. (2021). Modeling the structural relationship among primary students' motivation to learn artificial intelligence. *Computers & Education: Artificial Intelligence, 2*, 1–7.

Lo, C. K. (2023). What is the impact of ChatGPT on education? A rapid review of the literature. *Education Sciences, 13*(4), 410–425.

Ng, D. T., Leung, J. K., Chu, S. K., & M, Q. (2021). Conceptualizing AI literacy: An exploratory review. *Computers and Education: Artificial Intelligence, 2*.

Ng, D. T., Luo, W., Chan, H. M., & Chu, S. K. (2022). Using digital story writing as a pedagogy to develop AI literacy among primary students. *Computers and Education: Artificial Intelligence*.

Panigrahi, C. M. (2020). Use of artificial intelligence in education. *Management Accountant, 55*, 64–67.

Polat, Ö. (2024). A research on artificial ıntelligence literacy of public library employees: Izmir Public Libraries. *RumeliDE Journal of Language and Literature Studies* (15), 625–641.

Popenici, S. A., & Kerr, S. (2017). Exploring the impact of artificial intelligence on teaching and learning in higher education. *Research and Practice in Technology Enhanced Learning, 22*(12), ttps://doi.org/10.1186/s41039-017-0062-8.

Türel, E., Küçükvardar, M., Mayatıyev, O., & Avcı, Y. (2024). a research on artificial ıntelligence literacy in the process of education and communication. I*nternational European Journal of Managerial Research Dergisi (EUJMR), 8*(15), 198–212.

Uçar, A. K. (2024). The loss of truth and artificial ıntelligence literacy. *Iv International Symposium On Communication: Disinformation In The Digital Age* (s. 179–186). Bishkek: Kyrgzy-Turkish Manas University.

Wang, B., Rau, P., & Yuan, T. (2023). Measuring user competence in using artificial intelligence: validity and reliability of artificial intelligence literacy scale. *Behaviour & Information Technology, 42*(9), 1324–1337.

Yetişensoy, O., & Rapoport, A. (2023). Artificial intelligence literacy teaching in social studies education. *Journal of Pedagogical Research, 7*(3), 100–110. https://doi.org/10.33902/JPR.202320866.

Zhao, L., Wu, X., & Luo, H. (2022). Developing ai literacy for primary and middle school teachers in china: based on a structural equation modeling analysis. *Sustainability, 14*(21).

Digital Storytelling and Cultural Heritage

Seda ÇİÇEKLİ AYYILDIZ[1], *İsa BAYHAN*[2]

Introduction

Cultural heritage is the whole of the tangible and intangible values that societies have formed throughout history and transferred to the next generations. Cultural heritage has particular importance in terms of strengthening societies' sense of identity and belonging. It is categorized into two main categories: tangible and intangible cultural heritage (Song et al., 2025). Storytelling has been recognized as an essential method for the preservation, transmission and reconstruction of cultural heritage (Roque, 2022). Traditional storytelling, as a transmission model based on oral culture, has established emotional and cognitive ties between individuals and communities, while these narrative practices have evolved depending on changing socio-cultural dynamics over time (Diego, 2020). Due to the development of digital technologies, storytelling is being reformatted through new media tools, gaining an interactive and multi-layered narrative form and becoming accessible to wider groups of people. Thanks to digital storytelling, multimedia components, interactive structures and user-centered content production, it has provided the opportunities to share cultural heritage in a more inclusive, accessible and sustainable manner (Zort et al., 2023). In this section of the book, cultural heritage, storytelling, digital storytelling and the impact of digital storytelling on cultural heritage elements have been examined. The theoretical and conceptual perspectives of the information obtained from the literature review within the scope of the research have been presented and the results have been evaluated.

---

[1] Res. Asst. Bolu Abant İzzet Baysal University, Tourism Faculty, Department of Tourism Guidance, seda.cicekliayyildiz@ibu.edu.tr

[2] Asst. Prof. Bolu Abant İzzet Baysal University, Tourism Faculty, Department of Tourism Guidance, isabayhan@ibu.edu.tr

Cultural Heritage

Culture can be regarded as a dynamic structure that a society develops in the historical process, shaping individuals' behaviors, ways of thinking and practices of perception of the world. Culture comprises not only language, art, traditions and belief systems, but also the ways in which society produces knowledge, its engagement with technology and its ability to respond to changing situations. In this regard, culture should be considered as a process that is constantly evolving and reshaped by different social and historical contexts, rather than a stable structure (Göçer, 2012; Akova, 2024). Cultural heritage refers to the protection of certain elements of this dynamic cultural structure in time, conveying them to the future by transferring and making sense of them (Blake, 2000). Cultural heritage is not a limited notion only to the preservation of the past. On the contrary, it also contains the problematic of how the past is related to the present and how it is transferred to the future (Nilson & Thorell, 2018). While contributing to the preservation of cultural diversity, this heritage, which forms the identities of individuals and communities, also has a critical role in transferring the collective experience of humanity to future generations (Akyurt Kurnaz, 2024).

The first international initiatives for the protection of cultural conservation were expressed at the Athens Conference organized by the International Organization of Museums in 1931. This conference stressed the necessity of aesthetic preservation of historical values related to cultural heritage (Vecco, 2010). Another crucial factor for the preservation of cultural heritage is the Second World War. As a natural consequence of the war, it became necessary to repair the natural, historical and cultural destruction and to rebuild historical buildings and cities. In this respect, efforts to establish general principles and rules for the protection of natural, historical and cultural diversity have accelerated. These developments have formed the basis of the United Nations Educational, Scientific and Cultural Organization (UNESCO) in 1946 (Dönmez & Yeşilbursa, 2014).

UNESCO recognizes the habitats that human communities have valued over time and the artefacts they have left in these areas as the common heritage of all humanity. The Convention on the Protection of the World Cultural and Natural Heritage was adopted at the 17th General Conference in Paris in 1972 to preserve, promote and transfer this heritage to future generations. With this convention, for many years, when cultural heritage was defined, tangible elements such as movable cultural heritage (sculptures, inscriptions, paintings, archaeological artefacts, etc.) and immovable cultural heritage (architectural structures, archaeological sites, etc.) were at the forefront (UNESCO, 2025). Nevertheless, UNESCO's efforts on cultural heritage have shown that the concept of heritage is not only limited to physical structures and led to the adoption of the concept of Intangible Cultural Heritage. In 2003, the 32nd General Assembly of UNESCO adopted the Convention for the Safeguarding of Intangible Cultural Heritage as a precaution against the cultural erosion due to globalization, aiming to protect traditional knowledge, life practices and cultural expressions. Through this convention, UNESCO has emphasized the necessity of interaction with nature and the heritage of past cultures for human development and balance in the face of fast-changing social structures (UNESCO, 2025a). Cultural heritage is generally considered in two main categories as tangible and intangible cultural heritage. However, cultural heritage can also be found as underwater cultural heritage and natural heritage (Nilson & Thorell, 2018; Akyurt Kurnaz, 2024).

Tangible cultural heritage can be observed directly in the living space owing to its physical existence and can provide valuable insights into the identity and past of a society. These heritage components are divided into two main categories: movable and immovable cultural heritage. Sculptures, inscriptions, paintings and archaeological artefacts are among movable cultural heritage, while the monumental architectural structures, geological sites and archaeological sites are regarded as immovable cultural heritage (Hua, 2010; Dönmez & Yeşilbursa, 2014). When cultural heritage is defined only in terms of tangible cultural heritage, it means fragmenting cultural heritage and disregarding some of it. Whereas tangible cultural heritage objects appear more like a result, the process of formation around these objects, the narratives, practices and performances in the process are also complementary elements of cultural heritage (Vecco, 2010; Nilson & Thorell, 2018). These elements are generally named as intangible cultural heritage elements. Intangible cultural heritage is a set of knowledge, skills, traditions and narratives that form the identity of a community and are transmitted from generation to generation. This heritage is preserved through oral narratives, rituals, handicrafts and social practices. Although it changes over time, it forms the cultural memory of society and provides it to be moved to the future (Qiu et al., 2022).

UNESCO conducts various projects and programs to protect and digitalize cultural heritage. In this context, the Memory of the World program has been developed with the aim of digitalizing important documents and archives from around the world and transferring them to future generations (UNESCO, 2025b). Cultural heritage development is as concerned with the identification and conservation of heritage assets as it is with their effects on communities, regions, economic and social benefits, the provision of financial resources, marketing and promotion. Within this context, one of the prominent elements is storytelling. Stories create bridges between the past and the present by establishing bonds between individuals and thus keeping the cultural heritage alive (Lussetyowati, 2015; Podara et al., 2021; Yaralı & Baloğlu, 2023). In this context, storytelling is an essential tool for the preservation and sustainability of cultural heritage.

Digital Storytelling

Stories have a central role in human life. In general, human beings are characterized as 'storytellers' (Bassano et al., 2019). Stories are formed by individuals and societies, they are shaped over time and passed down from generation to generation. Communities use stories to communicate and transmit their myths, legends and histories to preserve and transmit them. Stories have a dynamic structure; they change and develop by time. Storytelling is one of the oldest communication strategies used by humanity. Definitions of storytelling may vary depending on the context and purpose of use (Kent, 2015). Fundamentally, stories are rich narratives, both emotional and symbolic content. For this reason, storytelling is an effective method in many fields such as distribution, marketing and communication (Moscardo, 2017; Mei et al., 2020; Moscardo, 2020; Zerman, 2023).

Storytelling has been one of the main instruments for the development of humanity. The oral narrative tradition has continued its existence in different forms from myths to tales, from epics to legends (Bassano et al., 2019). Technological advances have transformed the forms and methods of storytelling, leading to digital storytelling. After reaching large masses with the invention of the printing press, written stories became audio and visual narratives with the development of radio, television and cinema. In the 21st century, new forms of narration such as digitalization, interactive media, virtual reality (VR), augmented reality (AR) and artificial intelligence-supported narratives have become possible (Xiang et al., 2017; Aydemir & Fetah, 2023). Digital storytelling provides a more inclusive experience for the audience by combining traditional storytelling techniques with multimedia tools. Digital storytelling can create a link between the past and the present with various elements such as visuals, audio, interactive maps and documentary elements (De Jager et al., 2017; Quah & Ng, 2017). Digital storytelling is basically defined as storytelling by using digital technologies. Digital stories are regarded as narratives consisting of cyber culture elements (Alexander, 2017).

Digital storytelling is a narrative form that is formed by adapting traditional storytelling techniques to the digital environment and is attracting growing interest. Thanks to the widespread use of the Internet and social media, individuals and organizations have resorted to digital storytelling methods to achieve stronger ties with their target groups and to make their narratives more effective (Quah & Ng, 2017; Anadolu, 2019; Yılmaz, 2023). Tourism is one of the primary sectors that utilize this method. Storytelling is an effective factor in attracting tourists. Storytelling is considered a co-creative experience that has the potential to encourage visitor engagement and sense-making, leaving lasting memories (Moscardo, 2017; Moscardo, 2020). Storytelling is a key element in the design of tourist experiences and a common creative tool that engages audiences and translates to memorable tourist experiences (Bassano et al., 2019; Mei et al., 2020; Campos et al., 2023; Zins & Abbas Adamu, 2024). Storytelling and digital support play an important role both for the preservation and transmission of cultural heritage to future generations. Digital technologies enhance the effectiveness of storytelling by providing visitors with richer and more interactive experiences in cultural heritage sites, thereby enhancing the sustainability of cultural heritage (Bonacini, 2019; Roque, 2022; Yaralı & Baloğlu, 2023).

Digital Storytelling and Cultural Heritage

Cultural heritage sites are home to unique stories that capture and capture the imagination of visitors. Storytelling in the context of cultural heritage is defined as an experience centered around a continuous and coherent narrative that develops the basic elements of storytelling through the interpretation of existing cultural heritage objects (Campos et al., 2023). At this point, this method has been an essential tool for competing in the tourism sector and conveying cultural knowledge and values (Roque, 2022; Kasemsarn et al., 2023). Museums and archaeological sites are increasingly recognizing that digital storytelling is an area where investment is needed to attract and engage audiences. Katifori et al. (2020) have argued that digital storytelling applications engage visitors and encourage their learning and reflection. Furthermore, Bonacini's (2019) study has examined the impact of digital storytelling on cultural heritage communities and found that this method mobilizes communities to promote their cultural heritage and promotes cooperation with local heritage organizations.

Whereas traditional preservation methods are limited to physical spaces, contemporary preservation methods such as digital storytelling stand out as an effective and innovative method for both preserving and communicating cultural heritage to large masses (Alexander, 2017). Digital storytelling not only helps to document cultural heritage but also to transmit it in an interactive and dynamic manner. Through this method, cultural heritage assets become more recognizable and accessible on a global scale than in local populations (Quah & Ng, 2017; Anadolu, 2019; Yılmaz, 2023).

Traditional heritage elements are vulnerable to threats such as physical wear and tears and natural disasters. However, with the help of digital storytelling, these values can be recorded and passed on to future generations (Lawal, 2023). Use of multimedia elements such as video, audio recordings, etc. provides an in-depth understanding and experience of historical and cultural factors. Particularly technologies such as virtual reality (VR) and augmented reality (AR) significantly increase the effect of digital storytelling. These technologies transform users from passive observers into actors in cultural heritage (Remondino & Cowley, 2011; Podara et al., 2021; Kum, 2022).

The digital storytelling is becoming more and more popular for the preservation and dissemination of cultural heritage. Especially international organizations, museums and cultural heritage organizations actively use this method (Kum, 2022; Yaralı & Baloğlu, 2023). The other important occupational group that actively maintains storytelling is influencers with large following and strong interaction power. Influencers who are interested in cultural heritage and who want to raise awareness on this subject make it more popular by producing content related to cultural heritage. In order to remind a cultural heritage that is about to be forgotten, sometimes the authorities bring together social media phenomena with high followers to revive it (Magno & Cassia, 2018; Paslanmaz & Narmanlıoğlu, 2019; Mele et al., 2021). In this regard, influencers promoting cultural heritage on platforms such as YouTube, Instagram and TikTok have an important role in accessing youth generations as digital storytellers.

Conclusion

Digital storytelling is a method that offers new horizons in the preservation, dissemination and transfer of cultural heritage to the future. Societies have constructed their historical memories and identities through narratives from past to present and have transmitted their cultural values from generation to generation via these narratives. Due to the opportunities offered by digitalization, these narratives can now reach wider masses, transcend geographical boundaries and thus become part of a collective heritage. The combination of traditional storytelling with digital technologies enables cultural heritage not only to be preserved but also to be reinterpreted and reproduced in various perspectives.

Heritage narratives, which are generally defined by specific institutions and authorities, can be shaped by people and communities by means of digital platforms. Nevertheless, the difficulties that may be faced during this period should not be ignored. There are ethical and technical responsibilities for the sustainability, accessibility and accuracy of digital cultural heritage. The risk of commercialization of digitalized cultural heritage and issues such as data security are fundamental issues that should be considered with attention.

As a result, digital storytelling is a strong instrument that enables cultural heritage to be preserved not only as an object but also as a narrative. Digital storytelling should be used consciously and responsibly to achieve the sustainability of cultural heritage. Although digital storytelling has huge capacity to revitalize and disseminate cultural heritage, there is a need to establish certain standards for the historical accuracy of narratives. Artificial intelligence generated or manipulated content can lead to misinterpretation of cultural heritage and disruption of social memory. Therefore, it is crucial that digital storytelling projects are managed within the framework of ethical principles and conducted under expert guidance. In the future, it is necessary to develop digital sustainability policies for the protection of cultural heritage, to ensure the active participation of local communities in the process and to establish principles for digital storytelling.

References

Akova, S. (2024). Intercultural communication and creation of cultural awareness: to the dialogue between local and global municipality, the example of Istanbul Maltepe municipality. *The Journal of Academic Social Science*, *26*(26), 416–428.

Akyurt Kurnaz, H. (2024). Kültürel mirasın korunması [Protection of cultural heritage]. A.A. Aktaş & Kökmen Seyirci (Eds.) in Multidisipliner Perspektiften Kültürel Miras [Cultural Heritage from a Multidisciplinary Perspective] (1st edition, pp. 108–132). Detay Publishing.

Alexander, B. (2017). *The new digital storytelling: Creating narratives with new media--revised and updated edition*. Bloomsbury Publishing USA.

Anadolu, B. (2019). The impact of artificial intelligence on cinema in the context of digital storytelling: an analysis of sunspring and it's no game. *Erciyes İletişim Dergisi*, (1), 39–56. https://doi.org/10.17680/erciyesiletisim.483510

Aydemir, M., & Fetah, V. (2023). The use of artificial intelligence in digital storytelling and scenario design: a short film applied research. *Pamukkale Üniversitesi Sosyal Bilimler Enstitüsü Dergisi*, (58), 255–275. https://doi.org/10.30794/pausbed.1290986

Bassano, C., Barile, S., Piciocchi, P., Spohrer, J. C., Iandolo, F., & Fisk, R. (2019). Storytelling about places: tourism marketing in the digital age. *Cities*, *87*, 10–20. https://doi.org/10.1016/j.cities.2018.12.025

Blake, J. (2000). On defining the cultural heritage. *International & Comparative Law Quarterly*, *49*(1), 61–85.

Bonacini, E. (2019). Engaging participative communities in cultural heritage: using digital storytelling in Sicily (Italy). *International Information & Library Review*, *51*(1), 42–50. https://doi.org/10.1080/10572317.2019.1568786

Campos, A. C., Guerreiro, M. M., & Beevor, M. C. (2023). Storytelling in heritage tourism: an exploration of co-creative experiences from a

tourist perspective. *Museum Management and Curatorship*, 1–26. https://doi.org/10.1080/09647775.2023.2230194
De Jager, A., Fogarty, A., Tewson, A., Lenette, C., & Boydell, K. M. (2017). Digital storytelling in research: A systematic review. *The Qualitative Report*, *22*(10), 2548–2582. https://doi.org/10.46743/2160-3715/2017.2970
Diego, R. (2020). Heritage storytelling, community empowerment and sustainable development. *Правоведение*, *64*(1), 57–67.
Dönmez, C., & Yeşilbursa, C. C. (2014). The effect of cultural heritage education on students' attitudes toward tangible heritage. *İlköğretim Online (elektronik)*, *13*(2), 425–442.
Göçer, A. (2012). On language-culture relationship and interaction. *Türk Dili*, *729*(1), 50–57.
Howard, P. (2003). *Heritage: Management, interpretation, identity*. A&C Black.
Hua, S. (2010). World heritage classification and related issues - A case study of the "convention concerning the protection of the World cultural and natural heritage." *Procedia-Social and Behavioral Sciences*, *2*(5), 6954–6961.
Kasemsarn, K., Harrison, D., & Nickpour, F. (2023). Applying inclusive design and digital storytelling to facilitate cultural tourism: A review and initial framework. *Heritage*, *6*(2), 1411–1428. https://doi.org/10.3390/heritage6020077
Katifori, A., Tsitou, F., Pichou, M., Kourtis, V., Papoulias, E., Ioannidis, Y., & Roussou, M. (2020). Exploring the potential of visually-rich animated digital storytelling for cultural heritage: the mobile experience of the Athens university history museum. *Visual Computing for Cultural Heritage*, 325–345.
Kent, M. L. (2015). The power of storytelling in public relations: Introducing the 20 master plots. *Public Relations Review*, *41*(4), 480–489.
Kum, Ö. (2022). 3D modeling of cultural heritage by photogrammetric method: Tokat wood block print scarf. *Türkiye Bilimsel Araştırmalar Dergisi*, *7*(1), 237–252.
Lawal, B. (2023). Challenges Associated with the Documentation and Digital Preservation of Traditional Medicine by Traditional Health Practitioners in Katsina State, Nigeria. *International Journal for Multidisciplinary Research*, *5* (5), 1–9.
Lenzerini, F. (2011). Intangible cultural heritage: The living culture of peoples. *European Journal of International Law*, *22*(1), 101–120.

Lussetyowati, T. (2015). Preservation and conservation through cultural heritage tourism. Case study: Musi Riverside Palembang. *Procedia-Social and Behavioral Sciences*, *184*, 401–406.
Magno, F., & Cassia, F. (2018). The impact of social media influencers in tourism. *Anatolia*, *29*(2), 288–290.
Mei, X. Y., Hågensen, A. M. S., & Kristiansen, H. S. (2020). Storytelling through experiencescape: Creating unique stories and extraordinary experiences in farm tourism. *Tourism and Hospitality Research*, *20*(1), 93–104. https://doi.org/10.1177/1467358418813410
Mele, E., Kerkhof, P., & Cantoni, L. (2021). Analyzing cultural tourism promotion on Instagram: a cross-cultural perspective. *Journal of Travel & Tourism Marketing*, *38*(3), 326–340. https://doi.org/10.1080/10548408.2021.1906382
Metin Basat, E. (2013). Safeguarding the tangible and intangible cultural heritage together. *Milli Folklor*, *25*(100).
Moscardo, G. (2017). Stories as a tourist experience design tool. *Design science in tourism: Foundations of destination management*, 97–124.
Moscardo, G. (2020). Stories and design in tourism. *Annals of tourism research*, *83*, 102950. https://doi.org/10.1016/j.annals.2020.102950
Nilson, T., & Thorell, K. (2018). Cultural heritage preservation: The past, the present and the future. Halmstad University Press ISBN: 978-91-87045-95-0.
Paslanmaz, İ., & Narmanlıoğlu, H. (2019). Presentatıon of self on instagram storıes: a research about influencers. *AJIT-e: Academic Journal of Information Technology*, *10*(39), 23–51.
Podara, A., Giomelakis, D., Nicolaou, C., Matsiola, M., & Kotsakis, R. (2021). Digital storytelling in cultural heritage: Audience engagement in the interactive documentary new life. *Sustainability*, *13*(3), 1193. https://doi.org/10.3390/su13031193
Qiu, Q., Zuo, Y., & Zhang, M. (2022). Intangible cultural heritage in tourism: Research review and investigation of future agenda. *Land*, *11*(1), 139. https://doi.org/10.3390/land11010139
Quah, C. Y., & Ng, K. H. (2022). A systematic literature review on digital storytelling authoring tool in education: January 2010 to January 2020. *International Journal of Human–Computer Interaction*, *38*(9), 851–867. https://doi.org/10.1080/10447318.2021.1972608
Remondino, F., & Cowley, D. (2011). 3D recording for cultural heritage. In *Remote Sensing for Archaeological Heritage Management, D. Cowley (ed.), Proc. EAC Symposium* (pp. 107–116).

Roque, M. I. (2022). Storytelling in cultural heritage: Tourism and community engagement. In *Global perspectives on strategic storytelling in destination marketing* (pp. 22–37). IGI Global.

Song, H., Chen, J., & Li, P. (2025). Decoding the cultural heritage tourism landscape and visitor crowding behavior from the multidimensional embodied perspective: Insights from Chinese classical gardens. *Tourism Management*, 110, 105180. https://doi.org/10.1016/j.tourman.2025.105180

UNESCO (2025). Convention on the protection of the world cultural and natural heritage https://whc.unesco.org/en/conventiontext/ (Erişim tarihi: 05.03.2025).

UNESCO (2025a). Convention on the protection of intangible cultural heritage https://ich.unesco.org/en/convention (Erişim tarihi: 05.03.2025).

UNESCO (2025b). Memory of the World programı https://www.unesco.org/en/memory-world (Erişim tarihi: 05.03.2025).

Vecco, M. (2010). A definition of cultural heritage: From the tangible to the intangible. *Journal of cultural heritage*, *11*(3), 321–324.

Xiang, Z., Du, Q., Ma, Y., & Fan, W. (2017). A comparative analysis of major online review platforms: Implications for social media analytics in hospitality and tourism. Tourism Management, 58, 51–65. https://doi.org/10.1016/j.tourman.2016.10.001

Yaralı, C., & Baloğlu, Ö. Ö. (2023). The impact of digital processes on the development of natural and cultural heritage tourism. Nevşehir Hacı Bektaş Veli Üniversitesi SBE Dergisi, 13(İhtisaslaşma), 245–264. https://doi.org/10.30783/nevsosbilen.1320911

Yılmaz, M. S. (2023). Storytelling transforming in the context of media studies: transmedia storytelling. *Middle Black Sea Journal of Communication Studies*, *8*(1), 49–66. https://doi.org/10.56202/mbsjcs.1192909

Zerman, S. (2023). Storytelling in Tourism. In: Kalemci Tüzün, İ. & Öksüz, M. (eds.), *Multifaceted Research in The Field of Tourism II*. Özgür Publications. https://doi.org/10.58830/ozgur.pub143.c1043

Zins, A. H., & Abbas Adamu, A. (2024). Heritage storytelling in destination marketing: cases from Malaysian states. *Journal of Heritage Tourism*, *19*(5), 669–681. https://doi.org/10.1080/1743873X.2023.2232476

Zort, Ç., Karabacak, E., Öznur, Ş., & Dağlı, G. (2023). Sharing of cultural values and heritage through storytelling in the digital age. *Frontiers in Psychology*, *14*, 1104121. https://doi.org/10.3389/fpsyg.2023.1104121

Experiential Marketing Strategies in Gastronomy Tourism

İbrahim Akın ÖZEN[1]*, Ahmet Emre DERİNKÖK*[2]*, Hüseyin DENİZ*[3]

Introduction

In recent years, gastronomy has become an important tourism element that increases the attractiveness of destinations, going beyond being just a means of nutrition. In this context, when the reasons why tourists prefer a destination are examined, it is seen that travel motivation is a determining factor. Travel motivation is among the basic answers given to the question of 'Why do people travel?' and it is seen that the majority of the research conducted on this subject is addressed within the framework of the theory of push and pull factors. According to this theory, push factors are defined as socio-psychological drives that reveal an individual's desire to travel, while pull factors are expressed as elements that activate this motivation and make a particular destination attractive. While pull factors include natural and cultural elements specific to the destination; push factors include the individual's internal drives such as getting away from daily life, discovering oneself, relaxing, gaining prestige, establishing new relationships, developing close friendships, seeking innovation and learning (Dann, 1977).

While tourists are pushed by their internal needs, they are also attracted by the opportunities and appealing features offered by a destination. Therefore, when these internal needs are sufficiently satisfied through the advantages provided by the destination, the desire to travel is effectively formed. Within

1 Assoc. Prof. Dr., Nevşehir Hacı Bektaş Veli University, Tourism Faculty, Department of Tourism Guidance, akin@nevsehir.edu.tr

2 Asst. Prof. Dr., Malatya Turgut Ozal University, Kale Tourism and Hotel Management Vocational School, Travel Tourism and Entertainment Services Department Tourist Guide Program, ahmet.derinkok@ozal.edu.tr

3 Lecturer, Malatya Turgut Ozal University Kale Tourism and Hotel Management Vocational School, Hotel, Restaurant and Catering Services Department, Tourism Management Program, huseyin.deniz@ozal.edu.tr

this framework, gastronomy also occupies a significant place among the factors that shape travel motivations in tourism, and gastronomy-oriented experiences have become one of the primary pull factors influencing tourists' destination choices.

Gastronomy tourism has emerged as a rapidly developing field in recent years, driven by the growing interest in experiencing local culinary cultures and the support of these experiences through strategic marketing approaches. The presentation and marketing of gastronomic elements are among the key factors that enhance the attractiveness of destinations and directly influence tourists' destination preferences. Indeed, the motivation to experience flavors unique to particular destinations is a significant driver of travel decisions, which in turn underscores the growing importance of location-based marketing strategies (İlbay, 2020; Erbay & Sabur, 2022). In this context, the presentation of destination-specific food and beverages plays a crucial role in the success of gastronomy tourism.

The presentation of local flavors belonging to a destination not only appeals to the sense of taste but also enables the discovery of cultural elements unique to the region (İlbay, 2020). Accordingly, gastronomy tours and routes offer tourists the opportunity to experience the destination's cultural and gastronomic identity more deeply, while also creating significant opportunities for local businesses in terms of economic sustainability (Ercan et al., 2024; Erbay & Sabur, 2022). At this point, the success of gastronomy tourism depends not only on the presentation of local flavors but also on experiential marketing strategies that provide tourists with opportunities to taste these flavors and participate in their production processes. In gastronomy tourism, such strategies play a key role in encouraging tourists' active participation. These strategies are reinforced through components such as tasting events and culinary workshops, which provide deeper engagement with the local culinary culture (Çalhan, 2023; Düzgün, 2023). Thus, tourists are not only able to taste traditional flavors but also to take part in their preparation processes. These interactive experiences are reported to positively influence tourists' overall satisfaction and the quality of their experiences at the destination (Serçek & Serçek, 2015). Experiential marketing strategies not only enhance tourists' engagement with the destination but also contribute to the strengthening and branding of its gastronomic identity.

Experiential Marketing and Gastronomy Experience

Experiential marketing has emerged as a sub-field of experience economy studies. The first studies on experiential marketing were conducted by Pine and Gilmore (1998). Pine and Gilmore (1999) described experiential marketing as follows: "when a consumer buys a product, they also buy a series of abstract activities that the product carries along with it." According to Schmitt (1999), experiential marketing focuses on understanding customers' experiences rather than the functional features and benefits of the product. Unlike traditional marketing, experiential marketing evaluates customers not only as rational decision-makers but also as thinking and feeling guests, focusing on concepts such as imagination, emotions, and entertainment, which symbolize experiential consumption. Experiential marketing strategies are based on the principle of delivering and enhancing experiences for customers before, during, and after the purchase. In the context of experiential marketing, various studies have sought to identify and define experiences and experiential dimensions from different perspectives. The concept has often been examined in relation to the phenomena it is associated with. The concept of experience has been addressed from various perspectives in the literature. From the consumer's perspective, experience is defined as fun, enjoyable and unforgettable moments that occur during consumption (Oh, Fiore & Jeoung, 2007). From the perspective of product and service providers, experience is defined as a product that is difficult to replicate and, in this sense, offers a competitive advantage to the party delivering it. In addition, it is argued that the experience, shaped by consumers' interaction with the environment and their level of participation, consists of four fundamental components. Known as the "4Es," these components include entertainment, education, esthetics, and escapism. When a consumer experience incorporates all four dimensions, the experience becomes richer and delivers greater value to the customer (Pine & Gilmore, 1998; 1999). From the perspective of experience components, it can be observed that experiential marketing has emerged as a key strategy in the delivery of touristic products and services.

Considering that the tourism sector is service-based and experience-oriented, the idea that providing services to consumers alone is not enough has emerged. Therefore, tourism businesses and destination marketing organizations have adopted the concept of providing unique experiences to customers as a way to differentiate themselves from competitors and create customer satisfaction. It is observed that studies on the conceptualization of the tourist experience have predominantly been conducted from both social science and management-marketing perspectives (Cohen, 1979; Mossberg, 2007). Research on the tourism experience and its components varies due to several factors, including the unique characteristics of the tourism industry and tourism products, the simultaneous use of a wide range of services and products during the experience, and the multitude of elements that may directly or indirectly influence the tourist experience.

According to Quan and Wang (2004), tourist experiences are considered an organic whole in which peak experiences (art, culture, heritage) and supporting experiences (accommodation, transportation, shopping, entertainment, food and beverage establishments) complement each other.

Gastronomic Experience

While food has proven to be an integral element of the tourist experience (Hall & Sharples 2003) and a primary travel motivation (Quan & Wang 2004), it remains a subject of tourism research. Food experiences are no longer viewed as fleeting moments for travelers, but are instead believed to stem from tourists' interest in novelty, uniqueness, and the distinctive culture of the host country (Björk & Kauppinen-Räisänen 2016).

The gastronomy tourism experience is a form of tourism in which tourists engage in tasting local cuisine at restaurants, sampling street food, attending cooking classes, and visiting local food markets, culinary exhibitions, or food festivals in foreign destinations (Dixit, 2019).

Gastronomy Tourist Typologies

In gastronomy tourism, tourists' motivation to visit a destination is classified according to the degree of importance they give to gastronomy. According to this approach, tourists are considered in four different types (Boyne et al. 2003). Type I: These tourists give importance to gastronomy in their travel plans. They want to learn about the local products and gastronomic heritage of the places they visit. Type 2: Although these types of tourists give importance to gastronomy, they do not look for destinations specifically for gastronomy. They participate in and appreciate gastronomic presentations in the region they visit. Type 3: Gastronomy is not important in their travel experiences. However, they welcome gastronomic experiences if they are offered during their travels. Type 4: These types of tourists have no interest in gastronomy. They ignore gastronomic experiences during their travels.

From the tourist perspective, gastronomic experiences, which are a part of the experiences experienced in the destination, depend on the tourists' interest in gastronomy. Destination development strategies can be determined by taking these approaches into consideration. When implementing these strategies, it is essential to ensure that the food experience forms a significant part of the overall experience at the destination.

Memorable Gastronomic Experience

Gastronomic values have become an increasingly important element in destination marketing. In addition to contributing to brand image formation, they are among the key factors in enhancing the attractiveness of a destination (Kovalenko et al., 2023).

Memorable gastronomic experiences can be defined as distinctive experiences that remain vivid in a tourist's memory after their travel, leaving sensory, emotional, cognitive, behavioral, and relational traces (Nowacki et al., 2023). An examination of the relevant literature reveals that many researchers have concluded that memorable gastronomic experiences differ significantly from actual experiences (Richards, 2002; Braun-LaTour et al., 2006; Morgan, 2006; Huang & Hsu, 2009; Ritchie et al., 2011; Adongo et al., 2015).

Researchers suggest that participants tend to suppress disappointing elements of an experience from memory and reconstruct it based on the information presented afterward, or reinterpret the experience in a way that aligns with their initial expectations (Klaaren et al., 1994; Mitchelle et al., 1997).

Experiential marketing in gastronomy tourism aims to offer tourists a memorable experience by creating environments that stimulate all the senses and foster emotional connections through activities such as cooking classes, street food tours, grape harvest events, food festivals, or gastrodiplomacy initiatives (Gün, 2024).

Gastronomy festivals are not only occasions where local culinary traditions are showcased, but also offer participants memorable experiences on sensory, cultural, and social levels. These festivals are often enriched with music, sculpture, theater, workshops, and interactive tasting sessions, enabling visitors to engage not only with flavors but also in meaningful cultural interaction (Björk & Kauppinen-Räisänen, 2019).

Gastrodiplomacy refers to the strategy of leveraging a country's soft power to promote its cultural identity and national image internationally through culinary culture (Rockower, 2014). Local flavors presented to tourists through gastronomic experiences foster an emotional bond with the destination, positioning gastrodiplomacy as a facilitator in the creation of memorable experiences. For instance, initiatives such as Korea's "Kimchi Diplomacy," Thailand's "Global Thai" program, and Turkey's "Turkish Cuisine Week" strengthen perceptions of national identity through the global promotion of traditional cuisine (Chapple-Sokol, 2013; Kaya, 2023).

So what are the elements that make a gastronomic experience memorable? Recent studies highlight several factors that address this question. For instance, Nowacki et al. (2023) identified eight key components that characterize a memorable experience. These include:

- **Service and Staff Quality:** Attentive, friendly, and professional service is one of the most essential elements that contribute to a memorable experience.
- **Atmosphere:** The venue's ambiance—including its decoration, scents, music, and overall environment—enriches the experience by appealing to the guests' senses.
- **Food (Taste) Quality:** The originality, presentation, and flavor of the food lie at the heart of the memorable experience. Distinctive tastes linger on the palate and form the foundation of culinary memories.
- **Beverages:** The quality and harmony of beverages served with the meal enhance and complete the overall experience.
- **Local Flavors:** Regionally distinctive ingredients and elements of local cuisine foster authenticity and promote a deeper cultural connection.
- **Venue Location and Setting:** The view, authenticity, or historical character of the restaurant provides a memorable backdrop to the dining experience.
- **Price and Perceived Value:** When the cost is aligned with the overall quality of the experience, it enhances the perceived value and contributes to a more positively remembered visit.

- **Reservation and Convenience:** Smooth and efficient handling of reservations and waiting times plays a supportive role in shaping the overall experience.

As evidenced, a memorable gastronomic experience is a multidimensional phenomenon. Clearly, elements such as service quality, the physical setting, and authenticity are just as influential as taste itself.

Cooking Experience in Destinations

Tourists' growing interest in learning about the culinary culture of the destinations they visit and experiencing the preparation of traditional dishes has added a new dimension to gastronomy tourism (Ellis et al., 2018). Such gastronomic experiences provide valuable opportunities for tourists to learn about the local eating habits, customs, regional ingredients, and cooking techniques of the destination. Additionally, by participating in hands-on workshops and cooking alongside local chefs, tourists gain insights into the destination's history, cultural background, and social life (Sharples, 2004). In the relevant literature, this term is commonly referred to as a "cooking class." These classes are considered a type of participatory activity that enables tourists to directly engage with regional culinary culture (Agyeiwaah et al., 2019).

Cooking classes play multiple roles within the context of gastronomy tourism. These are;

- ***Authenticity-seeking effect:*** Tourists are often motivated by a desire to engage in authentic and meaningful experiences. Cooking workshops foster a sense of belonging and offer opportunities to satisfy this interest in authenticity, while also enabling tourists to gain cultural and social insights (Richards, 2002).
- ***Impact on sustainability and cultural heritage preservation:*** Cooking workshops raise awareness about the preparation of region-specific dishes and their introduction to tourists, contributing to their preservation and transmission to future generations. As a result, interest in traditional gastronomic values increases. Moreover, tourists' acquisition of culinary knowledge plays a vital role in the international promotion of local food cultures (Bessière, 1998).

- ***Personalized touristic experience effect:*** Cooking workshops allow individuals to actively engage in the process, making the experience more personal and meaningful. This enhances tourist satisfaction and fosters loyalty (Fields, 2003).
- ***Impact on the local economy:*** Cooking workshops are typically organized by small-scale businesses, local chefs, or home cooks. In this way, tourism revenue benefits not only large enterprises and restaurants but also the local community and small producers (Sims, 2009).

Conclusion

As a tourist attraction, destinations offer a variety of goods and services. Food and beverages are very important to attract tourists to a place. Tourists value culinary experiences as part of their larger travels. Traditional foods enhance the tourism of a region and increase tourist demand. This is because they play an important role in providing visitors with a unique and authentic experience. Therefore, they greatly influence tourists' intention to revisit a place (Grbac and Milohanovic, 2008; Alonso and Krajsic, 2013).

Food and beverages are related to the five senses. Thanks to this feature, memories of gastronomic travels leave both an emotional and sensory impression (Hall and Mitchell, 2005; Henderson 2009). Many countries choose food as the main product of tourism due to the increasing tourism interest and high-income potential (Jones and Jenkins, 2002; Okumuş et al., 2007).

Gastronomy, unlike other travel and tourist attractions, is not seasonal and is accessible throughout the year (Kivela and Crotts, 2006). This leads to longer stays and higher expenditures (Ignatov and Smith, 2006). It also plays an important role in increasing the demand for special tourism for a destination. Thanks to gastronomic values, a tourist sees eating during the day not as a basic necessity but as an important culinary experience. In this way, gastronomy provides the value that organizes a tourist's day (Richards, 2012).

To enhance competitiveness through culinary experiences, it is recommended that destinations implement the following strategies as part of their development efforts.

- Because gastronomic experiences are inherently unique and rooted in local culture, they are difficult to replicate. This uniqueness allows destinations to distinguish themselves from competitors. An inventory of local dishes specific to the destination can be created and local food businesses can be encouraged to include these dishes in their menus.

- Destination management can ensure that tourists actively participate in gastronomy by organizing various activities such as gastronomy-related food competitions, festivals and workshops. Thus, a memorable gastronomic experience can be created for tourists. In this way, destinations can increase their revenues by providing competitive advantage.
- By sharing informative content reflecting gastronomy culture in destination promotions via social networks, the desire to travel to the destination can be triggered.
- Informative environments and applications can be developed for tourists regarding local cuisine, such as the history of the food, ingredients, and cooking techniques.

References

Adongo, C. A., Anuga, S. W., and Dayour, F. (2015). 'Will they tell others to taste? International tourists' experience of Ghanaian cuisines', Tourism Management Perspectives, 15: 57–64.

Agyeiwaah, E., Otoo, F. E., Suntikul, W., & Huang, W. J. (2019). Understanding Culinary Tourist Motivation, Experience, Satisfaction, and Loyalty Using a Structural Approach. Journal of Travel & Tourism Marketing, 36(3), 295–313.

Alonso, A. D., & Krajsic, V. (2013). Food heritage down under: olive growers as Mediterranean 'food ambassadors'. Journal of Heritage Tourism, 8(2–3), 158–171.

Bessière, J. (1998). Local development and heritage: traditional food and cuisine as tourist attractions in rural areas. Sociologia ruralis, 38(1), 21–34.

Björk, P., & Kauppinen-Räisänen, H. (2016). 'Local food: A source for destination attraction', International Journal of Contemporary Hospitality Management, 28 (1): 177–94.

Björk, P., & Kauppinen-Räisänen, H. (2019). Destination foodscape: A stage for travelers' food experience. Tourism Management, 71, 466–475.

Boyne, S., Hall, D. R., and Williams, F. (2003). 'Policy, support and promotion for food-related tourism initiatives: A Marketing approach

to regional development', Journal of Travel and Tourism Marketing, 14 (3–4): 131–154.
Braun-LaTour, K. A., Grinley, M. J., and Loftus, E. F. (2006). 'Tourist memory distortion', Journal of Travel Research, 44 (4): 360–367.
Chapple-Sokol, S. (2013). Culinary diplomacy: Breaking bread to win hearts and minds. The Hague Journal of Diplomacy, 8(2), 161–183.
Çalhan, H. (2023). Kentsel inovasyon ve robotik teknolojiler üzerine bir değerlendirme. https://doi.org/10.58830/ozgur.pub71.c179
Cohen, E. (1979). A Phenomenology of tourist experience. The Journal of the British Sociological Association 13(1), 179–201.
Dann, G. (1977). Anomie, ego-enhamcement and tourism, Annals of Tourism Research, 4(4): 184–194
Dixit, S. K. (2019). Gastronomic tourism: A theoretical construct. In The Routledge handbook of gastronomic tourism (pp. 13–23). Routledge.
Düzgün, M. (2023). Elektronik ağızdan ağıza pazarlama bağlamında müşteri değerlendirmeleri ile gastronomi imajı arasındaki ilişki: gaziantep üzerine bir araştırma. Journal of Tourism and Gastronomy Studies. https://doi.org/10.21325/jotags.2023.1285
Ellis, A., Park, E., Kim, S., & Yeoman, I. (2018). What is food tourism?. Tourism management, 68, 250–263.
Erbay, M. and Sabur, D. (2022). Gastronomi turizmi kapsamında pazarlama stratejileri: türkiye ve avrupa örneği. Journal of Tourism and Gastronomy Studies. https://doi.org/10.21325/jotags.2022.1009
Ercan, M., Kaya, S., Mutlubaş, I., & Erdoğan, T. (2024). Türk mutfak kültüründe yer alan yöresel yoğurtlar ile ilgili gastronomi rotasi önerisi. 3023–7823. https://doi.org/10.69527/agats.2024.3
Fields, K. (2003). Demand for the gastronomy tourism product: motivational factors. Tourism and gastronomy, 50–64.
Gün, S. (2024). Gastronomi turizmine ilişkin deneyimsel pazarlamada arttırılmış gerçeklik uygulama kullanımının deneyim değerine etkisi: arttırılmış gerçeklik menü uygulaması üzerine bir araştırma. Safran Kültür ve Turizm Araştırmaları Dergisi, 7(3), 510–522.
Grbac, B., & Milohanovic, A. (2008). Contribution of food products in creating cultural identity of tourist destination. In V. Gekas, N. E. Mastorakis, & E. Stamatiou (Eds.), WSEAS International Conference. Proceedings. Mathematics and Computers in Science and Engineering. WSEAS.
Hall, C. M. and Sharples, L. (2003). 'The consumption of experiences or the experience of consumption? An introduction to the tourism of

taste'. In Food Tourism Around the World (pp.1–24). Oxford: Elsevier Butterworth-Heinemann.
Hall, C. M., & Mitchell, R. (2005). Gastronomic Tourism-Comparing Food and Wine Tourism Experiences. In M. Novelli (eds.), Niche Tourism. Contemporary Issues, Trends and Cases, Butterworth-Heinemann, p.73–88.
Henderson, J. C. (2009). Food tourism reviewed. British food journal 111(4), p.317-326
Huang, S., & Hsu, C. H. C. (2009). 'Effects of travel motivation, past experience, perceived constraint, and attitude on revisit intention', Journal of Travel Research, 48(1): 29–44.
Ignatov, E. (2003). The Canadian Culinary Tourists: How Well Do We Know Them? (Master's Thesis, University Of Waterloo).
İlbay, B. (2020). Determination of "çibörek" consumption motives of domestic tourists traveling to eskişehir. https://doi.org/10.5281/zenodo.3688767
Jones, A., & Jenkins, I. (2002). "A taste of wales—Blas Ar Gymru":Institutional malaise in promoting welsh food tourism products. InA. M. Hjalager, & G. Richards (Eds.),Tourism and gastronomy(pp. 113–115). London: Routldge
Kaya, F. (2023). Yabancı öğrencilerin gastrodiplomasi aracı olarak Türk mutfağı haftasına ilişkin deneyimleri. GSI Journals Serie A: Advancements in Tourism Recreation and Sports Sciences, 6(1), 152-167.
Kivela, J. & Crotts, J. C. (2006). Tourism and Gastronomy: Gastronomy's Influence on How Tourists Experience a Destination. Journal of Hospitality & Tourism Research 30 (3), 354-377.
Klaaren, K. J. Hodges, S. D. Wilson, T. D. (1994). The role of affective expectations in subjective experience and decisionmaking, Social Cognition, 12, 77-101.
Kovalenko, A., Dias, Á., Pereira, L., & Simões, A. (2023). Gastronomic experience and consumer behavior: Analyzing the influence on destination image. Foods, 12(2), 315.
Mitchelle, T. R. Thompson, L. Peterson, E. Cronk, R. (1997). Temporal adjustments in the evaluation of events: The "rosy view," Journal of Experimental Social Psychology, 33, 421–448.
Morgan, M. (2006). 'Making space for experience', Journal of Retail and Leisure Property, 6 (4): 305–313.
Mossberg, L. (2007). A Marketing Approach to the Tourist Experience. Scandinavian Journal of Hospitality and Tourism, 7(1), 59–74.

Nowacki, M., Stasiak, A., & Niezgoda, A. (2023). Exploring memorable gastronomic experiences: Automatic topic modelling of TripAdvisor reviews. Geographia Polonica, 96(3).

Oh, H., Fiore, A. M., and Jeoung, M. (2007). Measuring experience economy concepts: Tourism applications. Journal of travel research, 46(2), 119–132.

Okumus, B., Okumus, F., & McKercher, B. (2007). Incorporating local and international cuisines in the marketing of tourism destinations: The cases of Hong Kong and Turkey. Tourism management, 28(1), 253–261.

Pine, J., and Gilmore, J. H. (1998). Welcome to the Experience Economy. Harvard Business Review, 97–105.

Pine, J., and Gilmore, J. H. (1999). The Experience Economy: Work is theatre and every business a stage. Boston: Harvard Business School Press.

Quan, S. and Wang, N. (2004), 'Towards a structural model of the tourist experience: An illustration from food experiences in tourism', Tourism Management, 25 (3): 297–305.

Richards, G. (2002). 'Gastronomy: An essential ingredient in tourism production and consumption?' In A.-M. Hjalager and G. Richards (eds.), Tourism and Gastronomy (pp.3–20). London: Routledge.

Richards, G. (2003). Gastronomy: an essential ingredient in tourism production and consumption?. In Tourism and gastronomy (pp. 17–34). Routledge.

Ritchie, J. R. B., Tung, V. W. S., and Ritchie, R. J. B. (2011). 'Tourism experience management research: Emergence, evolution and future directions', International Journal of Contemporary Hospitality Management, 23 (4): 419–438.

Richards, G. (2012) Food and the tourism experience: major findings and policy orientations. In Dodd, D. (ed.) Food and the Tourism Experience. OECD, Paris, pp. 13–46.

Rockower, P. (2014). The state of gastrodiplomacy. Public Diplomacy Magazine, 11(14), 434–435.

Schmitt, B. (1999). Experiential marketing. Journal of marketing management, 15(1–3), 53–67.

Serçek, G. & Serçek, S. (2015). Destinasyon markalaşmasında gastronomi turizmin yeri ve önemi. Journal of Tourism Theory and Research, 1(1), 22–22. https://doi.org/10.24288/jttr.202819

Serçek, S. (2018). Turistlerin yerel yemek tüketimindeki motivasyon faktörlerinin incelenmesi (investigation of the motivational factors of local food consumption by tourists). Journal of Tourism and Gastronomy Studies, 6(4), 463–481. https://doi.org/10.21325/jotags.2018.319

Sharples, L. (2004). The world of cookery-school holidays. In Food tourism around the world (pp. 102–120). Routledge.

Sims, R. (2009). Food, place and authenticity: local food and the sustainable tourism experience. Journal of sustainable tourism, 17(3), 321–336.

www.ingramcontent.com/pod-product-compliance
Ingram Content Group UK Ltd.
Pitfield, Milton Keynes, MK11 3LW, UK
UKHW020833290926
15470UKWH00004B/1328